Sean Clark (Ed.)

Sustainable Agriculture–Beyond Organic Farming

MDPI

This book is a reprint of the Special Issue that appeared in the online, open access journal, *Sustainability* (ISSN 2071-1050) in 2016, available at:

http://www.mdpi.com/journal/sustainability/special_issues/BOrganicFarming

Guest Editor
Sean Clark
Department of Agriculture and Natural Resources
Berea College
USA

Editorial Office
MDPI AG
St. Alban-Anlage 66
Basel, Switzerland

Publisher
Shu-Kun Lin

Managing Editor
Guoshui Liu

1. Edition 2016

MDPI • Basel • Beijing • Wuhan • Barcelona • Belgrade

ISBN 978-3-03842-304-1 (Hbk)
ISBN 978-3-03842-305-8 (electronic)

Table of Contents

Section 1: Performance of Organic Farming

Section 2: Farming to Food Systems

Section 3: Beyond Organic: Shaping Future Farming and Food Systems

V

List of Contributors

Luis J. Belmonte A. R. College of Pharmacy and G H Patel Institute of pharmacy, Vallabh Vidyanagar, Dist. Anand (Gujarat) – 388120, India.

Eddie A. M. Bokkers Animal Production Systems Group, Wageningen University, P.O. Box 338, 6700 AH Wageningen, The Netherlands.

Natalia Brzezina Sustainable Food Economies Research Group, KU Leuven, Leuven 3001, Belgium.

Ian Christie Centre for Environmental Strategy, University of Surrey, Guildford GU2 7XH, UK.

Roland Clift Centre for Environmental Strategy, University of Surrey, Guildford GU2 7XH, UK.

Xiaoyi Cui Department of Mathematics and Statistics, University of Guelph, 50 Stone Rd E, Guelph, ON, N1G 2W1, Canada.

Imke J. M. de Boer Animal Production Systems Group, Wageningen University, P.O. Box 338, 6700 AH Wageningen, The Netherlands.

Evelien M. de Olde Department of Engineering, Aarhus University, Inge Lehmanns Gade 10, 8000 Aarhus, Denmark; Animal Production Systems Group, Wageningen University, P.O. Box 338, 6700 AH Wageningen, The Netherlands.

Seri Donnuea Center of Excellence for Marine Biotechnology, Department of Marine Science, Faculty of Science, Chulalongkorn University, Payathai Road, Pathumwan, Bangkok 10300, Thailand; National Center for Genetic Engineering and Biotechnology, National Science and Technology Development Agency, Pathum Thani 12120, Thailand.

Alfredo J. Escribano Independent Researcher and Consultant, C/Rafael Alberti 24, Cáceres 10005, Spain.

Eihab Fathelrahman Julius Kühn-Institut, Federal Research Centre for Cultivated Plants, Institute for Plant Protection in Field Crops and Grassland, Messeweg 11-12, 38104 Braunschweig, Germany.

Daniela Fiorito Department of Agricultural and Forest Sciences (SAF), University of Palermo, Viale delle Scienze, 11-90128 Palermo, Italy.

Vincenzo Girgenti Department of Agricultural, Forest and Food Sciences (DISAFA), University of Torino, Largo Paolo Braccini 2, Grugliasco, 10095 Torino, Italy.

Thomas Hahn Stockholm Resilience Centre, Stockholm University, SE-106 91 Stockholm, Sweden.

Carlos Herrero-Sánchez A. R. College of Pharmacy and G H Patel Institute of pharmacy, Vallabh Vidyanagar, Dist. Anand (Gujarat) – 388120, India.

Lifen Huang Department of Mathematics and Statistics, University of Guelph, 50 Stone Rd E, Guelph, ON, N1G 2W1, Canada.

Thomas Koellner Faculty of Biology, Chemistry and Geosciences, University of Bayreuth, BayCEER, Bayreuth 95440, Germany.

Birgit Kopainsky System Dynamics Group, University of Bergen, Bergen 5020, Norway.

Markus Larsson Division of Environmental Strategies Research (fms), KTH Royal Institute of Technology, SE-100 44 Stockholm, Sweden.

Saem Lee Faculty of Biology, Chemistry and Geosciences, University of Bayreuth, BayCEER, Bayreuth 95440, Germany.

Sarah Taylor Lovell 1009 Plant Sciences Laboratory, Department of Crop Sciences, University of Illinois at Urbana-Champaign, Urbana, IL 61801, USA.

George Martin Centre for Environmental Strategy, University of Surrey, Guildford GU2 7XH, UK.

Erik Mathijs Sustainable Food Economies Research Group, KU Leuven, Leuven 3001, Belgium.

Rebecka Milestad Division of Environmental Strategies Research (fms), KTH Royal Institute of Technology, SE-100 44 Stockholm, Sweden.

Safdar Muhummad Julius Kühn-Institut, Federal Research Centre for Cultivated Plants, Institute for Plant Protection in Field Crops and Grassland, Messeweg 11-12, 38104 Braunschweig, Germany.

Trung Thanh Nguyen Institute for Environmental Economics and World Trade, Leibniz University of Hannover, Hannover 30167, Germany.

Frank W. Oudshoorn Department of Engineering, Aarhus University, Inge Lehmanns Gade 10, 8000 Aarhus, Denmark; SEGES, Agro Food Park 15, 8200 Aarhus N, Denmark.

Prasert Pavasant School of Energy Science and Engineering, Vidyasirimedhi Institute of Science and Technology, Rayong 21210, Thailand; Chemical Engineering Research Unit for Value Adding of Bioresources, Department of Chemical Engineering, Faculty of Engineering, Chulalongkorn University, Bangkok 10330, Thailand.

Cristiana Peano Department of Agricultural, Forest and Food Sciences (DISAFA), University of Torino, Largo Paolo Braccini 2, Grugliasco, 10095 Torino, Italy.

Patrick Poppenborg Faculty of Biology, Chemistry and Geosciences, University of Bayreuth, BayCEER, Bayreuth 95440, Germany.

Sorawit Powtongsook Center of Excellence for Marine Biotechnology, Department of Marine Science, Faculty of Science, Chulalongkorn University, Payathai Road, Pathumwan, Bangkok 10300, Thailand; National Center for Genetic Engineering and Biotechnology, National Science and Technology Development Agency, Pathum Thani 12120, Thailand.

Hio-Jung Shin Department of Agricultural and Resource Economics, Kangwon National University, 1 Kangwondaehak-gil, Chuncheon-si, Gangwon-do 24341, Republic of Korea.

Francesco Sottile Department of Agricultural and Forest Sciences (SAF), University of Palermo, Viale delle Scienze, 11-90128 Palermo, Italy.

Claus A. G. Sørensen Department of Engineering, Aarhus University, Inge Lehmanns Gade 10, 8000 Aarhus, Denmark.

Puchong Sri-uam Chemical Engineering Research Unit for Value Adding of Bioresources, Department of Chemical Engineering, Faculty of Engineering, Chulalongkorn University, Bangkok 10330, Thailand.

Anke Stubsgaard SEGES, Agro Food Park 15, 8200 Aarhus N, Denmark.

Rafi Ullah Tasbih Ullah Julius Kühn-Institut, Federal Research Centre for Cultivated Plants, Institute for Plant Protection in Field Crops and Grassland, Messeweg 11-12, 38104 Braunschweig, Germany.

Nadia Tecco Department of Agricultural, Forest and Food Sciences (DISAFA), University of Torino, Largo Paolo Braccini 2, Grugliasco, 10095 Torino, Italy.

Juan Torres A. R. College of Pharmacy and G H Patel Institute of pharmacy, Vallabh Vidyanagar, Dist. Anand (Gujarat) – 388120, India; Department of Applied Pharmacy, Faculty of Pharmacy, Medical University of Lublin, 1 Chodzki Str. 20-093, Lublin, Poland.

Diego L. Valera A. R. College of Pharmacy and G H Patel Institute of pharmacy, Vallabh Vidyanagar, Dist. Anand (Gujarat) – 388120, India.

Paul Vincelli Department of Plant Pathology, 207 Plant Science Building, College of Agriculture, Food and Environment, University of Kentucky, Lexington, KY 40546, USA.

Jacob von Oelreich Division of Environmental Strategies Research (fms), KTH Royal Institute of Technology, SE-100 44 Stockholm, Sweden.

Shouhong Wang College of Information Science and Technology, Chengdu University of Technology, Chengdu 610059, China.

Matthew Heron Wilson 1105 Plant Sciences Laboratory, Department of Crop Sciences, University of Illinois at Urbana-Champaign, Urbana, IL 61801, USA.

Huozhong Yang Department of Mathematics and Statistics, University of Guelph, 50 Stone Rd E, Guelph, ON, N1G 2W1, Canada.

Jie Yang Department of Mathematics and Statistics, University of Guelph, 50 Stone Rd E, Guelph, ON, N1G 2W1, Canada.

Hengyang Zhuang Department of Mathematics and Statistics, University of Guelph, 50 Stone Rd E, Guelph, ON, N1G 2W1, Canada.

About the Guest Editor

Sean Clark is Professor of Agriculture and Natural Resources at Berea College and Director of the Berea College Farm. He regularly teaches undergraduate courses in farm management, food systems, horticulture, beekeeping, and aquaculture and occasionally offers international courses in farming, field ecology, and entomology. He facilitates a summer undergraduate internship program on the college's student farm. The farm, which includes a student-run campus farm store, serves both as an educational laboratory and a model of sustainable regional agriculture. The farm internship program emphasizes critical inquiry, practical experience and innovative problem-solving. His research activities have included studies of soil biology and chemistry in organic and conventional farming systems, waste management and nutrient cycling, evaluations of appropriate technology and production methods for small-scale organic farming, and assessments of the environmental impacts of agriculture. He currently serves as a technical advisor for the Organic Association of Kentucky. A list of publications can be found at: https://www.berea.edu/anr/faculty-and-staff/dr-sean-clark/

Preface to "Sustainable Agriculture–Beyond Organic Farming"

The current conception of organic farming—as an agricultural production system based on ecological understanding in contrast to one reliant upon external inputs, particularly synthetic agrichemicals and fertilizer—is the result of nearly a century of intellectual thought and dialogue, field observations and experiences, systematic experimentation, and codification of rules. Debates on the future viability of organic farming often focus on its capacity to produce sufficient food to meet the demands of a growing human population. Yet any thorough examination of the pros and cons of alternative farming approaches should consider much more—for example the side effects on soil, water and air; energy and land-use efficiency; global warming potential; conservation of biodiversity; waste minimization and recycling; farmer and community well-being; animal welfare; and the capacity to function and meet demands long into the future.

Today, organic farming is widely acknowledged as a viable alternative to conventional production under many conditions and certified organic foods are increasingly sought out by consumers concerned about environmental issues and human health. Considerable research supports the validity of such consumer choices. Organic products typically contain lower levels of pesticides and antibiotics, soil quality is generally improved and water pollution reduced on organic farms, and biodiversity is often greater in organic production systems compared to their conventional counterparts. Crops once thought to be impossible to raise organically are now widely found on grocery store shelves.

Research findings also show the trade-offs in productivity and efficiency that accompany the adoption of organic farming. Such systems often do not match conventional agriculture in measures such as in yield per unit of land or per unit of labor, as well as in costs of production (not including externalities). As a consequence, price premiums are often needed to make production economically viable for farmers. Higher prices, of course, put some products out of reach for lower income families and communities, prompting criticism by some that organic farming is elitist. Organic standards today also may not explicitly or sufficiently address important public concerns about climate change, animal welfare and the quality of life provided to farmers, farm workers and others in the supply chain.

It is important to recognize that organic agriculture is evolving. Rather than a static set of rules, the requirements, technologies, inputs and management practices comprising organic farming change as our understanding improves, new technologies become available, and the stakeholders involved with the political process that governs certification amend the rules. Research and debate today will help to shape organic farming in the future.

The chapters in this book represent perspectives on organic farming and food systems from widely different academic disciplines and different regions of the world. They include replicated field experiments, modelling, systems analyses, case studies and literature reviews. They address issues from the field-plot scale, such as resource-use efficiency in crop production, to the resilience of entire national food systems. Some chapters tackle controversial topics in organic farming, such as aquaculture and the use of genetically modified organisms. Some authors focus on the challenges to producers, while others examine consumer behavior and the education of the next generation of global citizens and decision-makers in the food system—today's children.

This book is composed of three sections. Authors in the first section—Performance of Organic Farming—examine how well organic and transitional production systems meet environmental, economic and social objectives and how performance could be improved. In the second section—Farming to Food Systems—the focus expands beyond crop and livestock production to consider other stakeholders forming food supply chains. The last section—Beyond Organic: Shaping Future Farming and Food Systems—delves into some vigorously debated topics that have the potential to substantially change future farming and food systems.

Two decades ago, few would have envisioned the expansion of organic farming and the dramatic growth in organic food sales that followed. Likewise, predictions about the state of organic food and farming decades from now would be fraught with the same uncertainty. Still, the improved understanding we gain from the contributions of researchers and thinkers today will influence production practices and food-system policies tomorrow. The findings, interpretations, and ideas shared in this book will likely generate as many questions as answers, but asking the relevant and difficult questions is as critical to progress as finding the right answers. This impressive and diverse group of authors makes interesting and useful contributions to our ongoing conversations about food, agriculture and the evolution of organic farming.

Sean Clark
Guest Editor

Section 1:
Performance of Organic Farming

Assessing the Sustainability Performance of Organic Farms in Denmark

Evelien M. de Olde, Frank W. Oudshoorn, Eddie A. M. Bokkers,
Anke Stubsgaard, Claus A. G. Sørensen and Imke J. M. de Boer

Abstract: The growth of organic agriculture in Denmark raises the interest of both producers and consumers in the sustainability performance of organic production. The aim of this study was to examine the sustainability performance of farms in four agricultural sectors (vegetable, dairy, pig and poultry) using the sustainability assessment tool RISE 2.0. Thirty seven organic farms were assessed on 10 themes, including 51 subthemes. For one theme (water use) and 17 subthemes, a difference between sectors was found. Using the thresholds of RISE, the vegetable, dairy and pig sector performed positively for seven themes and the poultry sector for eight themes. The performance on the nutrient flows and energy and climate themes, however, was critical for all sectors. Moreover, the performance on the economic viability theme was critical for vegetable, dairy and pig farms. The development of a tool, including decisions, such as the selection of themes and indicators, reference values, weights and aggregation methods, influences the assessment results. This emphasizes the need for transparency and reflection on decisions made in sustainability assessment tools. The results of RISE present a starting point to discuss sustainability at the farm-level and contribute to an increase in awareness and learning about sustainability.

Reprinted from *Sustainability*.   Cite as: de Olde, E.M.; Oudshoorn, F.W.; Bokkers, E.A.M.; Stubsgaard, A.; Sørensen, C.A.G.; de Boer, I.J.M. Assessing the Sustainability Performance of Organic Farms in Denmark. *Sustainability* **2016**, *8*, 957.

1. Introduction

A large number of sustainability assessment tools have been developed to gain insight into the sustainability performance of farms [1,2]. These tools generally integrate a wide range of themes and indicators to develop a holistic view on farm-level sustainability and are used for different purposes, such as monitoring, certification, consumer information, farm advice and research [3]. Applying sustainability assessment tools can help to identify challenges, related to environmental, economic and social impact, in the development of sustainable food production systems in conventional and organic agriculture [4–6]. On-farm assessment tools, however, show a large diversity in, for example, data, time and budget requirements, measurement and aggregation methods, output accuracy and complexity [2,7]. These differences should become more explicit when choosing a tool [7,8]. Moreover, after a sustainability assessment, additional efforts are needed to discuss the assessment outcomes with farmers and other stakeholders and translate them into meaningful decisions for change [2,9].

Studies on organic agriculture provide divergent views on its sustainability and potential to contribute to global food security [10–13]. Especially yield differences between conventional and organic agriculture are a topic of discussion [12,14–16]. Differences in yields are highly dependent on system and site characteristics (e.g., available nutrients and technology) [11,14]. Yields in organic farming are generally lower compared to conventional yields. Lower yields, on the one hand, are associated with a higher land use and, for example, higher global warming potential per kg live weight of pigs [17]. On the other hand, the restricted use of pesticides and mineral fertilizers in organic agriculture can have a positive effect on biodiversity and enhance ecosystem services [18,19]. From an economic and social perspective, organic agriculture is often associated with the use of local resources (i.e., local seed varieties, manure), benefits for animal welfare and opportunities to increase farmers' income and livelihood [10,13,20,21].

Organic agriculture in the European Union has increased over the past decades and currently accounts for about 5.7% of the agricultural area [22]. The consumption of organic products is increasing as well, and is worldwide the highest per capita in Switzerland, Luxembourg and Denmark [22]. Market shares of organic retail sales are highest in Denmark (7.6%), Switzerland (7.1%) and Austria (6.5%) [22]. Denmark, therefore, can be considered a pioneer in organic food production, with an expected ongoing growth in organic food consumption [23]. At the same time, producers and consumers are increasingly interested in getting insight in, and the development of, the sustainability performance of organic production [24,25].

The sustainability assessment tool RISE (Response-Inducing Sustainability Evaluation) [26] is used in Denmark to assess the sustainability performance of organic farms and to guide farmers in producing more sustainably [24,27,28]. This tool was selected based on the European project STOAS (Sustainability Training for Organic Advisors), in which experiences with different sustainability assessment tools were gathered [29]. The objectives of the present study were to analyze the sustainability performance of organic farms in Denmark using the RISE 2.0 tool and to analyze differences in the performance among a diversity of agricultural sectors (i.e., vegetable, dairy, pigs and poultry production). First, we describe the RISE tool and elaborate on the methods for data collection and assessment. Second, we present the RISE assessment results of Danish organic farms and discuss differences between sectors. Third, we reflect on the approaches in RISE to assess sustainability performance and discuss the implications of our findings for organic agriculture in Denmark and, more generally, for assessing sustainability at the farm level.

2. Materials and Methods

2.1. RISE 2.0

RISE is an indicator-based sustainability assessment tool developed at the Bern University of Applied Sciences (School of Agricultural, Forest and Food Sciences, HAFL) [30]. The aim of the tool developers is to provide a holistic evaluation of sustainability at the farm level and support the dissemination of sustainable practices [30]. Since its start in 1999, RISE has been applied in over 2500 farms in 56 countries [31]. Experiences with RISE 1.0 have been extensively described in the literature [9,26,32–34]. Studies describing the application of the updated Version 2.0, launched in 2011, however, are limited [2].

RISE 2.0 assesses the sustainability performance of a farm for 10 themes and 51 subthemes (Table 1). Although RISE defines the subthemes as indicators, we prefer to call them subthemes as they include the evaluation of various indicators and align the terminology with other sustainability assessment tools and publications, such as, for example, the Sustainability Assessment of Food and Agriculture systems (SAFA) guidelines [1,2]. The sustainability performance of each subtheme is based on an aggregation of various indicators. These indicators are normalized (i.e., converted to a 0–100 scale) differently for each subtheme and can include comparisons between farm and reference data. The score at the theme level is based on the average of the scores of the 4–7 subthemes included in each theme. Scores on theme and subtheme level range from 0–100 and are visualized in a polygon. According to RISE, a performance between 0 and 33 is considered problematic, between 34 and 66 critical and between 67 and 100 positive. RISE results are presented in a farm report, which includes the farm's sustainability polygon, a table with the theme and subtheme scores and an explanation of the calculation and scores. Based on this report, a farmer and auditor define the measures for improvement. The RISE software is available on a license and requires training.

To compute the sustainability performance of a farm, four types of data are used: points allocated to farm practices, quantitative farm data, regional data and master data (global reference data). Information on farm practices and quantitative farm data are gathered through a questionnaire-based interview with the farmer and farm workers, conducted by a trained auditor. For the themes working conditions and quality of life, the farmer decides whether the employees may be interviewed, and if so, who. A certain amount of points (positive or negative) are given based on the answers of the farmer, farm worker and/or auditor to questions on farm management, activities and the on-farm situation (e.g., animal welfare conditions). This way, qualitative information is translated into a quantitative score (see Box 1). The majority of subthemes (40) integrate this type of data to compute the performance of the farm on the subtheme. Of these subthemes, 19 subthemes are exclusively based on points allocated to certain measures, activities or situations on-farm. These subthemes are related to quality of life (6), farm management (5), animal husbandry (3), soil use (2), water use (1), nutrient flows (1) and working conditions (1). For the remaining 21 subthemes, this type of data is combined with one or more of the other data types.

Table 1. Themes and subthemes in RISE 2.0.

Theme	Subthemes	Theme	Subthemes
1. Soil use	1.1. Soil management 1.2. Crop productivity 1.3. Soil organic matter supply 1.4. Soil reaction 1.5. Soil pollution 1.6. Soil erosion 1.7. Soil compaction	6. Biodiversity	6.1. Plant protection management 6.2. Ecological priority areas 6.3. Intensity of agricultural production 6.4. Landscape quality 6.5. Diversity of agricultural production
2. Animal husbandry	2.1. Herd management 2.2. Livestock productivity 2.3. Possibility for species-appropriate behavior 2.4. Quality of housing 2.5. Animal health	7. Working conditions	7.1. Personnel management 7.2. Working times 7.3. Safety at work 7.4. Salaries and income level
3. Nutrient flows	3.1. Nitrogen balance 3.2. Phosphorus balance 3.3. N and P self-sufficiency 3.4. Ammonia emissions 3.5. Waste management	8. Quality of life	8.1. Occupation and education 8.2. Financial situation 8.3. Social relations 8.4. Personal freedom and values 8.5. Health 8.6. Further aspects of life
4. Water use	4.1. Water management 4.2. Water supply 4.3. Water use intensity 4.4. Risks to water quality	9. Economic viability	9.1. Liquidity reserve 9.2. Level of indebtedness 9.3. Economic vulnerability 9.4. Livelihood security 9.5. Cash flow-turnover ratio 9.6. Debt service coverage ratio
5. Energy and climate	5.1. Energy management 5.2. Energy intensity of agricultural production 5.3. Share of sustainable energy carriers 5.4. Greenhouse gas balance	10. Farm management	10.1. Farm strategy and planning 10.2. Supply and yield security 10.3. Planning instruments and documentation 10.4. Quality management 10.5. Farm cooperation

Quantitative farm data (e.g., energy consumption, crop yields and income) are used in 28 subthemes, especially in combination with other types of data (23 subthemes) (see Box 2). In five subthemes, quantitative farm data are used exclusively and compared to regional reference values. These subthemes are related to economic viability (4) and biodiversity (1).

Box 1. Example points-based subtheme: farm strategy and planning (10.1).

The score on this subtheme is based on the average score on four questions:
1   Is there a clear long-term farm development strategy?
2   Are there any short to medium-term measures for improvement on economy, social or ecology? (answered seperately for each dimension)
Each question can be anwered with yes (100 points), partly (50 points) or no (0 points).

Box 2. Example of a subtheme combining points, quantitative farm data, regional data and regional reference values: livestock productivity (2.2).

The subtheme is calculated in four steps.
1   The livestock units, per animal category (i) and in total (t), are calculated and corrected for temporarily absent or present animals. The livestock units are derived from regional data (livestock unit factors).
2   The productivity of each animal category (e.g., annual milk yield, growth rate, egg production) is compared to regional reference values. The score on the productivity for each animal category is calculated using this formula: productivity/regional productivity $\times$ 100 − 33.
3   For each animal category, the farmer is asked to give an estimation of the product quality (q1) and of the development of the performance and quality over the last 5 years (q2). For both questions, the farmer can select the answer from five options: significantly above average/improvement (20 points), slightly above average/improvement (10 points), average/stagnation (0 points), slightly below average/decline (−10 points), significantly below average/decline (−20 points).
4   The results of Steps 2 and 3 are added and corrected for the share of the animal category in the total livestock units on the farm: sum ((result step 2i + q1i + q2i) $\times$ (LUi/LUt)).

Regional data are specific to the respective region, but are not assessed or available at the farm level e.g., nitrogen losses from farm and storage facilities, livestock unit factors and water demand of crops. The regional data can be from a country, in this study Denmark, or from a smaller region. Master data are provided by RISE and cover, for example, the composition of feedstuffs, the toxicity and persistency of pesticides, the energy consumption of machine work, energy density (i.e., energy contained in MJ), the emissions of energy carriers (e.g., coal, wood, natural gas, petroleum) the and nutrient contents of organic fertilizers. Regional and master data are integrated in the calculations of 11 and 14 subthemes, respectively, always in combination with points and/or quantitative farm data. Five subthemes integrate all four data types.

Next to farm, regional and master data, regional reference values are used in 11 subthemes to compare the performance of the farm to the regional average or target (e.g., crop yields, livestock production, share of ecological priority areas, working hours and days per week). In the RISE software, a standard set of crops (i.e., yields, water content and cultivation period) and livestock (i.e., productivity and livestock units) is given that can be adjusted to the region and extended. What should be considered as 0 and 100 points is defined by the tool, except for six subthemes in which a regional reference value is used. In each subtheme, different calculations are used to aggregate data and compute a score. Decisions regarding these calculations, for example on indicators, units (i.e., hectares, MJ), weights and the use of an average or minimum score of the indicators, influence the result on the subtheme. These calculations are mostly fixed within RISE, except for quality of life-related subthemes, in which the interviewee determines the weight of each indicator within the subtheme and can include an additional subtheme.

2.2. RISE Assessments of Danish Organic Farms

The sustainability performance of organic farms in Denmark was assessed and analyzed in three phases (Figure 1). In the first phase, the RISE software had to be prepared for application in the Danish context. This preparation included entering regional data and regional reference values for Danish agriculture in RISE and translating the tool and questionnaire to Danish. These data were gathered from different sources, including databases and software on Danish farm management (e.g., Mark Online, Farmtal Online (SEGES)), expert consultation and discussions with the RISE tool developers. Regional reference values are based on Danish standards (e.g., weather, income levels and working hours) and the performance of Danish agriculture (not specifically organic agriculture).

In the second phase, the assessment process was prepared by training auditors, selecting farms, contacting farmers and entering available data in RISE. Ten consultants from Danish advisory services were trained as RISE 2.0 auditors by the tool developers from Switzerland. This training included a joint assessment and discussion of assessment procedures. Six food processing companies were involved in the selection of farmers for the assessments. This transdisciplinary approach, in which stakeholders from farming practice (i.e., farmers, advisors and processing companies) and research collaborate, can help to address sustainability challenges [35–37]. Farms out of four sectors (vegetables, dairy, pigs and poultry) were selected by six food processing companies. These food processing companies were involved as stakeholders in communication on the sustainability of food products. Each company freely selected 7 or 8 of their supplying farmers to participate voluntarily in the sustainability assessment. Although this selection of farms is not a representative sample of the Danish organic farmers, an analysis of the results might give insight into generic sustainability challenges in Danish organic farming and differences in the sustainability performance of four agricultural sectors.

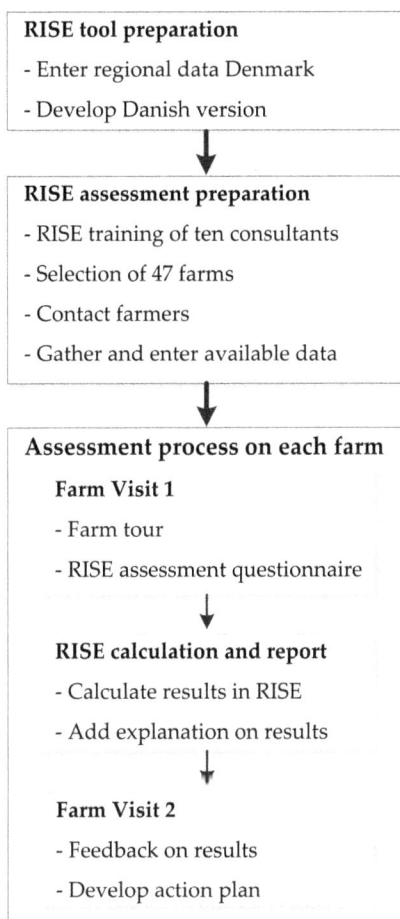

Figure 1. Phases in the sustainability assessment of organic farms in Denmark, using RISE 2.0.

Before the actual assessment, each farmer was contacted and asked to provide available data (i.e., farm accounts, crop rotation plan, fertilization plan). These data were entered in RISE beforehand to reduce on-farm assessment time.

Finally, assessments of organic farms were carried out and involved two farm visits, calculation and reporting. Each farm assessment started with a short farm tour. After this introduction, the questionnaire-based interview with the farmer was carried out by one or two auditors. In case a farmer did not have all data needed available at the moment of assessment, these data were emailed later to the auditor and entered in the RISE software. When all data needed for the assessment were gathered, the outcomes were calculated in RISE, and a report was made. This report included an explanation given by the auditor(s) on the outcomes and was discussed with the farmer during a second farm visit. Based on the outcomes of the tool and priorities of the farmer, a brief action plan for improvement was made.

2.3. Analysis of RISE Assessment Results

In total, 47 farms were assessed in the period 2013–2014. Six assessments had to be excluded from the data analysis due to insufficient data and errors in data storage. To compare the sectors, a farm was considered specialized in a particular sector if more than 50% of the total output and coupled subsidies resulted from that sector. Although the food processing companies selected supplying farmers, this branch was not in all cases the most important output of the farm. Four farms that were initially selected as poultry farms appeared specialized in other sectors and therefore were excluded from the analysis.

Data of 37 RISE assessments were analyzed. Seven farms were assessed based on data from 2012 and 30 with data from 2013. General characteristics of the farms are given in Table 2. Vegetable producers included in this study produce vegetables for vegetable and meal boxes. To compare farms with different species and ages of animals, livestock units (LU) are used as a reference unit. A dairy cow, for example, represents 1 LU; a heifer between 1 and 2 years old is 0.4 LU; while a fattening pig is 0.17 LU; and a laying hen is 0.01 LU [38].

Table 2. Number (N) of farms per sector, median and range of agricultural area on farm and livestock units (LU) and livestock density (LU/ha).

N	Sector	Agricultural Area (ha)	LU	LU/ha
5	Vegetables	77 (3–176)		
13	Dairy	153 (44–832)	168 (53–548)	1 (0.7–1.5)
8	Pigs	153 (25–351)	337 (55–1130)	1.8 (1.1–5.2)
11	Poultry	34 (5–81)	90 (44–610)	2.2 (1.4–18.1)

Assessment outcomes of the individual farms were analyzed in SPSS 22 to identify significant differences ($p < 0.05$) using nonparametric tests. Differences between sectors for themes and subthemes were analyzed using the Kruskal–Wallis test. In case of significant differences, additional analysis was carried out using the Mann–Whitney U test for pairwise comparisons [39,40].

3. Results

3.1. Soil Use

The score on the theme soil use is based on the average score of seven subthemes (Table 3). No difference between sectors was found for the theme soil use, while for the subtheme soil compaction, a difference was found (Table 3).

Table 3. Sustainability performance on the theme soil use and related subthemes for the vegetables, dairy, pig and poultry sector (median, min–max).

	Vegetables	Dairy	Pigs	Poultry	p-Value
1. Soil use	74 (60–95)	72 (64–81)	74 (62–85)	79 (71–88)	0.125
1.1. Soil management	84 (67–84)	83 (50–100)	84 (83–84)	67 (50–84)	0.224
1.2. Crop productivity	64 (15–100)	59 (15–98)	48 (16–63)	60 (31–75)	0.325
1.3. Soil organic matter supply	41 (22–100)	51 (43–75)	56 (31–93)	42 (33–100)	0.739
1.4. Soil reaction	100 (50–100)	95 (55–100)	100 (81–100)	100 (63–100)	0.233
1.5. Soil pollution	90 (70–100)	90 (60–100)	90 (60–90)	100 (70–100)	0.053
1.6. Soil erosion	98 (84–100)	93 (77–100)	94 (68–100)	96 (90–100)	0.871
1.7. Soil compaction	70 (25–100) [a,b]	55 (0–90) [a]	55 (0–100) [a,b]	90 (55–100) [b]	0.037

[a,b] Different superscripts indicate significant differences between sectors ($p < 0.05$).

The subtheme soil management combines quantitative farm data on the loss of agricultural land in the past ten years and points for knowledge and information about soil fertility. Erosion, salinization or building activity has caused losses in agricultural area on 24% of the farms (0.3%–1.4% of the farm area). Soil analyses for fertilization planning were applied regularly by 65% of the farmers, nutrient balances by 95% of the farmers, whereas soil organic matter balances were used by 5% of the farmers.

Crop productivity compares the farm yield of each crop per hectare to regional reference values. In addition, points are allocated based on the farmer's perception of product quality compared to the regional quality and the development of the quality over the past five years. The productivity differed strongly between farms and per crop, but was generally lower than the Danish reference values, which were not specifically for organic farms.

Soil organic matter supply determines the share of farm area with a high humus content and the soil organic matter balance in arable crops. It includes the share of permanent grassland and crops, removal and burning of crop residues and the use of organic fertilizer. The median share of farm area with a high humus content was 19%, with a range from 0%–100%.

Soil reaction focuses on the chemical condition and management of the soil (i.e., pH level, use of acidifying fertilizers, liming, irrigated soils without adequate drainage). Acidifying fertilizers were used by 14% of the farms, 80% of which apply liming.

The subtheme soil pollution evaluates farm practices to reduce the risk of chemical soil pollution. Organic fertilizers that may contain heavy metals were used by 11% of the farmers; residues (e.g., compost) without pollutant analyses were used by 3% of the farmers; and a risk of pollution from highways or industry was recognized by 8% of the farmers. The majority of farmers (60%) used farm manure (either from conventional or organic farms) that may contain antibiotic residues.

The subtheme soil erosion assesses the wind and water erosion risks and evaluates measures implemented to reduce soil erosion (e.g., ploughing, ground cover, hedges). Farmers observed water erosion on 5% of the farms and affected 1%–5% of the agricultural land. Wind erosion was observed by farmers on 16% of the farms and affected 1%–15% of their land.

Soil compaction evaluates practices that can positively or negatively affect soil compaction. Harmful soil compaction was observed by farmers on 35% of the farms. Heavy machines (i.e., machines with a wheel load above 2.5 tons) were used on 78% of the farms. Of the farms using heavy machines 22% also used them on arable land with clayey soils; 22% used them on wet soils; and 60% applied intensive cultivation of such soils (e.g., plowing, root crops). Soil conservation measures (e.g., dual tires, low tire pressure or controlled traffic farming) when using heavy machines were implemented by 65% of the farms, and 70% implemented measures to improve soil stability (e.g., liming, interim greening or reduced tillage). The score on soil compaction was higher for poultry farms compared to dairy farms. Of the poultry farms, 64% used heavy machines, compared to 92% of dairy farms. Moreover, poultry farmers did not observe harmful soil compaction at all, while 54% of the dairy farmers did observe harmful soil compaction.

3.2. Animal Husbandry

The theme animal husbandry consists of five subthemes (Table 4). Vegetable farmers included in this study did not have livestock; the scores are therefore based on 32 farms with livestock. No difference between sectors was found for this theme, while for the subthemes livestock productivity and quality of housing, a difference between sectors was found (Table 4).

Table 4. Sustainability performance on the theme animal husbandry and related subthemes for the dairy, pig and poultry sector (median, min–max).

	Dairy	Pigs	Poultry	p-Value
2. Animal husbandry	89 (73–95)	90 (84–96)	92 (84–95)	0.374
2.1. Herd management	100 (67–100)	100 (83–100)	100 (83–100)	0.612
2.2. Livestock productivity	65 (33–94) [a]	94 (75–100) [b]	98 (82–100) [b]	0.001
2.3. Possibilities for species-appropriate behavior	100 (58–100)	100 (100–100)	100 (100–100)	0.482
2.4. Quality of housing	100 (92–100) [a]	98 (91–100) [a]	90 (90–100) [b]	0.007
2.5. Animal health	75 (48–89)	67 (43–97)	73 (50–83)	0.287

[a,b] Different superscripts indicate significant differences between sectors ($p < 0.05$).

In the subtheme herd management, a farmer receives points for answers on questions related to information about livestock, health management and criteria for the selection of breeding animals. Of the farmers, 97% answered that they regularly observe the animals and information about the animals is documented and used for management. On health management, all farmers answered that they cleaned the barns properly, frequently and thoroughly and used preventive measures, such as separating animals with infectious diseases and regular claw trimming. Selection of breeding animals was generally made consciously considering robustness, adaptedness and expected life performance (78%).

The score on livestock productivity is computed by comparing the productivity of the animal category to the regional average and by taking into account farmers' perception of product quality and developments in performance and quality over the last five years (Box 2). Livestock productivity, without the scores on the perception of quality (development), was lower for dairy farms compared to pig and poultry farms (Table 4). The median livestock productivity at dairy farms (of all animal categories) was 90% of the regional reference values, while the productivity at pig and poultry farms was 108% and 105%.

The subtheme possibilities for species-appropriate behavior combines points for the possibility for the animal to express behavioral needs and the livestock density on the farm. Both aspects are evaluated by the auditor for each animal category separately. For each species, RISE defined certain behavioral needs to be scored on-farm (e.g., outdoor access, free moving space, clean floors).

Scores on the subtheme quality of housing are based on the auditor's observation of the cleanliness and amount of drinking places, protection from heat, light, air quality and protection from noise, for each animal category. Poultry farms had a lower score compared to dairy and pig farms on this subtheme. This difference is mainly due to an ammonia odor in the barns observed at 67% of the poultry farms.

The score on the subtheme animal health is based on the farm data of each animal category on animal treatment products used (i.e., share of animals treated curatively), mortality and mutilation (e.g., dehorning in cattle, debeaking in laying hens, castration in pigs).

3.3. Nutrient Flows

The theme nutrient flows consists of five subthemes (Table 5). No difference between sectors was found for the theme nutrient flows, while the scores differed between sectors for the subthemes N and P self-sufficiency and ammonia emissions.

The score on nitrogen balance is calculated by first calculating the nitrogen demand at the farm level (based on the nitrogen demand of each crop and, if relevant, exported organic material) and comparing this to the nitrogen supply (i.e., animal husbandry, organic material and crops (through nitrogen fixation)). An optimum nitrogen balance (100 points) is according to RISE between 90% and 110%, and a poor balance (zero points) is lower than 30% or more than 180% of the demand.

Table 5. Sustainability performance on the theme nutrient flows and related subthemes for the vegetable, dairy, pig and poultry sector (median, min–max).

	Vegetables	Dairy	Pigs	Poultry	p-Value
3. Nutrient flows	61 (37–71)	66 (54–79)	51 (27–71)	62 (39–75)	0.057
3.1. Nitrogen (N) balance	74 (0–95)	88 (52–100)	45 (0–100)	93 (41–100)	0.332
3.2. Phosphorus (P) balance	0 (0–99)	60 (0–100)	23 (0–64)	69 (0–100)	0.182
3.3. N and P self-sufficiency	36 (0–78) [a]	90 (68–96) [b]	48 (10–81) [a]	50 (27–82) [a]	0.000
3.4. Ammonia emissions	100 (85–100) [a]	57 (32–68) [b]	62 (33–76) [b]	47 (20–60) [c]	0.000
3.5. Waste management	60 (30–95)	65 (5–80)	80 (50–85)	60 (40–95)	0.213

[a–c] Different superscripts indicate significant differences between sectors ($p < 0.05$).

Phosphorus balance scores are calculated using a similar approach and the optimum as for nitrogen balance. RISE compares the demand of crops with the supply from animals and imported organic material.

The subtheme N and P self-sufficiency compares the nitrogen and phosphorus demand of livestock (i.e., feed) and crops to the on-farm supply. The scores of dairy farms were higher compared to all other sectors as a result of a high degree (90% median) of self-sufficiency in N and P in both feed and fertilizer in dairy farms (Table 5).

The subtheme ammonia emissions calculates emissions from animal husbandry, imported organic fertilizers and includes points for farm practices related to manure storage, manure spreading and slurry injection. The absence of livestock on vegetable farms resulted in lower ammonia emissions, hence a higher score compared to dairy, pig and poultry farms (Table 5). In addition, dairy and pig farms scored higher compared to poultry farms, which is related to a lower livestock density in dairy farms (Table 2).

In the subtheme waste management, the environmental risks of the disposal of twelve types of waste are assessed. Points are allocated to the different ways of disposing waste. The scores varied between farms indicating different approaches to dispose the various types of waste.

3.4. Water Use

The theme water use is based on the average score of four subthemes and differed between sectors with a lower score for poultry farms compared to dairy and pig farms (Table 6).

Table 6. Sustainability performance on the theme water use and related subthemes for the vegetable, dairy, pig and poultry sector (median, min–max).

	Vegetables	Dairy	Pigs	Poultry	p-Value
4. Water use	77 (69–83) [a,b]	83 (69–90) [a]	80 (72–84) [a]	72 (61–84) [b]	0.026
4.1. Water management	75 (50–80) [a]	75 (50–90) [a]	53 (45–70) [b]	45 (35–80) [c]	0.000
4.2. Water supply	100 (90–100)	100 (80–100)	100 (50–100)	100 (90–100)	0.918
4.3. Water use intensity	26 (26–76)	67 (34–76)	74 (64–76)	62 (26–76)	0.068
4.4. Risks to water quality	100 (91–100) [a]	96 (77–100) [a,b]	100 (94–100) [a]	84 (75–100) [b]	0.027

[a–c] Different superscripts indicate significant differences between sectors ($p < 0.05$).

The score on the subtheme water management is based on points received for farm practices related to water management (i.e., information on water availability and quality, technical water-storing measures and hygienic recycling of waste water) and the implementation of water-saving measures. Whereas 95% of the farmers had access to information on water availability and quality and 87% recycled waste water hygienically, only 11% implemented measures to increase the water storage capacity. Vegetable and dairy farms scored higher compared to pig and poultry farms, as a result of differences in the amount and type of water saving measures applied on the farms. Moreover, pig farms scored higher than poultry farms.

The subtheme water supply evaluates problems on the farm related to water supply (through minus points) and includes the regional value for water stress. A decrease of water availability was observed by 3% of the farmers; 8% observed a lowering of the ground water level; 5% observed a decrease of water quality; and 3% were confronted with water conflicts. Fossil groundwater was used by 14% of the farmers.

Water use intensity compares the water demand for agricultural production (i.e., crops, livestock and service) per hectare (in m^3/year) with the regional moisture index (a regional value for the availability of water, calculated from the FAO Moisture Index [41]). The median water use intensity was 82 (0–100), and the regional value for the moisture index was 51.

The subtheme risk to water quality evaluates risks to water quality caused by storage facilities and effluent disposal and risks of nutrient input into the water. Scores were lower for poultry farms compared to vegetable and pig farms (Table 6). Of the poultry farmers, 45% indicated that areas with high nutrient input (e.g., as a result of the outdoor run) are present, compared to 0% of farmers in the other sectors. Moreover, 82% of the poultry farms had buffer strips along open water, compared to 92% of dairy farms and 100% of vegetable and pig farms. Frequent (at least once a week) access of livestock to open water occurred on 9% of the poultry farms, 15% of the dairy farms and 0% on pig farms and could cause local eutrophication or water contamination.

3.5. Energy and Climate

The energy and climate theme covers four subthemes (Table 7). No difference between sectors was found for the theme energy and climate, while a difference was found for the subthemes energy intensity of agricultural production and greenhouse gas balance (Table 7).

Table 7. Sustainability performance on the theme energy and climate and related subthemes for the vegetable, dairy, pig and poultry sector (median, min–max).

	Vegetables	Dairy	Pigs	Poultry	p-Value
5. Energy and Climate	49 (7–67)	53 (42–61)	60 (28–69)	61 (21–75)	0.251
5.1. Energy management	95 (25–100)	60 (45–100)	60 (25–75)	65 (35–85)	0.526
5.2. Energy intensity of agricultural production	0 (0–79) [a]	81 (63–91) [b]	60 (39–100) [b,c]	69 (0–92) [a,c]	0.010
5.3. Share of sustainable energy carriers	25 (2–68)	17 (10–40)	16 (7–42)	30 (7–76)	0.158
5.4. Greenhouse gas balance	83 (0–100) [a,b]	42 (18–74) [a]	100 (0–100) [b]	100 (0–100) [b]	0.012

[a–c] Different superscripts indicate significant differences between sectors ($p < 0.05$).

The score on energy management is based on points for energy saving measures applied on the farm and monitoring of energy consumption. Energy consumption was monitored on all farms. The type and number of energy saving measures, however, varied strongly between farms.

Energy intensity of agricultural production is a comparison of the farm's energy consumption with the regional average. The energy consumption of the farm (in MJ per ha) is a sum of all energy carriers multiplied with the energy density in MJ given in the master data and is corrected for imported or exported contract machinery work. Farm energy consumption of 25% or less of the regional average results in a score of 100; a consumption of 175% or more results in a score of zero points. The Danish reference value was 11,000 MJ per hectare. The score of vegetable farms was lower compared to dairy farms and pig farms. In addition, poultry farms scored lower compared to dairy farms (Table 7).

The subtheme share of sustainable energy carriers determines the share of energy from renewable sources in comparison to the total energy consumption of the farm (in MJ). On average, 25% of the energy consumption on the farms was renewable. Although 44% of the electricity in Denmark is from renewable sources, non-renewable energy from diesel and gas represented a large share of the energy used on farms.

The greenhouse gas balance is calculated per hectare and includes emissions from livestock, fuel and fertilizer use (i.e., energy consumed, imported and exported machine work, N mineralization), carbon sequestration and afforestation and forest clearing. Greenhouse gas emissions were higher for dairy farms compared to pig and poultry farms, as a result of higher livestock related emissions in dairy farms (Table 7).

3.6. Biodiversity

The theme biodiversity consists of five subthemes (Table 8). No difference between sectors was found for score on theme-level, while for the subthemes intensity of agricultural production and diversity of agricultural production, differences were found (Table 8).

Table 8. Sustainability performance on the theme biodiversity and related subthemes for the vegetable, dairy, pig and poultry sector (median, min–max).

	Vegetables	Dairy	Pigs	Poultry	p-Value
6. Biodiversity	76 (69–91)	76 (56–92)	72 (57–85)	67 (49–82)	0.468
6.1. Plant protection management	100 (82–100)	100 (90–100)	100 (100–100)	100 (83–100)	0.142
6.2. Ecological priority areas	100 (45–100)	100 (23–100)	100 (41–100)	100 (40–100)	0.943
6.3. Intensity of agricultural production	81 (51–93) [a,b]	88 (81–95) [a]	69 (59–82) [b]	64 (47–72) [b]	0.000
6.4. Landscape quality	70 (50–90)	65 (20–100)	60 (10–100)	63 (15–100)	0.812
6.5. Diversity of agricultural production	68 (48–90) [a]	46 (29–71) [b,c]	48 (35–57) [b]	32 (18–52) [b,c]	0.005

[a–c] Different superscripts indicate significant differences between sectors ($p < 0.05$).

Plant protection management aggregates points for implemented biodiversity conservation practices and the degree of toxicity and persistency of plant protection products used. Crop rotation and selection of varieties based on resistance to pests were implemented on all farms. Of the farms, 73% participated in biodiversity programs. Plant protection products permitted in organic agriculture were used by 5% of the farmers.

Ecological priority areas is the share of land (including agricultural area, forest, courtyard, open water, unused land) with a high ecological quality. The share on-farm is compared to the regional target for the share of ecologically-protected areas (9% in Denmark). The median share of land with ecological quality was 10%, with a range from 2%–62%.

Intensity of agricultural production aggregates the calculated intensity of nitrogen fertilization, livestock density and intensity of plant protection products and points for biodiversity promoting measures applied on the farm. Dairy farms scored higher compared to pig and poultry farms, which is related to the number and type of biodiversity-promoting measures implemented and a higher livestock density in LU/ha in pig and poultry farms (Tables 2 and 8).

The score on landscape quality is based on points allocated to the development of ecological elements that structure the landscape and the share of agriculture areas in the vicinity of ecological landscape elements (i.e., within a buffer of 50 m around all ecologically-valuable habits, e.g., trees bushes, hedges, stone heaps and ecological priority areas). The farms' median share of areas in the vicinity of ecological landscape elements was 37% with a range of 10%–100%; the regional target was 100%.

In the subtheme diversity of agricultural production, a farm receives points for diversity in land use types, crop species and varieties in cultivation, old and endangered crop species, livestock breeds, old and endangered breeds and bee keeping. Vegetable farms scored higher compared to all other sectors. In addition, pig farms have a higher score compared to poultry farms. Old and endangered crop species were grown on 19% of the farms of which 57% was on vegetable farms. On 57% of the farms, bees were kept. For livestock farms, the presence of old and endangered livestock breeds is considered. As old and endangered livestock breeds were absent on 97% of the farms, the score of these livestock farms is lower.

3.7. Working Conditions

The theme working conditions covers four subthemes (Table 9). No differences were found in the scores between sectors for the theme working conditions or for the four subthemes (Table 9).

In the subtheme personnel management, fifteen aspects are evaluated by all interviewees (farmer and farm workers). Based on their answers, points are allocated to reflect the farm performance on aspects such as housing of employees, education of apprentices, working contracts, assurance of replacement of work forces, illness benefit, equality and forced labor. In the aggregation, the minimum scores on three aspects (equality (gender), equality (other e.g., age, religion or origin) and forced labor are taken, while for the other aspects, the average score is used. All farm workers interviewed had a work permit; 97% participated in ongoing training; and on 38% of the farms apprentices were trained.

Table 9. Sustainability performance on the theme working conditions and related subthemes for the vegetable, dairy, pig and poultry sector (median, min–max).

	Vegetables	Dairy	Pigs	Poultry
7. Working conditions	86 (63–90)	82 (43–95)	77 (46–92)	75 (59–85)
7.1. Personnel management	90 (87–100)	93 (80–100)	93 (80–100)	90 (80–100)
7.2. Working times	80 (29–98)	69 (1–100)	61 (9–87)	58 (17–98)
7.3. Safety at work	98 (95–100)	98 (72–100)	100 (71–100)	92 (61–100)
7.4. Salaries and income level	49 (24–95)	70 (17–100)	74 (22–100)	49 (25–73)

In the subtheme working times, the working time (i.e., hours, days, holidays) of employees and self-employed farm workers are compared to regional reference values. In addition, compensation of overtime is considered. Scores on working time for employees were 93 (39–100), whereas for family members, this was 34 (1–97). A cause for the lower score for family members is a higher number of working hours and working days per week for family workers.

Safety at work covers farm-related incidents (i.e., accidents and illnesses), implementation of a safety strategy, safety of pesticide use and veterinary treatments and child welfare. On 73% of the farms, no occupational accidents or illnesses occurred in the last five years, and 67% had implemented a professional safety strategy.

The subtheme salaries and income level determines the salaries and income level of employees and self-employed workers. It calculates the attractiveness of the hourly wage paid to employees and family members compared to minimum hourly wage. An attractive income in Denmark is set at double the minimum wage. Based on this regional target, an attractive hourly wage for both employees and family members was reached on 8% of the farms.

3.8. Quality of Life

Scores for the theme quality of life are derived from five subthemes. The farmer or farm-worker can define an additional sixth subtheme if he/she considers it important for his/her quality of life. In total, 15 of the 37 assessments included a score on this sixth subtheme further aspects of life. For the other assessments, the score was based on the five subthemes given in Table 10. No difference between sectors was found for the theme quality of life, while for the subtheme occupation and education, a difference was found (Table 10). The voluntary evaluation of the theme quality of life is based on one or more farm workers. The procedure for the assessment of each subtheme in the same; the interviewee is asked to rate the importance (i.e., weight) and his or her satisfaction on aspects related to the subtheme. The importance and satisfaction are both evaluated on a five-level Likert scale. The procedure for each subtheme is similar; therefore, we do not discuss all subthemes in detail, except for the one in which differences were found.

The subtheme occupation and education covers the importance and satisfaction on occupation, education and ongoing training. Scores were lower for vegetable farms compared to dairy and poultry farms. Vegetable farmers were less satisfied with regard to their ongoing training.

Table 10. Sustainability performance on the theme quality of life and related subthemes for the vegetable, dairy, pig and poultry sector (median, min–max).

	Vegetables	Dairy	Pigs	Poultry	p-Value
8. Quality of life	70 (64–91)	83 (32–96)	81 (52–89)	81 (71–92)	0.351
8.1. Occupation and education	71 (54–83) [a]	85 (38–100) [b]	82 (63–88) [a,b]	89 (72–100) [b]	0.047
8.2. Financial situation	81 (71–100)	75 (25–89)	84 (11–89)	75 (55–100)	0.633
8.3. Social relations	88 (64–100)	89 (25–100)	88 (38–100)	89 (70–100)	0.662
8.4. Personal freedom and values	71 (54–83)	78 (38–100)	75 (45–96)	75 (66–100)	0.549
8.5. Health	65 (62–88)	80 (38–89)	84 (63–100)	75 (55–100)	0.304
8.6. Further aspects of life	88 (75–100)	88 (25–100)	88 (60–100)	75 (50–100)	0.981

[a,b] Different superscripts indicate significant differences between sectors ($p < 0.05$).

3.9. Economic Viability

The theme economic viability covers six subthemes (Table 11). No difference was found between sectors for the theme; however, a difference was found between sectors for four out of six subthemes (Table 11). As not all farmers were able or willing to share their economic results, assessment results of 32 farms were included (Table 11).

Table 11. Sustainability performance on the theme economic viability and related subthemes for the vegetable, dairy, pig and poultry sector (median, min–max).

	Vegetables (n = 3)	Dairy (n = 12)	Pigs (n = 8)	Poultry (n = 11)	p-Value
9. Economic viability	65 (57–88)	47 (26–78)	55 (44–71)	69 (31–88)	0.093
9.1. Liquidity reserve	55 (30–100) [a]	62 (13–100) [a]	18 (7–31) [b]	47 (10–100) [a]	0.039
9.2. Level of indebtedness	98 (89–100) [a]	47 (0–97) [b]	75 (58–100) [a,b]	85 (29–99) [a,b]	0.037
9.3. Economic vulnerability	76 (68–76) [a]	76 (55–82) [a]	62 (52–71) [b]	69 (51–77) [a]	0.018
9.4. Livelihood security	13 (0–100)	63 (37–100)	81 (32–100)	50 (24–98)	0.479
9.5. Cash flow–turnover ratio	61 (23–100) [a]	7 (0–46) [b]	34 (0–100) [a]	48 (11–100) [a]	0.010
9.6. Debt service coverage ratio	91 (80–98)	29 (0–100)	66 (0–100)	85 (0–99)	0.131

[a,b] Different superscripts indicate significant differences between sectors ($p < 0.05$).

The score on the subtheme liquidity reserve is calculated by comparing the liquid assets to the total farm expenditure (including private expenditure) and is the average of the examined financial year. The liquidity reserve that can be used to meet financial obligations is expressed in weeks and compared to regional reference values. In Denmark, a minimum liquidity reserve is 15 weeks (33 points); optimal is 25 weeks (66 points); and ideal is 40 weeks (100 points). The liquidity reserve was lower for pig farms compared to vegetables, dairy and poultry farms (Table 11). This could be related to the generally lower economic results for pig farms in 2013 [42].

In the subtheme level of indebtedness, the total borrowed capital (short and long debts) and liquid assets are compared to the operational cash flow to determine the number of years required to pay off debts with the current cash flow. The Danish reference values consider a low level of indebtedness of 10 years (66 points); medium is 20 years (33 points); and a high level is 50 years (zero points). Dairy farms had a higher level of indebtedness (median 16 years) resulting in a lower score compared to vegetable farms (median one year).

Economic vulnerability evaluates the vulnerability of each revenue source (i.e., secondary activities, direct payments and operation branches) based on the farmers' perception of market trends, infrastructure condition and income security. In addition, the main income source (percentage of total business value) is determined to evaluate the concentration risk. Pig farmers evaluated the market trends, infrastructure condition and income security less positive, resulting in lower scores on economic vulnerability compared to vegetable, dairy and poultry farms.

The subtheme livelihood security compares the private household expenditure to the household expenditure target (poverty threshold; basic household needs of a single person or family in Denmark (e.g., food, clothes, health, housing, transport costs)). An income sufficient to meet the basic needs (100%) is awarded 34 points. An attractive income of 200% or more than the basic needs is awarded 100 points. Based on these reference values, 13% of the farmers did not have an income sufficient to meet the basic needs, while 28% of the farmers had an attractive income.

The subtheme cash flow-turnover ratio evaluates the profitability of the farm. It compares the operational cash flow to the business turnover (i.e., farming income, secondary activities and success of financial investments). A cash flow-turnover ratio of 20% results in a score of 33 points, 35% in 66 points and 55% in 100 points. The scores were lower for dairy farms compared to vegetable, pig and poultry farms.

The debt service coverage ratio compares the debt service (interest and mandatory amortization) to the operational cash flow. If the short-term debt service exceeds the operational cash flow (110%) zero points are awarded; a medium debt service ratio of 85% results in 33 points; and a low ratio of 50% in 66 points. On 16% of the farms, the short-term debt service exceeded the operational cash flow.

3.10. Farm Management

The theme farm management includes five subthemes (Table 12). No difference between sectors was found for the theme farm management, while for the subtheme planning instruments and documentation, a difference was found.

Table 12. Sustainability performance on the theme farm management and related subthemes for the vegetable, dairy, pig and poultry sector (median, min–max).

	Vegetables	Dairy	Pigs	Poultry	p-Value
10. Farm management	78 (72–98)	78 (60–96)	83 (70–95)	75 (62–95)	0.274
10.1. Farm strategy and planning	63 (50–100)	75 (25–100)	88 (38–100)	75 (25–88)	0.311
10.2. Supply and yield security	79 (64–100)	86 (71–100)	93 (79–100)	93 (86–100)	0.136
10.3. Planning instruments and documentation	91 (76–98) [a]	95 (78–100) [a,b]	98 (92–98) [b]	91 (69–98) [a]	0.040
10.4. Quality management	90 (83–99)	90 (51–95)	95 (70–96)	83 (65–99)	0.116
10.5. Farm cooperation	67 (50–100)	58 (17–100)	63 (17–92)	33 (17–92)	0.212

[a,b] Different superscripts indicate significant differences between sectors ($p < 0.05$).

In the subtheme farm strategy and planning, a farmer receives points for having a farm strategy and formulated measures for improvements on economic, environmental and social aspects (Box 1). Although the majority of farmers had a farmer strategy (65% yes, 32% "partly"), not all farmers defined short- or medium-term measures for improvements on economic (14% "no"), environmental (16% "no") and social aspects (35% "no").

In the subtheme supply and yield security, a farmer evaluates aspects that negatively affected the supply and yield in the last five years (i.e., shortages in energy, work force, water, nutrients or problems related to marketing, equipment or diseases and pests). The main problems that negatively affected yield and farm income were "diseases, pests, weeds or fungi" (8% "yes", 27% "partly") and "marketing" (8% "yes", 19% "partly").

The subtheme planning instruments and documentation evaluates the use of planning tools and documentation for personnel management and production. It covers documentation on farm performance, personnel management, livestock production, soil management, energy, water, as well as risk insurance and expert consultation. Scores were higher in pig farms compared to poultry farms and vegetable farms as a result of higher scores on personnel management and product (Table 12).

Quality management evaluates the control of product quality, bookkeeping quality, work safety and waste management. Product quality and bookkeeping quality were fully and regularly controlled on all farms. Scores on the subthemes waste management and work safety were lower, as discussed in Sections 3.3 and 3.7.

The score on the subtheme farm cooperation is based on the level of cooperation with other farmers on land use, machines, buildings, work force, collective purchase of inputs and sale of products. Cooperation was mainly focused on sales of products (78% "yes" and 19% "partly"), land use (54% "yes" and 27% "partly") and machines (51% "Yes" and 30% "partly") and least on buildings (49% "no").

3.11. RISE Polygon

To summarize, the median scores on the theme level per sector are presented in the RISE 2.0 polygon (Figure 2). The sustainability performance of vegetable, dairy and pig farms is positive (>67 points) for seven out of 10 themes and in poultry farms for eight themes. The performance on nutrients flows and energy and climate is critical (34–67 points) for all sectors, according to RISE. The performance of vegetable, dairy and pig farms is also critical for the theme economic viability. None of the sectors has problematic (<33) median scores at the theme level, while at the subtheme level, the scores of 6%–11% of the subthemes (dependent on the sector) can be considered problematic. The share of subthemes in the category critical ranges from 20%–31% and in the category positive 63%–70%. The sustainability performance on one theme (water use) and 33% of subthemes differed between sectors.

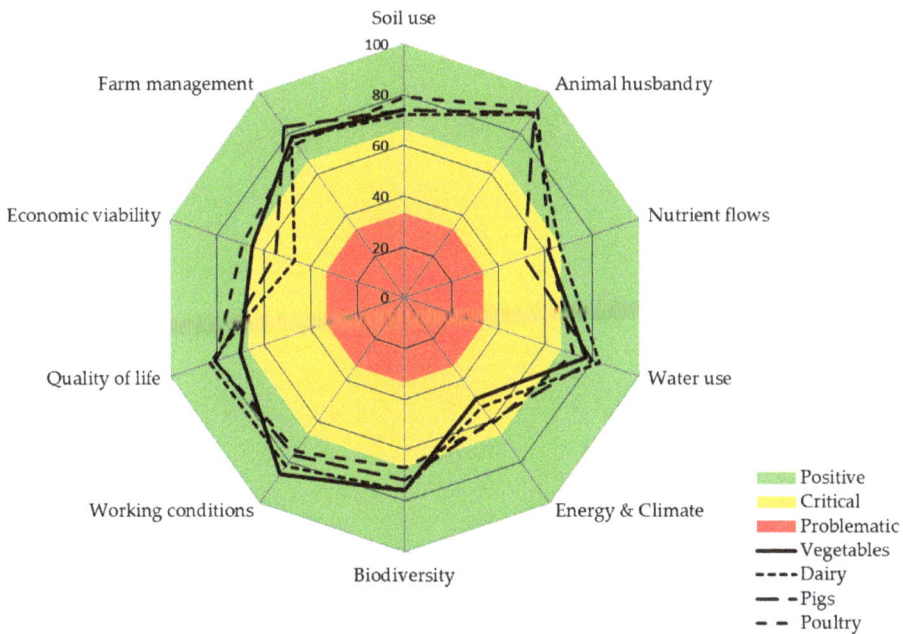

Figure 2. Median sustainability performance at the theme level per sector.

4. Discussion

We used RISE 2.0 to analyze the sustainability performance of organic farms in Denmark and to analyze differences in the performance between sectors. Participation of farms was based on a selection by processing companies and voluntary participation. Generalizations based on this not necessarily representative sample should be taken with care. Next to the sustainability performance of the farms, the results provided insight into the assessment approach of RISE. RISE facilitates the assessment of a wide variety of themes and subthemes and aggregates different types of data: points, quantitative farm data, regional and master data. By allocating points to possible answers, the tool facilitates qualitative data to be included in the assessment. This normalization process is needed to transform the data into units (e.g., 0–100) that can be integrated in the tool and aggregated with other indicators [43,44]. This procedure enables the assessment of issues that tend to be left out of sustainability assessments due to the challenge in assessing and quantifying such information [45]. Transforming such qualitative data into scores, however, requires decisions on the number of answers and the allocation of points to each answer. In the example mentioned in Box 1, the farmer has three possible answers: yes, partly or no, with 100, 50 and 0 points, respectively. This has an influence on the possible score a farmer can obtain. Moreover, for an auditor, it can be challenging to decide when an answer corresponds to, for example, yes or partly [2]. In RISE, 37% of the subthemes are exclusively based on points, and an additional 41% of subthemes use points in addition to other data types. In the interpretation and communication of the results, the role of decisions made in the development of a tool, such as the number of possible answers and allocated points, needs to be acknowledged [46].

The analysis of the assessment results showed differences in the sustainability performance between sectors on subthemes, such as N and P self-sufficiency, ammonia emissions, greenhouse gas emissions, energy intensity and diversity of agricultural production related to differences in farming systems (e.g., presence/absence of livestock). This raises the question of whether the use of a generic approach, without sector-specific comparisons, is valid. A sector-specific comparison would, however, disable a comparison between the performances of sectors. Differences in the performance on the subthemes water management, occupation and education and planning instruments and documentation could not be explained and might be related to the selection of farms and/or auditors involved in the assessment. Due to the relatively high number of auditors involved in this study, their role could not be assessed. Experiences with other tools, however, indicate that the auditor plays a role in the assessment result [2,47].

Next to decisions related to the point-based data and the assessment approach, also decisions made in the selection of indicators, reference values, weights and aggregation methods have an impact on the assessment results. In the selection of reference values, for example, we used Danish reference values that were not specific for organic farming. This had an impact on the scores because of the generally lower productivity in organic farming, as shown by results on crop productivity [11,14]. Whereas the comparison of farm productivity to regional reference values makes the tool more context specific, it also increases the dependency on the quality of regional reference values and reduces the possibility to compare the performance between regions [48].

To summarize the datasets and clarify the assessment results, different aggregation functions are used in sustainability assessment tools [49]. In RISE, the score at the theme level is determined by the average score (arithmetic mean) of the subthemes; as a result, all subthemes become equally important. Although this aggregation allows a quick overview of the sustainability performance, it can result in the compensation of poor scores. For example, in the subtheme crop productivity, the median scores of all sectors are considered as critical, whereas the scores at the theme level (soil use) are positive. Without the consideration of the underlying data, the aggregated scores can lead to simplistic conclusions [44]. At the subtheme level, this type of aggregation is used, as well; however, in several subthemes, a 'risk-based approach' is used, for example in the subtheme personnel management; instead of the average, the minimum score on indicators related to equality and forced labor is taken to reduce the compensation of poor scores.

To select indicators, different criteria can be used, such as sensitivity, precision, affordability and time demand [46,50]. How these criteria are prioritized is dependent on the context and perspectives of those involved in developing the tool [46]. A recent study showed that even amongst sustainability experts, a lack of consensus on what is most important in selecting sustainability indicators can be observed [46]. Once indicators are selected, methods for data collection and calculation are defined. Again, these decisions, for example to express greenhouse gas emissions per hectare instead of per kg product, have an influence and can lead to different conclusions [51]. In addition, the assessment of indicators related to, for example, the subthemes farm management, working conditions and quality of life, farm size is not taken into account, whereas this might play a role. This emphasizes the need for transparency and reflection on the role of decisions and value judgements in sustainability assessment tools and is important to be able to explain their implications for the assessment results [46,52–54].

As the concept of sustainable development is evolving, so are sustainability assessment tools. Some of the comments made in this paper may therefore already be addressed in new versions of the tool and reference values. For example, in the new version of RISE (RISE 3.0, available since 2016), direct input of data from other tools, such as greenhouse gas calculations, is enabled. This could allow more precise data to be entered and reduce time investments; yet, also here, transparency in how these data were calculated and what has been included and excluded is crucial.

Marchand, Debruyne, Triste, Gerrard, Padel and Lauwers [7] identified a continuum from rapid to full sustainability assessments. Rapid sustainability assessment tools are, for example, self-evaluations of farmers based on their knowledge and can be characterized by a limited time investment, low costs, low complexity, high subjectivity, transparency and user-friendliness. Full sustainability assessments are expert-based assessments with high time investment and costs, high complexity and scientifically underpinned output accuracy, with lower transparency and user-friendliness. Given these characteristics, rapid sustainability assessment tools are particularly applicable to raise awareness and interest in sustainability among larger groups of farmers [7]. Full sustainability assessment tools are more focused on monitoring and support farmers interested in sustainability and willing to invest time and money [7]. RISE can be positioned in between these extremes and is confronted with the challenge of balancing in between the characteristics of a rapid and full sustainability assessment tool, for example by combining subjective and more scientifically underpinned data, and combining precise and accurate measurements with user-friendliness and transparency. Similarly, Marchand, Debruyne, Triste, Gerrard, Padel and Lauwers [7] indicated that combining functions (i.e., learning and monitoring) in a sustainability assessment tool can cause tensions.

In a recent study comparing different sustainability assessment tools, RISE was considered by farmers as a relevant tool to gain insight into their sustainability performance [2]. The use of a context-specific approach using regional reference values, the input of quantitative farm data and the user-friendliness were aspects contributing to the perceived relevance. Additional efforts, however, are needed to support farmers in translating the sustainability assessment results into sustainable development at the farm level [2,54]. Further research is needed to evaluate the implementation of the sustainability assessment results and to reflect on the contribution of RISE to the learning and monitoring of sustainability at the farm level.

5. Conclusions

The sustainability performance of 37 organic farms in Denmark was assessed using RISE 2.0. RISE contributes to the diversity of sustainability assessment tools by providing its own perspective on how to assess sustainability. Decisions made in the development of a tool like RISE, such as the selection of themes and indicators, reference values, weights and aggregation methods, influence the assessment results. This emphasizes the need for transparency and reflection on decisions made in sustainability assessment tools. Although all decisions made in the development of RISE can be debated, the outcomes of RISE are a starting point to discuss sustainability at the farm level and to contribute to awareness and learning about sustainability.

Acknowledgments: The sustainability assessments were part of the KØB project (Kompetenceudvikling til Økologisk Bæredygtighed (in English: competence development for organic sustainability) funded by the Danish "Okologifremmeordning", which is an EU-funded scheme aimed at promoting the production and sale of organic products. We would like to thank the farmers and auditors for their participation in the sustainability assessments. We would like to acknowledge Jan Grenz (HAFL) for his constructive feedback on an earlier version of this paper.

Author Contributions: Anke Stubsgaard coordinated the data collection and sustainability assessments on the farms. Evelien M. de Olde analyzed the data and wrote the paper based on discussions with the other authors. All authors participated in reviewing the manuscript and discussing the results.

Conflicts of Interest: The authors declare no conflict of interest.

Abbreviations

The following abbreviations are used in this manuscript:

RISE   Response-Inducing Sustainability Evaluation
LU   Livestock Units

References

1. Food and Agricultural Organization. *Sustainability Assessment of Food and Agriculture Systems (SAFA): Guidelines*, version 3.0; Food and Agricultural Organization of the United Nations: Roma, Italy, 2013.
2. De Olde, E.M.; Oudshoorn, F.W.; Sørensen, C.A.G.; Bokkers, E.A.M.; de Boer, I.J.M. Assessing sustainability at farm-level: Lessons learned from a comparison of tools in practice. *Ecol. Indic.* **2016**, *66*, 391–404.
3. Schader, C.; Grenz, J.; Meier, M.S.; Stolze, M. Scope and precision of sustainability assessment approaches to food systems. *Ecol. Soc.* **2014**, *19*, 42.
4. Waas, T.; Hugé, J.; Block, T.; Wright, T.; Benitez-Capistros, F.; Verbruggen, A. Sustainability assessment and indicators: Tools in a decision-making strategy for sustainable development. *Sustainability* **2014**, *6*, 5512–5534.
5. Pope, J.; Annandale, D.; Morrison-Saunders, A. Conceptualising sustainability assessment. *Environ. Impact Assess. Rev.* **2004**, *24*, 595–616.
6. Schader, C.; Baumgart, L.; Landert, J.; Muller, A.; Ssebunya, B.; Blockeel, J.; Weisshaidinger, R.; Petrasek, R.; Mészáros, D.; Padel, S.; et al. Using the sustainability monitoring and assessment routine (smart) for the systematic analysis of trade-offs and synergies between sustainability dimensions and themes at farm level. *Sustainability* **2016**, *8*, 274.
7. Marchand, F.; Debruyne, L.; Triste, L.; Gerrard, C.; Padel, S.; Lauwers, L. Key characteristics for tool choice in indicator-based sustainability assessment at farm level. *Ecol. Soc.* **2014**, *19*, 46.
8. Coteur, I.; Marchand, F.; Debruyne, L.; Dalemans, F.; Lauwers, L. A framework for guiding sustainability assessment and on-farm strategic decision making. *Environ. Impact Assess. Rev.* **2016**, *60*, 16–23.

9. Häni, F.; Gerber, T.; Stämpfli, A.; Porsche, H.; Thalmann, C.; Studer, C. An evaluation of tea farms in southern India with the sustainability assessment tool rise. In Proceedings of the Symposium ID-105: The First Symposium of the International Forum on Assessing Sustainability in Agriculture (INFASA), Bern, Switzerland, 16 March 2006.

10. International Assessment of Agricultural Knowledge, Science and Technology for Development (IAASTD). *Agriculture at a Crossroads—IAASTD Synthesis Report*; Island Press: Washington, DC, USA, 2009. Availble online: http://www.unep.org/dewa/agassessment/reports/IAASTD/EN/Agriculture%20at%20a%20Crossroads_Synthesis%20Report%20(English).pdf (accessed on 20 September 2016).

11. Seufert, V.; Ramankutty, N.; Foley, J.A. Comparing the yields of organic and conventional agriculture. *Nature* **2012**, *484*, 229–232.

12. Badgley, C.; Moghtader, J.; Quintero, E.; Zakem, E.; Chappell, M.J.; Avilés-Vázquez, K.; Samulon, A.; Perfecto, I. Organic agriculture and the global food supply. *Renew. Agric. Food Syst.* **2007**, *22*, 86–108.

13. Sundrum, A. Organic livestock farming: A critical review. *Livest. Prod. Sci.* **2001**, *67*, 207–215.

14. De Ponti, T.; Rijk, B.; van Ittersum, M.K. The crop yield gap between organic and conventional agriculture. *Agric. Syst.* **2012**, *108*, 1–9.

15. Connor, D.J. Organic agriculture cannot feed the world. *Field Crop. Res.* **2008**, *106*, 187–190.

16. Reganold, J.P.; Wachter, J.M. Organic agriculture in the twenty-first century. *Nat. Plants* **2016**, *2*, 15221.

17. Dourmad, J.Y.; Ryschawy, J.; Trousson, T.; Bonneau, M.; Gonzàlez, J.; Houwers, H.W.J.; Hviid, M.; Zimmer, C.; Nguyen, T.L.T.; Morgensen, L. Evaluating environmental impacts of contrasting pig farming systems with life cycle assessment. *Animal* **2014**, *8*, 2027–2037.

18. Bengtsson, J.; Ahnström, J.; Weibull, A.C. The effects of organic agriculture on biodiversity and abundance: A meta-analysis. *J. Appl. Ecol.* **2005**, *42*, 261–269.

19. Merfield, C.; Manhire, J.; Moller, H.; Rosin, C.; Norton, S.; Carey, P.; Hunt, L.; Reid, J.; Fairweather, J.; Benge, J.; et al. Are organic standards sufficient to ensure sustainable agriculture? Lessons from New Zealand's argos and sustainability dashboard projects. *Sustain. Agric. Res.* **2015**, *4*, 158–172.

20. MacRae, R.J.; Frick, B.; Martin, R.C. Economic and social impacts of organic production systems. *Can. J. Plant Sci.* **2007**, *87*, 1037–1044.

21. Castellini, C.; Bastianoni, S.; Granai, C.; Bosco, A.D.; Brunetti, M. Sustainability of poultry production using the emergy approach: Comparison of conventional and organic rearing systems. *Agric. Ecosyst. Environ.* **2006**, *114*, 343–350.

22. Willer, H.; Lernoud, J. *The World of Organic Agriculture. Statistics and Emerging Trends 2016*; Research Institute of Organic Agriculture (FIBL): Frankfurt, Germany; IFOAM—Organics International: Bonn, Germany, 2016. Avaliable online: https://shop.fibl.org/fileadmin/documents/shop/1698-organic-world-2016.pdf (accessed on 20 September 2016).

23. Wier, M.; O'Doherty Jensen, K.; Andersen, L.M.; Millock, K. The character of demand in mature organic food markets: Great britain and denmark compared. *Food Policy* **2008**, *33*, 406–421.

24. Ministeriet for Fødevarer, Landbrug og Fiskeri. *Økologiplan Danmark. Sammen om Mere Økologi*; Ministeriet for Fødevarer, Landbrug og Fiskeri: Copenhagen, Danmark, 2015.

25. Arbenz, M.; Gould, D.; Stopes, C. *Organic 3.0 for Truly Sustainable Farming & Consumption*; ISOFAR International Organic EXPO 2015; SOAAN & IFOAM—Organics International: Goesan County, Korea, 2015.

26. Häni, F.; Braga, F.; Stämpfli, A.; Keller, T.; Fischer, M.; Porsche, H. RISE, a tool for holistic sustainability assessment at the farm level. *Int. Food Agribus. Manag. Rev.* **2003**, *6*, 78–90.

27. Landbrug & Fødevarer. *Danske Økologer får Papir på Bæredygtighed*; Landbrug & Fødevarer: København, Denmark, 2015.

28. Terkelsen, M. Medlemsbaseret landbrug inddrager bæredygtighed. *Økologi & Erhverv*, 4 March 2016.

29. ORC. Stoas. Avaliable online: http://www.organicresearchcentre.com/?go=IOTA&page=STOAS (accessed on 13 April 2016).

30. Grenz, J. *RISE (Response-Inducing Sustainability Evaluation)*, version 2.0; HAFL: Zollikofen, Switzerland, 2016. Available online: https://www.hafl.bfh.ch/en/research-consulting-services/agricultural-science/sustainability-and-ecosystems/sustainability-assessment/rise.html (accessed on 4 April 2016).

31. Thalmann, C.; Grenz, J. Factors affecting the implementation of measures for improving sustainability on farms following the rise sustainability evaluation. In *Methods and Procedures for Building Sustainable Farming Systems: Application in the European Context*; Marta-Costa, A.A., Soares da Silva, G.E.L.D., Eds.; Springer: Dordrecht, The Netherlands, 2013; pp. 107–121.

32. Grenz, J.; Thalmann, C.; Stämpfli, A.; Studer, C.; Häni, F. RISE—A method for assessing the sustainability of agricultural production at farm level. *Rural Dev. News* **2009**, *1*, 5–9.

33. Urutyan, V.; Thalmann, C. Assessing sustainability at farm level using rise tool: Results from Armenia. In Proceedings of the 2011 International Congress, Zurich, Switzerland, 30 August–2 September 2011.

34. Mobjörk, M. Consulting versus participatory transdisciplinarity: A refined classification of transdisciplinary research. *Futures* **2010**, *42*, 866–873.

35. Popa, F.; Guillermin, M.; Dedeurwaerdere, T. A pragmatist approach to transdisciplinarity in sustainability research: From complex systems theory to reflexive science. *Futures* **2015**, *65*, 45–56.

36. Baumgärtner, S.; Becker, C.; Frank, K.; Müller, B.; Quaas, M. Relating the philosophy and practice of ecological economics: The role of concepts, models, and case studies in inter-and transdisciplinary sustainability research. *Ecol. Econ.* **2008**, *67*, 384–393.

37. ADEME. *Guide des Valeurs Dia'terre*®, version 1.1; ADEME: Paris, France, 2010; pp. 1–130.

38. Ott, R.; Longnecker, M. *An Introduction to Statistical Methods and Data Analysis*; Nelson Education: Scarborough, ON, Canada, 2015.

39. Baarda, D.B.; de Goede, M.P.; van Dijkum, C. *Introduction to Statistics with SPSS: A Guide to the Processing, Analysing and Reporting of (Research) Data*; Wolters-Noordhoff; Taylor & Francis: Oxfordshire, UK, 2004.

40. Zomer, R.J.; Trabucco, A.; Bossio, D.A.; Verchot, L.V. Climate change mitigation: A spatial analysis of global land suitability for clean development mechanism afforestation and reforestation. *Agric. Ecosyst. Environ.* **2008**, *126*, 67–80.

41. Regnskabsstatistik for Økologisk Jordbrug. Available online: http://www.dst.dk/Site/Dst/Udgivelser/nyt/GetPdf.aspx?cid=20040 (accessed on 29 June 2016).

42. Andreoli, M.; Tellarini, V. Farm sustainability evaluation: Methodology and practice. *Agric. Ecosyst. Environ.* **2000**, *77*, 43–52.

43. Gómez-Limón, J.A.; Sanchez-Fernandez, G. Empirical evaluation of agricultural sustainability using composite indicators. *Ecol. Econ.* **2010**, *69*, 1062–1075.

44. Binder, C.R.; Schmid, A.; Steinberger, J.K. Sustainability solution space of the swiss milk value added chain. *Ecol. Econ.* **2012**, *83*, 210–220.

45. De Olde, E.M.; Moller, H.; Marchand, F.; McDowell, R.W.; MacLeod, C.J.; Sautier, M.; Halloy, S.; Barber, A.; Benge, J.; Bockstaller, C.; et al. When experts disagree: The need to rethink indicator selection for assessing sustainability of agriculture. *Environ. Dev. Sustain.* **2016**.

46. Gerrard, C.; Smith, L.G.; Pearce, B.; Padel, S.; Hitchings, R.; Measures, M. Public goods and farming. In *Farming for Food and Water Security*; Volume 10 of the Sustainable Agriculture Reviews; Springer: Dordrecht, The Netherlands; Heidelberg, Germany; New York, NY, USA; London, UK, 2012; pp. 1–22.

47. Gasso, V.; Oudshoorn, F.W.; de Olde, E.; Sørensen, C.A.G. Generic sustainability assessment themes and the role of context: The case of danish maize for german biogas. *Ecol. Indic.* **2015**, *49*, 143–153.

48. Pollesch, N.; Dale, V.H. Applications of aggregation theory to sustainability assessment. *Ecol. Econ.* **2015**, *114*, 117–127.

49. Niemeijer, D.; de Groot, R.S. A conceptual framework for selecting environmental indicator sets. *Ecol. Indic.* **2008**, *8*, 14–25.

50. De Boer, I.J.M. Environmental impact assessment of conventional and organic milk production. *Livest. Prod. Sci.* **2003**, *80*, 69–77.

51. Bell, S.; Morse, S. Breaking through the glass ceiling: Who really cares about sustainability indicators? *Local Environ.* **2001**, *6*, 291–309.

52. Triste, L.; Marchand, F.; Debruyne, L.; Meul, M.; Lauwers, L. Reflection on the development process of a sustainability assessment tool: Learning from a flemish case. *Ecol. Soc.* **2014**, *19*, 47.

53. Thorsøe, M.H.; Alrøe, H.F.; Noe, E. Observing the observers: Uncovering the role of values in research assessments of organic food systems. *Ecol. Soc.* **2014**, *19*, 46.

54. Binder, C.R.; Feola, G.; Steinberger, J.K. Considering the normative, systemic and procedural dimensions in indicator-based sustainability assessments in agriculture. *Environ. Impact Assess. Rev.* **2010**, *30*, 71–81.

Economic and Social Sustainability through Organic Agriculture: Study of the Restructuring of the Citrus Sector in the *"Bajo Andarax"* District (Spain)

Juan Torres, Diego L. Valera, Luis J. Belmonte and Carlos Herrero-Sánchez

Abstract: Over 1000 hectares of citrus fruits crops are grown in the Bajo Andarax district in Almeria (Spain). The withdrawal of EU subsidies for conventional production led to a drastic loss of economic profitability of the holdings and, consequently, the abandonment of most of the conventionally managed farms of the district. In this context, a restructuring of the citrus sector from conventional to organic farming was implemented as a strategic measure to achieve the long-term sustainable development of the holdings. This study examines the citrus sector of the district and performs a comprehensive evaluation of the economic sustainability of this shift from conventional to organic production. In addition, the impact of the restructuring of the sector on the social sustainability both at the farm level and at the municipality level is studied. The results of the study are of interest to other agricultural areas of compromised profitability in which a shift towards organic production can represent a viable alternative for the economic and social sustainability of the holdings.

Reprinted from *Sustainability*. Cite as: Torres, J.; Valera, D.L.; Belmonte, L.J.; Herrero-Sánchez, C. Economic and Social Sustainability through Organic Agriculture: Study of the Restructuring of the Citrus Sector in the *"Bajo Andarax"* District (Spain). *Sustainability* **2016**, *8*, 918.

1. Introduction

The ability of any sector to support a defined level of development is directly linked to the fulfillment of the principles of sustainability [1]. Sustainability is a three-dimensional concept that encompasses economic, environmental, and social aspects [2]. In this context, the shift from conventional to organic farming practices can contribute to the sustainability of those areas that would be otherwise at risk of abandonment, as reported by other authors [3].

Spain has a long tradition in citrus fruit agriculture and has increased its total production in recent years. The country's annual 7 million tons of citrus fruit production (3000 million €) is surpassed only by China, Brazil, the USA, India, and México [4].

In 2015, Spain dedicated 299,518 hectares to citric crops, 7020 of which were farmed organically. Oranges are the most commonly grown citrus fruit in the world, and this also holds true for Spain, where 148,777 hectares were grown last year [5].

The Bajo Andarax district of Almería accounts for 1080 hectares of citric fruit crops, over half of which are dedicated to the so-called "white" varieties of lesser organoleptic quality, which are largely unsuitable for fresh produce but are in great demand for juice (Figure 1). The lack of alternative sources of employment in the area means that the need to maintain these crops is not only a major economic concern, but also a social priority.

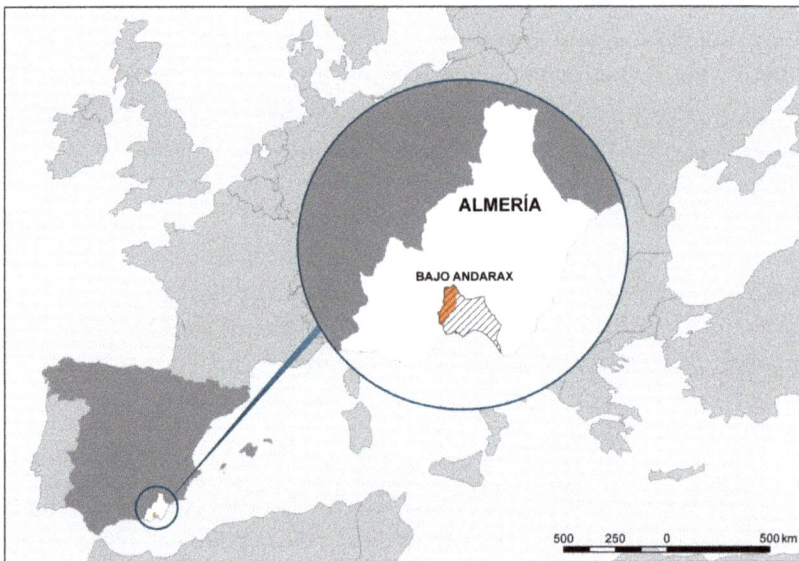

Figure 1. Intensive areas of organic citrus in the Bajo Andarax district (Almería). In red, the municipalities of Santa Fé de Mondújar and Gador, which are the object of the present study.

Most villages in this district enjoy limited options for economic growth, have a very aged population, and suffer from high unemployment rates [6]. From an environmental point of view, citrus crops also play an important role in the semi-arid surrounding landscape, since without them desertification would progress in the area.

Traditionally, citrus farming in the region received economic subsidies for transformation from the former common organization of the market (COM). However, the reform of the common organization of the citrus market led to a drastic change in the economic prospects of the farming sector of the Bajo Andarax district. Indeed, with aid awarded according to the area of cultivated land (350 €/ha) rather than the volume of production, a substantial gap between actual production and the aid received led to a sharp decline in the profitability of the orange plantations, with an overall drop of profitability of 60%. This decrease in profitability, in turn, resulted in the abandonment of the majority of the conventional "white" orange plantations and the socio-economic collapse of the district.

In this context, the Agricultural Processing Society (*Sociedad Agraria de Transformación*, or SAT, by its acronym in Spanish) "SAT Cítricos del Andarax" played a major role in the restructuring of the citrus sector in the Bajo Andarax district. A "*Sociedad Agraria de Transformación*" is a cooperative-type association of independent farmers with both economic and social objectives, and to which the totality of the production of the farmers is allocated. Since the further exploitation of conventional varieties was deemed economically unfeasible, the diversification into higher-value crops was considered the best alternative for the majority of the small and medium-sized farms of the district as a means to guarantee not only the socio-economic feasibility of the holdings, but also the reform of the sector in line with a climate-smart agricultural approach [7].

Purchasing preferences of European consumers have shifted in the last decades towards the consumption of natural products with little presence of chemical agents. In this context, sales of organically farmed produce have climbed to 3% of the total marketed produce from practically null, and are expected to double in the coming years [8]. Organic farming is increasing its share of the world food market and receives growing support from agricultural policies concerned with sustainability [9,10]. Despite the fact that the desire for sustainable agriculture is universal [11], there is no consensus on how to achieve such an ambitious goal [12]. Organic farming has been considered in prior research as an important means to ensure sustainable development [1]. In this context, organic farming is viewed as a means to produce food through the integration of cultural, biological, and mechanical practices aimed at preserving natural resources, biodiversity, animal welfare, and human health [13]. In addition, organic products are greatly appreciated by an increasing share of consumers, who consider them of higher quality, mainly due to the lack of chemical products used during the production or conservation phase, which, in turn, allows a more sustainable and environmentally friendly supply chain [14,15].

Taking into account that organic farming has been shown in other cases to result in higher economic and financial results than conventional farming, due to both reduced labor input and greater market appreciation [3,16–18], SAT Cítricos del Andarax performed a thorough analysis of the legislation regulating organic farming in Europe in order to study the feasibility of a shift of the conventionally-managed farms of the district, which were in a situation of semi-abandonment, towards organic farming. In general terms, organic farming requires the avoidance of GMO and ionizing treatments, as well as of synthetic chemical products (such as pesticides, herbicides, fertilizers, waxes, and preservatives) in the cultivation, handling, and commercialization of produce. In the context of a shift towards organic farming, the SAT plays a major role in regulating the use of numbered labels or seals of quality that certify the produce's organic production, which are awarded after the control by the Organic Farming Committee and its authorized control bodies has taken place.

The application of EU regulations regarding organic farming was a priority for the Bajo Andarax citric plantations. Interestingly, the previous state of semi-abandonment of the plantations due to the economic unfeasibility of the holdings facilitated the shift to organic farming, since it is mandatory in any organic certification scheme to verify that the holdings under consideration have not been subjected to the aggressive use of fertilizers, herbicides, and plant protection treatments.

In the Bajo Andarax district, the municipalities of Santa Fé de Mondújar and Gádor have the highest density of organic citrus farming (90% of the total) of the district. SAT Cítricos del Andarax, in turn, plays a considerable role in the citric sector at the district level. Indeed, SAT Cítricos del Andarax sells over 85% of the district's citrius production, and its associates manage 450 hectares of certified organic farming inside the district and another 240 hectares in other territories.

Figure 2. Organic oranges in the Bajo Andarax district of Almería (Spain).

The present work analyzes the suitability of conversion from conventional to a totally organic production scheme in the Bajo Andarax district of Almería (Figure 2). It also studies the recent changes in marketing focus, from selling "white" varieties to the fruit juice industry to the fresh marketing of organic farm oranges, which provide 40% more return than conventional citrus sales thanks to the increased demand for organically farmed citrus in the EU. This higher demand made it possible for SAT Cítricos del Andarax to start a new business line in organic orange juice, which boosts the added value of the members' crops by using the discards from both fresh market citrus and white varieties for juice.

2. Materials and Methods

Two municipalities were considered in our study, namely, Gádor and Santa Fe de Mondújar. The selection of these municipalities was based on three criteria. First, these municipalities account for more than 85% of the production of citrus in the Bajo Andarax district and are therefore highly representative of the citrus sector in this area. Second, both territories are neighboring municipalities, thus guaranteeing similar agronomic and geoclimatic conditions (Table 1). Finally, SAT Cítricos del Andarax accounts for 66.78% of the employment provided by private companies in these municipalities [19], which, in turn, provides ideal conditions for the study of the impact of the farmers' association on the economic and social sustainability of the territory.

Table 1. Agronomic and geoclimatic frame conditions of the study.

Municipality	Altitude (m)	Surface (km²)	Annual Rainfall (mm)	Climate	Main Crops	Watering Regime
Gádor	166	87.7	249	Continental Mediterranean	Orange	Flood Irrigation
Santa Fe de Mondújar	217	34.9	271			

2.1. Economic Sustainabilty

The analyzed data comprised 44 plantations producing the two main varieties of orange in the area (navelina and castellana) under both conventional and organic cultivation systems (Table 2). Due to the difficulties of carrying out an entirely random sampling of the farms, a stratified sampling that guarantees the validity of the sample [20] was performed according to the number of surveyed plots in the region and their typology.

Table 2. Sample distribution according to type of crops and variety.

Crop Type	Orange Variety	Sampled Surface (ha)	Number of Sampled Farms
Organic farming	Navelina	6.7	13
	Castellana	9.7	14
Conventional farming	Navelina	2.2	4
	Castellana	6.4	13
Total		**25**	**44**

The sample consisted of 44 plots which make up 25 of the 1080 orange crop hectares in the district and featured no newly-built farms, as the aim of the study is to analyze the changes triggered by the restructuring of the sector from the old plantations to organic farming. The sample was initially expected to cover 50 hectares, but only 25 of these were found to include adequate cost management mechanisms. Nonetheless the 44 plots surveyed represent a valid sample of the number of the surveyed plots, with a 12.09% margin of error and a 95% level of confidence. All of the farms under consideration had grown conventional orange crops for at least 10 years prior to the start of the study.

Both conventional and organic farms in this sample use conventional flood irrigation and have an average area of under one hectare. The predominant planting pattern is 6 m × 4 m between trees for the navelina variety and 6 m × 6 m for the castellana variety. All plantations combine "white" varieties of castellana oranges, which are intended for the processing industry, and navelina oranges for the fresh market.

The economic sustainability of the cropping systems has been assessed by means of appropriate indices, as previously employed in the literature [3,16,17]. Table 3 summarizes the indicators of economic sustainability employed in the study.

Table 3. Economic sustainability indicators employed in the study.

Indicator	Measure	Source
Net Present Value Internal Rate of Return Discounted Cost-Benefit Rate	Profitability of the investment	Testa et al. [3] Sgroi et al. [16] Sgroi et al. [17]
Discounted Pay-Back Time	Return period of an investment	

An economic analysis was performed in order to determine the Net Present Value (NPV), the Internal Rate of Return (IRR), the Discounted Cost-Benefit Rate (DCBR), and the Discounted Pay-Back Time (DPBT), in accordance with the methodology proposed by Sgroi et al. [16].

The Net Present Value (NPV) was calculated by the difference between the discounted gross income values generated during the investment life of the project or investment and the corresponding fixed costs [21] by means of the following formula:

$$NPV = \sum_{I=0}^{n} \frac{GI_i - FC_i}{(1+r)^i} \qquad (1)$$

where GI represents the gross income, FC are the fixed costs, n corresponds to the lifetime of the investment, and i and r are the year under consideration and the discount rate, respectively. In this formula, GI is calculated as the difference between gross production value and variable costs. In our study, the lifetime of the investment was 25 years and the discount rate is set to 5%, considering market conditions. By employing this criterion, an investment is deemed convenient if the NPV is positive; in the case of two alternative investment projects, the one providing the highest NPV is to be chosen [16,22].

The Internal Rate of Return (IRR) is the discount rate at which NPV equals zero, i.e., the discount rate at which the discounted benefits are equal to the discounted costs [16]. By using this criterion, an investment is deemed convenient if its IRR exceeds the chosen alternative discount rate [23].

In addition, the Discounted Cost-Benefit Rate (DCBR) was calculated to assess the economic sustainability of the cropping systems. The DCBR is defined as the ratio between the discounted gross income values generated during the investment life and the corresponding fixed costs. The following formula is employed to calculate the DCBR:

$$DCBR = \frac{\sum_{I=0}^{n} \frac{GI_i}{(1+r)^i}}{\sum_{I=0}^{n} \frac{FC_i}{(1+r)^i}} \qquad (2)$$

According to this economic indicator, a ratio greater than 1 reveals a financially convenient investment [24] since the sum of the gross revenue provided by the investment exceeds the sum of the fixed costs.

Finally, the economic indicator DPBT has been employed in the study. DPBT corresponds to the number of years for which the sum of the discounted gross income equals the sum of the fixed costs [25].

In order to determine these indicators, an analysis of the information from the representative sample of plots in the study was performed by identifying the structure of costs and revenues of each farm. To this end, the structure and quantification of costs, income, and timeframe based on the methodology proposed by Juliá and Server [26] was employed.

Income was defined as the average settlement price of conventional and organically farmed navelina and castellana varieties over the last two years, for which internal price data of SAT Cítricos del Andarax was used.

The timeframe to analyze the profitability of both organic and conventional farms was set at 25 years. All farms were farmed by using conventional farming methods for the first 11 years (years 1 to 11). Then, a two-year period was established for the conversion from conventional to organic farming (years 12 and 13); during this period, the farms were adapted to meet the administrative requirements for the certification of organic production, which was then obtained at the end of year 13. Finally, the farms were completely managed with organic production methods during the last 12 years of the study (years 14 to 25).

2.2. Social Sustainability

A substantial body of research has been developed in the last years with regard to the environmental and economic dimensions of sustainability. However, less attention has been paid in the literature to the social dimension of sustainability [2]. In addition, literature devoted to social sustainability is highly focused on specific research contexts, thus hindering the attainment of an integrative, all-encompassing framework of social sustainability [27].

Social sustainability was assessed in our study by selecting a number of indicators proposed in the literature for which relevant quantitative data was available. Table 4 summarizes the indicators of social sustainability employed in the study. The differentiation between internal (i.e., at organization level) and external (i.e., at the territory level) social sustainability dimensions proposed by Van Calker et al. [28] was employed as a first classification criterion. Farm-level data was collected and analyzed to evaluate those indicators related to internal social sustainability, whereas municipality-level data was employed for the assessment of external social sustainability.

The study of the impact of the restructuring of the sector led by SAT Cítricos del Andarax on the social sustainability of the municipalities was possible due to the very high degree of interdependence between the farmers' association and the socioeconomic conditions of the territory since the association accounts for 66.78% of the employment of private companies in these municipalities [19].

In order to determine the impact of the restructuring of the sector on social sustainability, the evolution of these indicators was assessed during a 10-year period between 2001, the last year in which conventional farming was practiced, and 2011, a representative year of full organic production for which statistical data was available. This methodology allowed a direct comparison between the indicators of social sustainability in the period of conventional production and those obtained during organic-only production.

Table 4. Social sustainability indicators employed in the study.

Scope	Indicator	Source	Measure
Internal social sustainability	Educational attainment	Dillon et al. [29]	Increase of the percentage of qualified personnel in the association
	Employment creation	Manara and Zabaniotou [30]	Increase of the number of workers in the association
	In-house training	Amaral and La Rovere [31] Veldhuizen et al. [32]	Increase of the number of on-the-job training hours per worker and year
	Workforce gender balance	Mani et al. [2]	Increase of the percentage of female personnel in the association
External social sustainability	Employment	Amaral and La Rovere [31]	Increase of unemployment rate
	Education level	Weingaertner and Moberg [27] Amaral and La Rovere [31]	Increase of the proportion of population with secondary or tertiary education

Data for the assessment of the internal social sustainability was obtained from the historical record of SAT Cítricos del Andarax. In addition, the evolution of the social sustainability indicators of the farmers' association during the period of study was compared to the evolution of the same indicators in the two immediate geographic aggregation levels, i.e., the province of Almería and the region of Andalusia. This provided a valuable comparison with the reference territories and allowed the drawing of meaningful conclusions with regard to the evolution of the social sustainability indicators in other reference territories in which the farmers' association had no influence.

Statistical datasets for these territories were obtained from the Multi-territory Information System of Andalusia (*Sistema de Información Multiterritorial de Andalucía*, SIMA, by its acronym in Spanish) published by the Andalusian Institute of Statistics and Cartography of the Regional Government of Andalusia [33]. Unemployment rates at the regional and provincial levels were obtained from the historical series of the National Institute of Statistics (*Instituto Nacional de Estadística*, INE, by its acronym in Spanish) of the Spanish Ministry of Economy and Competitiveness [34].

Among the internal social sustainability indicators, education attainment was assessed as the increase during the period of study of the percentage of qualified personnel, defined as the proportion of personnel with secondary or tertiary education in the association. This result was then compared to the same measure in the province of Almería and in the region of Andalusia. As a further indicator of internal social sustainability, employment creation was determined as the increase of the number of workers in the organization over the period of study. This, in turn, was compared to the same measure in the two immediate geographic aggregation levels, i.e., at the provincial and regional level. In addition, in-house training was evaluated as the increase of the number of on-the-job training hours per worker and year in the association, for which the most recent data until 2015 could be used. No statistical data was available for this indicator at the provincial and regional level. Finally, workforce gender balance was calculated as the percentage of female personnel in the association, which was then compared to the same measure at the two immediate geographic levels.

External social sustainability was assessed by evaluating the evolution of the employment and the education level between 2001, the last year of conventional production, and 2011, a representative year of full organic production. In order to evaluate the evolution of the employment in the municipalities under consideration, statistical datasets of the evolution of the unemployment rate during the period of study were processed and compared to the evolution of the unemployment rate at both the provincial and regional level. Finally, the education level was evaluated by determining the increase of the proportion of the population with secondary or tertiary education in the study area during the period from 2001 to 2011 and comparing it with the evolution of the same measure at the provincial and regional levels.

3. Results

3.1. Economic Sustainability

3.1.1. Cultivation Costs

The cost structure analysis of the sample shows that the average cultivation costs in the region are lower than those of other areas [9,35]. Despite this relative cost advantage, low selling prices and the lack of economic profitability after the withdrawal of the public subsidies resulted in the abandonment of the conventional citrus farms of the region.

Table 5 shows the cost structure of the surveyed plantations with detail of the actual costs, yields, and income as obtained from the internal datasets of SAT Cítricos del Andarax in the surveyed farms. To this end, the cost structure proposed by Caballero, de Miguel and Juliá [36] has been employed and adapted to represent the results for the two varieties under consideration. Marked differences can be observed for both fixed and variable cost structures.

Regarding variable costs, we should highlight the lower annual cost of conventional crops (2185 €/ha) compared to organic crops (4147 €/ha for navelina and 3470 €/ha for castellana). The variable cost of conventional farming is 47% and 37% lower than the organic farming of navelina and castellana varieties, respectively. As regards the type of cost, conventional farming proves to be more economical in terms of variable costs, especially with respect to irrigation and fertilizers. This is partly due to the fact that organic fertilizers are more expensive than conventional synthetic ones. Moreover, no phytosanitary products were used under conventional management practices due to the state of semi-abandonment of the surveyed plots.

On the other hand, plant-health treatments in organic farming are limited to mineral oils (mostly used in navelina crops destined for fresh consumption) and diammonium phosphate (used in fly traps).

Table 5. Cost structure of the orange crop in the Bajo Andarax (€/ha).

	Conventional Crops				Organic Crops			
	Navelina (Processing)	%	Castellana (Processing)	%	Navelina (Fresh Market)	%	Castellana (Processing)	%
Irrigation	489	18	489	18	794	15	794	18
Fertilizers	284	10	284	11	728	14	512	11
Phytosanitary products	-	-	-	-	337	6	120	3
Cropping practices	468	17	468	18	761	14	761	17
Workforce	893	33	893	34	1.464	27	1.220	27
Others	51	2	51	2	63	1	63	1
Variable Costs (Total)	2185	81	2185	82	4147	77	3470	77
Annual Cost of Working Capital	64	2	64	2	121	2	101	2
Plantation repayment instalments	91	3	54	2	91	2	54	1
Interest on the plantation	57	2	34	1	57	1	34	1
Repayment instalments on the capital facilities	38	1	38	1	38	1	38	1
Interest on the capital of the facilities	24	1	24	1	24	0	24	1
Costs of replacement and maintenance of the facilities	65	2	65	2	65	1	65	1
Rent of the land, taxes, and others	187	7	187	7	587	11	487	11
Quality certifications	-	-	-	-	225	4	225	5
Fixed Costs (Total)	462	17	402	15	1087	20	927	21
Total Annual Costs	2711 €/ha	100%	2651 €/ha	100%	5355 €/ha	100%	4498 €/ha	100%

Source: Internal data of SAT Cítricos del Andarax based on the cost structure proposed by Caballero, de Miguel and Juliá [36].

The remaining cost factors covering cropping practices and labor are higher for organic crops due to the special attention they require, in particular the navelina variety which is intended for fresh consumption.

In order to determine the annual cost of working capital, we have considered the volume of variable costs to be financed according to crop type and variety. We have assumed an average interest of 5% for seasonal loans and an average reimbursement period of seven months.

The fixed costs shown in this table are annual and do not depend on production volume. Such is the case of farming costs, which cover grafting, removal, or substitution of orange trees and associated labor costs. They also include investment in new irrigation channels or the maintenance of existing channels, average annual renting of the plantations (which varies depending on farming system and orange variety), and payment of immovable property tax.

3.1.2. Profitability and Expected Income

Table 6 illustrates the returns by crop type and variety as obtained from the actual results from surveyed plots during the period of study. Farms yielded similar volumes of produce during the transition and organic-only periods, despite the lack of synthetic fertilizers of the latter, due to the fact that the trees in the surveyed farms had reached their maturity and therefore full production capacity.

Table 6. Returns by crop type and variety by age of the plantation.

Variety: Navelina					
Period (Years)	Production Method	Target Market	Weighted Average Annual Production (kg/ha)	Weighted Average Annual Price (€/kg)	Weighted Average Annual Returns [1] (€/ha)
1 to 11	Conventional	Processing industry	27,038	0.0605	1636
12 to 13	Transition	Processing industry	33,200	0.0605	2009
14 to 25	Organic	Fresh market	33,200	0.2575	8549
Variety: Castellana					
Period (Years)	Production Method	Target Market	Weighted Average Annual Production (kg/ha)	Weighted Average Annual Price (€/kg)	Weighted Average Annual Returns [1] (€/ha)
1 to 11	Conventional	Processing industry	39,331	0.0605	2379
12 to 13	Transition	Processing industry	42,200	0.0605	2553
14 to 25	Organic	Processing industry	42,200	0.1265	5338

[1] Deducting harvesting costs, transportation, fees, and the SAT operational program.

Under conventional crop management, both navelina and castellana varieties are intended for the processing market, whereas the navelina variety managed under organic management techniques is targeted to the fresh market. Prices of those varieties destined for the organic market, both as fresh and processed products, is higher than the equivalent varieties for the conventional market.

Regarding sales-generated income, the price per kilogram is the same for organic and conventionally farmed oranges due to the fact that all farms grow conventionally for the first 11 years under consideration. After the shift to organic farming, significant differences in sales prices are observed.

In the case of organic farming, we have considered other income from aid and subsidies, which is usually geared towards the improvement of the quality of facilities, investigation, and counseling for producers. Such aid was specifically intended for organic production and usually had a maximum validity of five years, though in some cases this period could be extended. There are two kinds of financial aid. On the one hand, agro-environmental measures regulated by the decree of 24 March 2011, BOJA of the Junta de Andalucía under Regulation (CE) 1698/2005 [37], which grants payments of 510.40 €/ha for the first three years after shifting to organic production and 459.36 €/ha for the following two years. The second is the subsidy covering 80% of the costs of registration and renovation with organic produce Control Organisms, limited to a maximum of 3000 € over the five years of financial aid.

3.1.3. Financial Analysis of Conventional and Organic Farming

After determining the costs and income structure of the farms, their profitability was analyzed. It must be remembered that the aids and subsidies granted to organic farming clearly benefit its financial analysis. Bearing in mind that the subsidy is not a regular source of income, the analysis included current expiration dates of the aid, but it is possible that this institutional support to organic agriculture will continue in the future. However, this analysis includes neither direct aid that certified organic producers receive nor additional aid that all producers receive due to the mismatches in aid to production generated by the new common organization of the market (COM).

Table 7 summarizes the profitability indicators of conventional and organic farming according to variety (navelina or castellana). This analysis reveals better results during the period of organic farming, both for its fresh market variety (navelina) and its fruit juice industry variety (castellana). This result is consistent with other studies of the sector [16,38].

Table 7. Profitability indicators by crop type and variety.

	NPV		IRR		DCBR		DPBT	
	Organic	Conventional	Organic	Conventional	Organic	Conventional	Organic	Conventional
Navelina	12,024	−10,590	11%	-	1.32	0.77	16.46	-
Castellana	5222	−555	13%	2%	1.17	1.05	14.98	-

Specifically, we have obtained positive NPV values only in the organic varieties. 12,024 €/ha in case of navelina and 5222 €/ha in case of castellana variety. The analysis of the IRR provided returns of 11% and 13% for navelina and castellana, respectively. Moreover, growers who have opted for organic production have recovered their investment in 16.46 years, in the case of navelina, and 14.98 years for castellana.

Certain guidelines can be recommended to improve the profitability of organic plantations of the fresh market varieties in the Bajo Andarax district. One would be to bring the shift to organic methods forward to the third year of cultivation in new farms, once the seedling has developed, in order to recoup the investment sooner. It is also necessary to improve the quality of the fruit in all plantations, even at the expense of incrementing production costs in order to reduce the quantity of discarded produce, which is decisive to the producers' price settlement. Finally, given these results it is also advisable to valorize the discarded produce for organic fruit juice, thus increasing the added value of the varieties.

Finally, these results confirm that the restructuring to organic farming can be an economically sound alternative that can guarantee the economic sustainability of the holdings for those agricultural areas in which citrus production is still managed under conventional production schemes and the plantations are fully or partially abandoned due to the lack of economic profitability. This is the case, for example, of the Lecrín Valley in Granada, the Guadalquivir Valley in Seville and Cordoba, and the citrus-farming areas of the Almanzora Valley in Almería. The farms in these areas are similar in their varieties and crops to those analyzed in this study, and are also considered to be low-intensity agriculture. However, the shift to organic production would not be easily implemented in high-yield conventional holdings of the Spanish Levante regions (mainly Valencia and Murcia), where conventional farming is a more profitable alternative due to the production structure of the holdings and the use of conventional varieties of higher yield.

3.2. Social Sustainability

Social sustainability was assessed by differentiating between internal (i.e., at farm level) and external (i.e., at municipality level) sustainability. Table 8 summarizes the measures of internal social sustainability employed in the study.

Table 8. Internal social sustainability measures employed in the study.

Measure	Period	SAT Cítricos del Andarax	Geographic Reference Areas	
			Province of Almeria	Region of Andalusia
Increase of the percentage of qualified personnel in the SAT	2001–2011	−20.21%	14.04%	12.99%
Increase of the number of workers in the SAT	2001–2011	1168.75%	7.28%	6.99%
Increase of the number of on-the-job training hours per worker	2001–2015	20.23%	Not applicable	
Increase of the percentage of female personnel	2001–2011	6.47%	7.67%	8.62%

Educational attainment was assessed as an indicator of internal social sustainability by calculating the increase of qualified personnel in SAT Cítricos del Andarax during the period from 2001 to 2011. A decrease of 20.21% in the ratio of qualified personnel was observed during this period, thus indicating an average annual decrease of 2.02%. This significant decrease in the qualification profile of the farmers' association can be explained by the fact that most of the new employment took place in the areas of harvesting and processing, where qualified personnel are less prevalent. This decrease in the qualification profile of the association is in sharp contrast to the evolution of the qualification level in the in the two immediate reference geographic levels, i.e., the province of Almería and the region of Andalusia, where an increase of 14.04% and 12.99%, respectively, was registered for the same indicator in the period from 2001 to 2011.

As a further indicator of internal social sustainability, employment creation by the farmers' association was also evaluated in the study. In the period from 2001 to 2011, the number of workers in the association increased from 16 to 203, thus resulting in a total increase of 1168.75% and an average annual increase of 116.87% during this period. This drastic increase in employment was due to the rapid and consistent rise in the turnover of the association as a result of the conversion from low-yield conventional production to organic production. Indeed, the turnover of SAT Cítricos del Andarax increased by 495.18% during the period from 2001 to 2011. When considered in relation to the increase of employment during the same period in the province of Almería and in the region of Andalusia, it becomes apparent that the increase at the SAT surpassed more than significantly the increase in both territorial domains: the total increase of employment in the province of Almería amounted to 7.28% over the same period (average annual increase of 0.52%), whereas the total increase and the average annual increase of employment in Andalusia during the same period was 6.99% and 0.50%, respectively.

In a similar vein, professional training was also evaluated as a measure of internal social sustainability. To this end, the total number of hours of on-the-job training at SAT Cítricos del Andarax was computed and then the average number of training hours per worker and year were calculated. The results show a total increase of 20.23% in the period from 2001 to 2015 (average annual increase of 1.44%). This rise in the workforce training can be explained by the increasing regulatory qualification requirements in agricultural holdings, especially in the fields of work safety and quality assurance.

As an additional internal social sustainability measure, workforce gender balance in the association was also assessed. To this end, the increase in female personnel was calculated. In this case, an increase of 6.47% was observed in the period from 2001 to 2011. As in the case of educational attainment, this can be explained by the fact that the areas in which most of the new employment was created during this period are those in which traditionally mostly female personnel are hired. This result is in line with the increase in female occupation ratios during the period from 2001 to 2011 in the province of Almeria and in Andalusia, of 7.67% and 8.62%, respectively.

Lastly, two indicators of external social sustainability were considered: employment and education level. Table 9 summarizes the results of the external social sustainability indicators employed in the study.

Table 9. External social sustainability measures employed in the study.

Measure	Period	Municipalities of Study in the Bajo Andarax	Geographic Reference Areas	
			Province of Almería	Region of Andalusia
Increase of unemployment rate	2001–2011	−0.38%	21.14%	10.98%
Increase of the proportion of population with secondary or tertiary education	2001–2011	13.94%	14.04%	12.99%

As measures of external social sustainability, both measures were determined in those territories in which the farmers' association has a high social impact, i.e., in the municipalities of Gádor and Santa Fe de Mondújar. Indeed, the high proportion of employment directly accountable to the SAT in these municipalities results in a high degree of interdependence between the association and the social conditions in the territory.

Firstly, the increase of the unemployment rate in the studied municipalities was assessed. The analysis of the statistical data for these municipalities shows a decrease of 0.38% over the 10-year period from 2001 to 2011, i.e., an average annual decrease of 0.04%. This figure has to be put into the context by comparing it with the results of the same measure in the two reference territorial domains of Almería and Andalusia. Indeed, the increase of the unemployment rate in the same period amounted to 21.14% and 10.98% in the province of Almería and in the region of Andalusia, respectively. Hence, the municipalities under consideration have been able to counteract to a large extent the more than significant increase in unemployment rate in the immediate reference territories during the most severe years of the last financial crisis. Since the SAT is the largest employer in these municipalities and, as previously discussed, employment creation in the SAT during the same period has increased by 1168.75%, it becomes apparent that this considerable increase of employment in the farmers' association during this period has had a favorable social impact in the municipalities in terms of a reversal of the rise of the unemployment rate that has been experienced in other territories.

Finally, education level was studied as a measure of social sustainability. To this end, the increase of the population with primary or secondary education in the two municipalities under consideration was studied during the period of 2001 to 2011. An increase of 13.94% was registered during this period in the municipalities, i.e., an average annual increase of 1.39%. This result, in turn, has to be considered in relation to the same measure in the reference territorial aggregation levels of the province of Almería and Andalusia. Indeed, the population with primary or secondary education in the province of Almería and Andalusia in the same period increased by 14.04% and 12.99%, respectively, during the same period. We can therefore conclude that a similar variation has taken place in all three territorial domains.

4. Conclusions

The profitability of organic farming in the area is higher than that of conventional farming in both of the orange varieties under consideration. Crop production costs reveal the need for a high sale price for the farms to be profitable, and organic varieties reach higher prices than the conventional varieties: sale prices of organically grown navelina and castellana oranges are 425% and 209% higher, respectively, than those of their conventionally grown counterparts. Cultivation costs are 98% higher in organic navelina farming and 70% higher in organic castellana farming, mostly due to the fact that conventional orange farms are in a state of semi-abandonment, which brings down cultivation costs of the conventional varieties.

From a social sustainability perspective, the restructuring of the citrus sector in the Bajo Andarax district has resulted in a notable improvement of the employment indicators both at the farm level and at the municipality level in comparison to the reference territories of the province of Almería and the region of Andalusia. This improvement, however, has not resulted in an increase of the qualification level of the workforce of the farmers' association in comparison to the reference territories. Moreover, no significant differences have been found in terms of workforce gender equality in the association and of education level in the municipalities in comparison to the reference territories.

Future work will focus on the study of environmental sustainability as a result of the shift from conventional to organic production in the sector, thus complementing the results of this paper.

Acknowledgments: The authors wish to express their gratitude to SAT Cítricos del Andarax and the Research Centre CIAIMBITAL of the University of Almería (Spain), for all the support provided.

Author Contributions: This work is the result of the collaboration between all authors. All authors have equally contributed, reviewed, and improved the manuscript. All authors have revised and approved the final manuscript.

Conflicts of Interest: The authors declare no conflict of interest.

References

1. Aceleanu, M.I. Sustainability and competitiveness of Romanian farms through organic agriculture. *Sustainability* **2016**, *8*, 245.
2. Mani, V.; Agarwal, R.; Gunasekaran, A.; Papadopoulos, T.; Dubey, R.; Childe, S. Social sustainability in the supply chain: Construct development and measurement validation. *Ecol. Indic.* **2016**, *71*, 270–279.

3.	Testa, R.; Foderà, M.; Trapani, A.M.; Tudisca, S.; Sgroi, F. Choice between alternative investments in agriculture: The role of organic farming to avoid the abandonment of rural areas. *Ecol. Eng.* **2015**, *83*, 227–232.

4.	Food and Agriculture Organization of the United Nations (FAO). Available online: http://faostat3.fao.org/home/E (accessed on 28 May 2016).

5.	Ministerio de Agricultura, Alimentación y Medio Ambiente (MAGRAMA). Gobierno de España. Encuesta Sobre Superficies y Rendimientos de Cultivos. Available online: http://www.magrama.gob.es/es/estadistica/temas/estadisticas-agrarias/agricultura/esyrce/ (accessed on 28 May 2016). (In Spanish)

6.	Belmonte, L.J.; León, J.M. El Mercado de Trabajo en Almería: 1980–2004. In *La Economía de la Provincia de Almería*; Instituto de Estudios de Cajamar: Almería, Spain, 2005; pp. 533–560. (In Spanish)

7.	Tzouramani, I.; Liontakis, A.; Sintori, A.; Alexopoulos, G. Assessing organic cherry farmers' strategies under different policy options. *Mod. Econ.* **2014**, *5*, 313–323.

8.	Ministerio de Agricultura, Alimentación y Medio Ambiente (MAGRAMA). Gobierno de España. Agricultura Ecológica Estadísticas 2013. Available online: http://www.magrama.gob.es/es/alimentacion/temas/la-agricultura-ecologica/Estadisticas_AE_2013_tcm7-351187.pdf (accessed on 28 May 2016). (In Spanish)

9.	Beltran-Esteve, A.J.; Picazo-Tadeo, J.; Reig-Martínez, E. What makes a citrus farmer go "organic"? Empirical evidence from Spanish citrus farming. *Span. J. Agric. Res.* **2012**, *10*, 901–910.

10.	Alonso, A.M.; Gonzalez, R.; Foraster, L. Comparación Económica Entre Cultivos Ecológicos y Convencionales. In Proceedings of the Actas del VIII Congreso de la Sociedad Española de Agricultura Ecológica, Bullas, Spain, 16–20 September 2008; p. 11. (In Spanish)

11.	Domínguez, A. *La Citricultura Ecológica*; Servicio de Asesoramiento a los Agricultores y Ganaderos, Dirección General de Agricultura Ecológica, Consejería de Agricultura y Pesca de la Junta de Andalucía: Sevilla, Spain, 2008. (In Spanish)

12.	Rigby, D.; Cáceres, D. Organic farming and the sustainability of agricultural systems. *Agric. Syst.* **2001**, *68*, 21–40.

13.	Lombardo, L.; Zelasco, S. Biotech approaches to overcome the limitations of using transgenic plants in organic farming. *Sustainability* **2016**, *8*, 497.

14.	Govindan, K.; Azevedo, S.G.; Carvalho, H.; Cruz-Machado, V. Impact of supply chain management practices on sustainability. *J. Clean. Prod.* **2014**, *80*, 119–138.

15.	Mota, F.; Gomes, M.I.; Carvalho, A.; Barbosa-Povoa, A.P. Towards supply chain sustainability: Economic, environmental and social design and planning. *J. Clean. Prod.* **2015**, *105*, 14–27.

16.	Sgroi, F.; Candela, M.; Trapani, A.M.; Forderà, M.; Squatrito, R.; Testa, R.; Tudisca, S. Economic and financial comparison between organic and conventional farming in Sicilian lemon orchards. *Sustainability* **2015**, *7*, 947–961.

17. Sgroi, F.; Foderà, M.; Trapani, A.M.; Tudisca, S.; Testa, R. Cost-benefit analysis: A comparison between conventional and organic olive growing in the Mediterranean area. *Ecol. Eng.* **2015**, *82*, 542–546.

18. Liontakis, A.; Tzouramani, I. Economic sustainability of organic *Aloe Vera* farming in Greece under risk and uncertainty. *Sustainability* **2016**, *8*, 338.

19. Sistema de Análisis de Balances Ibéricos (SABI). Available online: https://sabi.bvdinfo.com/ (accessed on 27 July 2016). (In Spanish)

20. Calatrava, J.; Sayadi, S. Permanencia de la actividad agraria y políticas de desarrollo rural: Un análisis a partir de un seguimiento (1981–2001) a explotaciones agrarias en zonas de montaña del sureste español. *Rev. Estud. Agrosoc. Pesq.* **2004**, *204*, 207–218. (In Spanish)

21. Prestamburgo, M.; Sccomandi, V. (Eds.) *Economia Agraria*; Etaslibri: Milano, Italy, 1995. (In Italian)

22. Tudisca, S.; Di Trapani, A.M.; Sgroi, F.; Testa, R.; Squatrito, R. Economic analysis of PV systems on buildings in Sicilian farms. *Renew. Sustain. Energy Rev.* **2013**, *28*, 691–701.

23. Kelleher, J.C.; MacCormack, J.J. Internal Rate of Return: A Cautionary Tale. Available online: http://www.cfo.com/printable/article.cfm/3304945 (accessed on 29 June 2016).

24. Zunino, A.; Borget, A.; Schultz, A.A. The integration of benefit-cost ratio and strategic cost management: The use on a public institution. *Espacios* **2012**, *33*, 1–2.

25. Bedecarratz, P.C.; López, D.A.; López, B.A.; Mora, O.A. Economic feasibility of aquaculture of the giant barnacle *Austromegabalanus psittacus* in southern Chile. *J. Shellfish Res.* **2011**, *30*, 147–157.

26. Julià, J.F.; Server, R.J. Evaluación Económico-Financiera de los Sistemas de Cultivo en Cítricos Biológicos (Orgánicos) versus Convencionales. Depósito de Documentos de la FAO. Available online: http://www.fao.org/3/a-y2746s.pdf (accessed on 29 June 2016). (In Spanish)

27. Weingaertner, C.; Mober, A. Exploring social sustainability: Learning from perspectives on urban development and companies and products. *Sustain. Dev.* **2014**, *22*, 122–133.

28. Van Calker, K.; Berentsen, P.; Giese, G.; Huirne, R. Identifying and ranking attributes that determine sustainability in Dutch dairy farming. *Agric. Hum. Values* **2005**, *22*, 53–63.

29. Dillon, E.; Hennessy, T.; Buckley, C.; Donnellan, T.; Hanrahan, K.; Moran, B.; Ryan, M. Measuring progress in agricultural sustainability to support policy-making. *Int. J. Agric. Sustain.* **2016**, *14*, 31–44.

30. Manara, P.; Zabaniotou, A. Indicator-based economic, environmental, and social sustainability assessment of a small gasification bioenergy system fuelled with food processing residues from the Mediterranean agro-industrial sector. *Sustain. Energy Technol. Assess.* **2014**, *8*, 159–171.

31. Amaral, S.; La Rovere, E. Indicators to evaluate environmental, social, and economic sustainability: A proposal for the Brazilian oil industry. *Oil Gas J.* **2003**, *101*, 30–35.

32. Veldhuizen, L.; Berentsen, P.; Bokkers, E.; de Boer, I. A method to assess social sustainability of capture fisheries: An application to a Norwegian trawler. *Environ. Impact Assess. Rev.* **2015**, *53*, 31–39.

33. Sistema de Información Multiterritorial de Andalucía (SIMA). Pabellón de Nueva Zelanda. Available online: http://www.juntadeandalucia.es/institutodeestadisticaycartografia/sima/index2.htm (accessed on 27 July 2016). (In Spanish)

34. Instituto Nacional de Estadística (INE). Available online: http://www.ine.es/inebmenu/indice.htm (accessed on 27 July 2016). (In Spanish)

35. Peris, E. Estudio de las diferencias de costes de producción de cultivo del naranjo convencional, ecológico o integrado en la Comunidad Valenciana mediante el análisis vectorial discriminante. *Econ. Agrar. Recur. Nat.* **2005**, *5*, 69–87. (In Spanish)

36. Caballero, P.; de Miguel, M.D.; Juliá, J.F. *Costes y Precios en Horto-Fruticultura*; Mundi-Prensa: Madrid, Spain, 1992; p. 768. (In Spanish)

37. Boletín Oficial de la Junta de Andalucía (BOJA). Orden de 24 de Marzo de 2011, Por la Que Se Aprueban en la Comunidad Autónoma de Andalucía las Bases Reguladoras Para la Concesión de Subvenciones a las Submedidas Agroambientales en El Marco del Programa del Desarrollo Rural de Andalucía 2007–2013, y Se Efectúa Su Convocatoria Para El Año 2011. Available online: http://www.juntadeandalucia.es/boja/2011/66/2 (accessed on 29 June 2015). (In Spanish)

38. Scuderi, A.; Zarbà, A.S. Economic analysis citrus fruits destined to markets. *Ital. J. Food Sci.* **2011**, *23*, 34–37.

Synergy and Transition of Recovery Efficiency of Nitrogen Fertilizer in Various Rice Genotypes under Organic Farming

Lifen Huang, Jie Yang, Xiaoyi Cui, Huozhong Yang, Shouhong Wang and Hengyang Zhuang

Abstract: Despite the growing demand for organic products, research on organic farming (OF) such as genotype screening, fertilizer application and nutrition uptake remains limited. This study focused on comparisons of the apparent recovery efficiency of nitrogen fertilizer (REN) in rice grown under OF and conventional farming (CF). Thirty-two representative conventional *Japonica* rice varieties were field grown under five different treatments: control check (CK); organic farming with low, medium and high levels of organic fertilizer (LO, MO and HO, respectively); and CF. Comparisons of REN between OF and CF classified the 32 genotypes into four types: high REN under both OF and CF (type-A); high REN under OF and low REN under CF (type-B); low REN under OF and high REN under CF (type-C); and low REN under both OF and CF (type-D). Though the yield and REN of all the rice varieties were higher with CF than with OF, organic N efficient type-A and B were able to maintain relatively high grain yield under OF. Physiological activities in flag leaves of the four types from booting to maturity were subsequently investigated under OF and CF. Under OF, high values of soil and plant analyzer development (SPAD) and N were observed in type-B varieties, while in contrast, both indexes slowly decreased in type-C varieties under CF. Moreover, the decline in N content in type-C and D varieties was greater under OF than CF. The decrease in glutamine synthetase (GS), glutamic-pyruvic transaminase (GPT) and glutamic oxaloacetic transaminase (GOT) activity in flag leaves was smaller under OF than CF in type-A and B varieties, while in contrast, type-C and D varieties showed an opposite trend. The findings suggest that OF slows the decline in key enzymes of N metabolism in organic N-efficient type rice, thus maintaining a relatively high capacity for N uptake and utilization and increasing yield during the late growth period. Accordingly, we were able to screen for varieties of rice with synergistically high REN and high grain yield under OF.

Reprinted from *Sustainability*. Cite as: Huang, L.; Yang, J.; Cui, X.; Yang, H.; Wang, S.; Zhuang, H. Synergy and Transition of Recovery Efficiency of Nitrogen Fertilizer in Various Rice Genotypes under Organic Farming. *Sustainability* **2016**, *8*, 854.

1. Introduction

As a consequence of the negative impacts of conventional farming (CF) on the environment and human health [1,2], organic agriculture is becoming increasingly widespread. Moreover, due to increasing consumer demand and political support, the popularity of organic food is also on the rise [3,4]. According to the United Nations Conference on Trade and Development [5], the global organic food market has expanded by 10%–15% in the past 10 years, with conventional markets growing by only 2%–4%. In addition to the safety aspects, organic farming is deemed beneficial to the environment and the biodiversity of a wide range of taxa including birds and mammals, invertebrates and arable flora [6,7]. In China, organic agriculture and the production of organic products was introduced in 1990 [8]. Since then, both the international and domestic market for organic products has continued to grow, with experts predicting huge market potential in the future [9].

During the past three decades, farming practices and management systems have intensified in many rice-producing countries [10–12]. Organic rice, as an important organic product, is thought to represent approximately 3 million hm^2 and is becoming increasingly popular [8]. However, due to its late start, there is a lack of related research and system technology for organic rice production, especially in terms of variety selection and fertilizer management [13]. Moreover, since nitrogen (N) is among the most important elements in agriculture systems, efficient organic fertilizer application that minimizes negative effects on the environment also needs to be examined. Moll et al. [14] defined nutrient use efficiency (NUE) as being the yield of grain per unit of available N in the soil (including the residual N present in the soil and the fertilizer). This NUE can be divided into two processes: uptake efficiency (the ability of the plant to remove N from the soil as nitrate and ammonium ions) and utilization efficiency (the ability to use N to produce grain yield) [15]. Although high amounts of organic fertilizer can maintain or increase rice yield, there is a risk of non-point source N pollution [16]. Under organic farming (OF), the gradual and continuous application of nutrients differs from that under CF [17]. Moreover, evaluation of yield and NUE in rice is commonly established on the basis of chemical N fertilizer [18], not the conditions of OF. Thus, there is an urgent need for research aimed at a thorough understanding of potential varieties that show high yield, quality and NUE under OF.

Nitrogen is usually available to plants in an inorganic form such as nitrate, nitrite or ammonia. Nitrate is a major source of inorganic nitrogen utilized by higher plants [19]. In *Arabidopsis*, the process of nitrate assimilation involves the initial uptake of nitrate ions from the soil, followed by reduction to nitrite and then to ammonia. The reduction reactions are catalyzed by the enzymes nitrate reductase and nitrite reductase, respectively. Ammonium ions enter the amino acid pool primarily by way of the action of glutamine synthetase [20]. Generally, in higher plants, there are some key N metabolism enzyme involved in assimilating intracellular ammonium into organic compounds, such as GS, GPT, GOT and so on. Glutamine synthetase (GS; EC 6.3.1.2) is a key enzyme for the assimilation of ammonium into glutamine. Numerous studies of the GS enzyme have emphasized the importance of this enzyme in plant nitrogen metabolism [21]. Leaf glutamic-pyruvic transaminase (GPT) and glutamic-oxaloacetic transaminase (GOT) activities, which in the presence of their co-substrates pyruvate and oxaloacetate, convert L-Glutamate into α-ketoglutarate, were determined to gain a better understanding of the nitrogen assimilation response [22]. However, there has been very little study of the influence of organic farming on nitrogen metabolism of rice plants.

In the present study, 32 representative conventional *Japonica* rice varieties were field-grown under CF and OF in Gaoyou city, Jiangsu Province, East China. The objectives were to screen for genotypes showing high yield and NUE under OF. Subsequently, key enzymes of N metabolism were investigated to determine the biochemical mechanism underlying effective N utilization in OF. The results provide a scientific basis for organic rice cultivation and nutrient management.

2. Materials and Methods

2.1. Plant Materials and Growth Conditions

Thirty-two representative *Japonica* rice varieties, commonly field-grown in the middle and lower reaches of Yangtze district, were selected (Table 1) and grown by organic farming and conventional farming during the rice growing season (from May to October) of 2013–2014. Two experiments were carried out, one in 2013 and a second one in 2014. In 2013, 32 varieties were screened under CK, CF and OF (including three different dosage of organic fertilizer) (Tables 1 and 2). In 2014, we repeated the same experiment as 2013, eight rice varieties with synergy and transition of REN were chosen for thorough analysis under CF and medium dosage of OF.

Experiments were carried out on an eco-agricultural farm in Gaoyou county, Jiangsu Province, China (32°47′ N, 119°25′ E, 2.1 m altitude). In the farm, half of the planting area had been verified by the China Organic Food Certification Center in 2009, where the organic farming of rice was produced. Conventional farming was carried on the other part of the farm. The farm was under continuous cultivation with green manure (*Astragalus sinicus* L.)-rice rotation. During the study period of 2013 and 2014, the study site experienced annual mean precipitation of 968 mm and 992 mm, annual mean evaporation of 1118 mm and 1082 mm, an annual mean temperature of 15.0 °C and 14.8 °C, annual total sunshine hours of 2124 h and 2076 h, and a frostless period of 223 day and 217 day, respectively. The soil on the farm is clay with 22.4 g·kg^{-1} organic matter, 80.8 mg·kg^{-1} alkali hydrolysable N, 23.3 mg·kg^{-1} Olsen-P, 105.3 mg·kg^{-1} exchangeable K, and 1.03 g·kg^{-1} total N. In both years, seedlings were sown in a seedbed on May 1st then transplanted on 20 May at a hill spacing of 0.30×0.15 m with three seedlings per hill.

Table 1. Rice varieties used in this experiment.

Type	Rice Variety	Growth Period (Days)	No. of Cultivars
Late-maturing medium *Japonica* rice	Huaidao 5, Nangjing 49, Nangjing 9108, Wulingjing 1, Wuyunjing 24	150–155	5
Early-maturing late *Japonica* rice	Changjing 09-6, Changjing 09-10, Changnongjing 4, Changyou 2, Changyou 5, Nangjing 5055, Suxiangjing 1, Wu 28105, Wu 2817, Wu 28181, Wuxiangjing 19, Wuxiangjing 9, Wuyunjing 7, Wuyunjing 23, Xiangjing 20-18, Xiangjing T31, Yangjing 4227, Yangjing 4038, Yinyu 2084, Yongyou 8, Zhendao 158, Zhendao 11, Zhendao 15, Zhendao 210	155–165	24
Medium-maturing late *Japonica* rice	Nangjing 46, Ningjing 1, Sujing 8	165–170	3

2.2. Treatments

The study was laid out in a split-plot design with farming system as the main plot and varieties as the split plot factor. Each block was 6 m^2, with four replicates per treatment.

A zero N control (control check) was designated as CK to evaluate the background of N supply in soil fertility.

As OF treatments, three rates of organic fertilizer (low, medium and high) were applied (designated LO, MO and HO, respectively). Among the OF treatments, the MO treatment were adopted following the protocols of organic rice produced by the farm. Treatment details are shown in Table 2. Contents of N, phosphorus, potassium, and organic matter in each type of organic fertilizer were as follows. Rapeseed cake fertilizer (a byproduct of rapeseed after pressing oil): 4.60%, 1.03%, 1.21%, and 81.60%, respectively; Sanan organic fertilizer: 4.00%, 0.84%, 1.16%, and 45.00%, respectively (confirmed by measuring); and clover grass manure: 0.33%, 0.03%, 0.19%, and 0.21%, respectively. Sanan organic fertilizer, which has been certified as a biological organic fertilizer [23], was supplied by Beijing Sanan Agricultural Science and Technology Corporation (Beijing, China). The average fresh weight yield of clover grass was 12,000 kg·ha^{-1}. Grass manure was plowed into the soil 15 days before cultivation as a basal fertilizer then rapeseed cake and specified Sanan organic fertilizer were applied evenly one day before transplanting. In early August, Sanan organic fertilizer was reapplied as top-dressing. All the OF treatments involving organic manure were managed according to the standards for organic rice cultivation, and weeds were controlled by hand weeding [9].

As CF treatments, it followed local high-yield agricultural management practices (Table 2). Nutrients were delivered throng chemical fertilizer (i.e., Urea as a source of N, P_2O_5 as a source of P and K_2O as a source of K). Water, insects, and disease were controlled as required to avoid yield losses, and weeds were controlled with chemical herbicide treatment.

Table 2. Fertilizers applied to rice grown under organic and conventional farming.

Treatment		Basic and Tiller Fertilizer (kg·ha^{-1})			Panicle Fertilizer (kg·ha^{-1})	Total N (kg·ha^{-1})
		Rapeseed Cake	Sanan Organic Fertilizer	Grass Manure	Sanan Organic Fertilizer	
CK		0	0	0	0	0
OF	LO	600	630	12,000	270	103.2
	MO	1200	1155	12,000	495	160.8
	HO	1800	1680	12,000	720	218.4
CF		Nitrogen was supplied as urea ($n = 46\%$): 40% of the total as basal dressing on 20 May; and 10%, 20%, and 30% as top-dressing on 27 May, 10 June, and 6 August, respectively. Phosphorus (150 kg·ha^{-1} as P_2O_5) and potassium (150 kg·ha^{-1} as K_2O) were applied before transplanting				270.0

CK, control check; OF, organic farming; CF, conventional farming. LO, MO, and HO: low, medium, and high application of organic fertilizer, respectively.

2.3. Plant Sampling and Measurements

Plants were hand-harvested on 25 October 2013, and 26 October 2014. Measurements of grain yield and yield components followed the procedures described in Yoshida et al. [24]. Plants in the two rows on each side of the plot were discarded to avoid border effects. Grain yield in each plot was thereby determined from a harvested area of 2.0 m^2 and adjusted to a 14% moisture content.

Above-ground biomass and yield components; that is, the number of panicles per square meter, number of spikelets per panicle, percentage of filled grains and grain weight, were determined in a 1 m^2 area (excluding borders) sampled randomly in each plot. The percentage of filled grains was defined as the number of filled grains expressed as a percentage of the total number of spikelets. The dry weight of each component was determined by oven-drying at 70 °C to a constant weight prior to weighing [25].

Tissue N content was determined by micro Kjeldahl digestion, distillation and titration to calculate aboveground N uptake [24]. The apparent recovery efficiency of N fertilizer (REN; the percentage of N fertilizer recovered in aboveground plant biomass at the end of the cropping season) was determined according to Zhang et al. [26]. In this paper, we chose the REN as the variable to estimate NUE.

A chlorophyll meter (SPAD-502) was used to determine soil and plant analyzer development (SPAD) values in flag leaves of 10 randomly selected plants of each variety. Average values of upper, middle and bottom portions of each flag leaf were determined. For soluble protein analysis, three biological replicated roots, rice leaves under different treatments were harvested at each growth stage.

Glutamine synthetase (GS), glutamic-pyruvic transaminase (GPT) and glutamic oxaloacetic transaminase (GOT) activities in leaves were measured at booting, heading, middle-filling, late-filling and maturity. The activity of GS was measured with the biosynthetic reaction assay, using NH_2OH as artificial substrate, by measuring the formation of glutamyl hydroxamate. Leaf GS activity was measured in pre-incubated assay buffer (37 °C) consisting of 70 mm MOPS (pH 6.80), 100 mm Glu, 50 mm $MgSO_4$, 15 mm NH_2OH, and 15 mm ATP. The reaction was terminated after 30 min at 37 °C by addition of an acidic $FeCl_3$ solution (88 mm $FeCl_3$, 670 mm HCl, and 200 mm trichloroacetic acid) at booting, heading, middle-filling, late-filling and maturity. One unit (U) of activity was defined as the increase in glutamylhydroxamate per g protein per hour [21]. Glutamic-oxaloacetic transaminase (GOT) and glutamic-pyruvic transaminase (GPT) activity were assayed by the method of Wu et al. [27]; one unit (U) of activities was defined as the increase in pyruvic acid content per g protein per hour.

2.4. Statistical Analysis

Analysis of variance was performed using the SAS/STAT statistical analysis package (version 6.12, SAS Institute, Cary, NC, USA). Data from each sampling date were analyzed separately. Four replicates were used to calculate the means. Differences between means were tested by least significant difference test at the 0.05 level of probability. For the mean comparisons, the Tukey test was the chosen method. Unweighted pair-group method with arithmetic means (UPGMA) of agglomeration method were used for the hierarchical clustering. Hierarchical cluster analysis was applied in the present work. After determining the Euclidean distance and sum of squared residuals, the dendrogram was built. Accordingly, the varieties were classified into three types as low, medium and high NUE under different farming systems respectively [28].

3. Results

3.1. Effects of OF and CF on REN

3.1.1. Statistical Analysis of REN under OF and CF

Table 3 shows the computed F-values of the differences in grain yield, REN and N content in the 32 rice varieties between years. All measurements showed significant differences among varieties and organic fertilizer treatments (LO, MO and HO, respectively), and in the interaction between variety and organic fertilizer treatment. Variations with year and the interactions between year and variety as well as between year and organic fertilizer treatment were not significant. Similar results were obtained in the measurements of SPAD and enzyme activity of GS, GPT and GOT (data not shown). Because year was not a significant factor, the data from both years were averaged.

Table 3. Analysis of variance of F-values of grain yield, REN, and N content under organic N fertilizer.

Source	Degrees of Freedom	Grain Yield $(t \cdot hm^{-2})$	REN (%)	N Content $g \cdot kg^{-1}$
Year (Y)	1	1.49	9.8	12.46
Organic fertilizer (O)	3	38,756.05 **	2589.42 **	1645.38 **
Variety (V)	31	760.01 **	586.83 **	561.54 **
Y × O	3	0.54	Ns	Ns
Y × V	31	10.96	Ns	Ns
O × V	93	155.22 **	143.54 **	201.47 **

** Indicates a significant difference at 0.01; Ns, not significant.

3.1.2. Effects of OF on REN and Classification of Different Rice Varieties

REN showed an initial increase followed by a decrease with increasing organic fertilizer. Under LO, MO and HO treatment, the average REN of all varieties was 20.18%, 24.39% and 22.33%, respectively (Table 4). The highest variable coefficient (31.40%) was recorded with LO treatment and the lowest (18.88%) with MO. Thus, MO treatment resulted in the highest REN value, and was therefore chosen as the optimal OF treatment for the remainder of the analysis.

Table 4. Variation in the REN of different rice varieties at different organic fertilizer levels under organic farming (%).

Organic Fertilizer Levels	Mean	Std	Max	Min	Range	CV
LO	20.18C	6.34	31.47	13.35	18.12	31.40
MO	24.39A	4.60	33.66	14.36	19.30	18.88
HO	22.33B	5.05	34.14	13.58	20.56	22.61

Std, standard deviations. Means followed by the same capital letters in a column are not significantly different at the 0.01 probability levels.

To clarify the REN attributes of the different rice varieties under OF-MO, after determining the Euclidean distance and sum of squared residuals, the dendrogram was built. Accordingly, the varieties were classified into three types as follows (Figure 1A): organic high REN (7 varieties; REN: 27.04%–33.66%, mean REN: 29.96%, variation coefficient: 7.93%), organic medium REN (18 varieties; 20.09%–24.37%, 21.88% and 6.11%, respectively) and organic low REN (7 varieties; 13.35%–18.86%,

17.15% and 11.06%, respectively). The mean REN value of all varieties was 22.61%, with the highest value (33.66%) recorded in Yongyou 8 and lowest (13.35%) in Suxiangjin 1.

3.1.3. Difference in REN between Varieties under CF

Under CF, the average REN was 37.66%, with the highest value (45.83%) recorded in Yongyou 8 and the lowest (31.44%) in Suxiangjing 1. Based on these results, the varieties were classified into three types by Euclidean distance (Figure 1B): high REN under CF (8 varieties; average REN of 43.34% and variation coefficient of 3.58%); medium REN under CF (15 varieties; 37.39% and 4.64%, respectively); and low REN under CF (9 varieties; 33.06% and 3.16%, respectively). Comparisons of REN under CF and OF revealed extremely significant higher efficiency under CF.

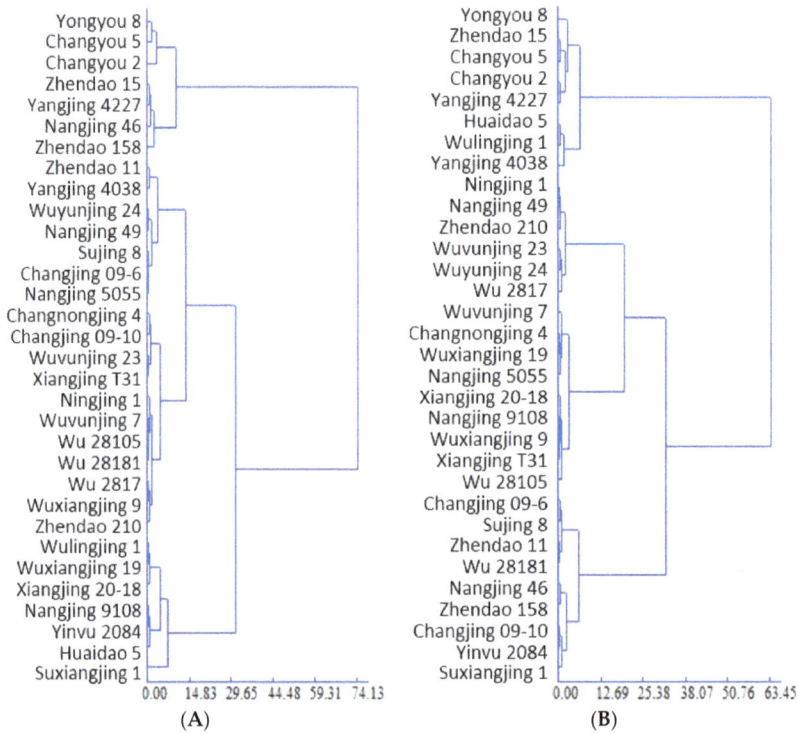

Figure 1. Dendrogram showing the *Japonica* rice varieties based on recovery efficiency of nitrogen under organic farming (**A**) and conventional farming (**B**).

3.1.4. Genotype Screening of Transformation and Synergy in REN under OF and CF

Comparisons of the differences in REN between OF and CF (Table 5) resulted in classification of the genotypes into four types: high REN under both OF and

CF (synergistically high REN, type-A); high REN under OF and low REN under CF (high-low REN transition, type-B); low REN under OF and high REN under CF (low-high REN transition, type-C); and low REN under both OF and CF (synergistically low REN, type-D).

All varieties were regular *Japonica* rice varieties except for three hybrids, Changyou 5, Yongyou 8 and Changyou 2, which have been proven to be a particular germplasm resource with significantly increasing yield and excellent quality traits [29] despite having high REN under OF. Of the regular *Japonica* varieties, Zhendao 15 and Yangjing 4227 were typical type-A varieties, Nangjing 46 and Zhendao 158 were type-B, Huaidao 5 and Wulingjing 1 were type-C, and Suxiangjing 1 and Yinyu 2084 were type-D.

Table 5. Representative varieties showing REN Synergism and transformation under organic (OF) and conventional farming (CF).

Type	REN Characteristics	Synergism and Transformation	Typical Varieties
Type-A	High REN under both OF and CF	High REN Synergism	Zhendao 15, Yangjing 4227
Type-B	High REN under OF whereas low REN under CF	Transformation from high to low REN	Nangjing 46, Zhendao 158
Type-C	Low REN under OF whereas high REN under CF	Transformation from low to high REN	Huaidao 5, Wulingjing 1
Type-D	Low REN under both OF and CF	Low REN Synergism	Suxiangjing 1, Yinyu 2084

3.2. Grain Yield and Yield Components in the Different Rice Varieties under OF and CF

In all the typical rice varieties, the number of panicles per unit area was significantly higher under CF than OF (Table 6). The OF/CF ratio of panicle number per unit area was highest in type-B followed by type-D, and lowest in type-C. Compared to CF, the number of spikelets per panicle in type-A and B rice showed a small but insignificant increase under OF. In contrast, type-C and D rice showed a significant decrease. The average OF/CF ratio of percentage of filled grains was 90.77%, 86.51%, 88.34% and 90.04% in type-A, B, C and D rice, respectively, with a significant decrease under OF. In contrast, the average OF/CF ratio of 1000-grain weight was greater under OF than CF in all varieties. Compared with CF treatment, grain yield significantly decreased under OF by 20.67%, 17.46%, 41.84% and 26.09% in type-A, B, C and D rice, respectively. Type-C and D rice showed higher grain yield losses than type-A and B rice under OF, suggesting that organic N efficient type-A and B were able to maintain relatively high grain yield under OF.

Table 6. Grain yield and yield components in the different types of rice under organic (OF) and conventional farming (CF).

Type *	Farming System	No. of Panicles ($\times 10^4$ hm^{-2})	Spikelets per Panicle	Percentage of Filled Grains (%)	1000-Grain Weight (g)	Yield (t·hm^{-2})
Type-A	OF	231.04B	166.06A	81.29B	25.87A	8.07B
	CF	281.4A	161.28A	89.55A	25.02A	10.17A
	OF/CF (%)	82.1	102.96	90.77	103.38	79.33
Type-B	OF	280.31B	137.74A	77.99B	26.00A	7.83B
	CF	308.55A	135.38A	90.14A	25.19A	9.48A
	OF/CF (%)	90.85	101.75	86.51	103.21	82.54
Type-C	OF	232.38B	121.55B	79.89B	25.21A	5.69B
	CF	320.92A	136.3A	90.43A	24.73A	9.78A
	OF/CF (%)	72.41	89.18	88.34	101.95	58.16
Type-D	OF	218.3B	118.13B	72.67B	25.14A	4.66B
	CF	241.13A	130.29A	80.71A	24.85A	6.30A
	OF/CF (%)	90.53	90.67	90.04	101.17	73.91

* See Table 5 for a description of each type. The REN characteristics of type-A is high REN under both OF and CF, type-B is high REN under OF whereas low REN under CF, type-C is low REN under OF whereas high REN under CF, type-D is low REN under both OF and CF, respectively. In a column, letter A and B labeled after figures are statistically significant at the 0.01 probability levels. Means comparisons were assessed independently for each type.

3.3. SPAD and N Values in Flag Leaves during the Key Growth Periods under OF and CF

3.3.1. SPAD Values

SPAD values report on the leaf chlorophyll content, which indicates the leaf N remobilization status. A dynamic change in SPAD values was detected in the flag leaves, the most important functional leaf in rice. SPAD values first increased then decreased with growth under both CF and OF (Table 7), and were significantly higher in all varieties under CF compared to OF. In type-A rice, SPAD values under OF were 95.35%, 96.83%, 96.06%, 95.85% and 95.50% of those under CF at booting, heading, middle-filling, late-filling and maturity, respectively. Type-D rice presented a similar trend; however, in type-B rice values under OF were 96.26%, 97.23%, 98.64%, 98.70% and 99.29% of those under CF, respectively. Moreover, the decreasing trend in type-B rice under OF was slow and smooth compared to that under CF. In contrast, SPAD values in type-C rice showed an opposite trend to type-B. These data demonstrate that farming mode, organic versus conventional, had only a small effect on SPAD values in type-A and D rice, but a greater effect on type-B and C.

Table 7. Effects of organic (OF) and conventional farming (CF) on SPAD values in flag leaves during key growth periods.

Type *	Farming System	Growth Period				
		Booting	Heading	Middle-Filling	Late-Filling	Maturity
Type-A	OF	38.55Ab	43.62Ab	39.48Ab	30.48Ab	17.62Ab
	CF	40.43Aa	45.05Aa	41.10Aa	31.80Aa	18.45Aa
	OF/CF (%)	95.35	96.83	96.06	95.85	95.50
Type-B	OF	35.47Ab	41.03Ab	43.4Ab	34.05Ab	20.95Aa
	CF	36.85Aa	42.20Aa	44.00Aa	34.50Aa	21.10Aa
	OF/CF (%)	96.26	97.23	98.64	98.70	99.29
Type-C	OF	38.72Ab	42.03Ab	43.30Ab	23.75Bb	15.00Bb
	CF	40.80Aa	44.35Aa	45.75Aa	25.80Aa	16.35Aa
	OF/CF (%)	94.90	94.77	94.64	92.05	91.87
Type-D	OF	40.93Ab	43.53Ab	45.05Ab	38.23Ab	23.45Ab
	CF	43.25Aa	45.90Aa	47.65Aa	40.35Aa	24.73Aa
	OF/CF (%)	94.64	94.84	94.54	94.75	94.82

* See Table 5 for a description of each type. The REN characteristics of type-A is high REN under both OF and CF, type-B is high REN under OF whereas low REN under CF, type-C is low REN under OF whereas high REN under CF, type-D is low REN under both OF and CF, respectively. In a column, letter A and B, letter a and b labeled after figures are statistically significant at the 0.01, 0.05 probability levels, respectively. Means comparisons were assessed independently for each type.

3.3.2. N Content

During growth, the N content of the flag leaves first increased then decreased similarly to SPAD values, reaching a maximum at heading (Table 8). The N content of all varieties was higher under CF than OF. Moreover, the decline in N content in type-A and D rice was relatively smaller than that in type-B and C. In type-B, the range of N content was much smaller under OF, while in type-C, the range was smaller under CF. Thus, OF slowed down the decline in N content in varieties with a high REN under OF, whereas CF did not.

Table 8. Effects of organic (OF) and conventional farming (CF) on N content in flag leaves during key growth periods $(g \cdot kg^{-1})$.

Type *	Farming System	Growth Period				
		Booting	Heading	Middle-Filling	Late-Filling	Maturity
Type-A	OF	22.28Aa	22.69Bb	22.56Ab	10.99Ab	6.04Ab
	CF	24.52Aa	25.45Aa	24.82Aa	12.09Aa	6.75Aa
	OF/CF (%)	90.88	89.12	90.91	90.88	89.48
Type-B	OF	21.10Bb	21.98Bb	18.83Ab	15.01Ab	9.87Ab
	CF	22.94Aa	23.64Aa	20.06Aa	15.74Aa	10.21Aa
	OF/CF (%)	91.99	92.98	93.86	95.33	96.71
Type-C	OF	21.23Ab	22.15Ab	17.01Bb	8.55Bb	6.10Bb
	CF	23.02Aa	24.26Aa	21.41Aa	11.26Aa	8.12Aa
	OF/CF (%)	92.22	91.29	79.44	75.89	75.14
Type-D	OF	26.32Ab	28.07Ab	24.20Ab	18.04Ab	11.63Ab
	CF	28.28Aa	30.07Aa	25.93Aa	19.27Aa	12.55Aa
	OF/CF (%)	93.06	93.34	93.30	93.59	92.64

* See Table 5 for a description of each type. The REN characteristics of type-A is high REN under both OF and CF, type-B is high REN under OF whereas low REN under CF, type-C is low REN under OF whereas high REN under CF, type-D is low REN under both OF and CF, respectively. In a column, letter A and B, letter a and b labeled after figures are statistically significant at the 0.01, 0.05 probability levels, respectively. Means comparisons were assessed independently for each type.

3.4. Effect on Key Enzymes of N Metabolism under OF and CF

3.4.1. GS Activity

GS catalyzes the adenosine triphosphate (ATP)-dependent condensation of ammonia (NH_3) with glutamate to yield glutamine. Many studies of the GS enzyme justify the importance of this enzyme in metabolism of plant N [18,21]. GS activity in all varieties showed the biggest drop from booting stage to heading (Figure 2). Under both OF and CF, GS activity decreased with growth in all varieties, except at maturity. Under OF, GS activity in type-A rice was 82.71%, 83.85%, 85.88%, 86.16% and 117.74% that under CF at booting, heading, middle-filling, late-filling and maturity, respectively. In contrast, in type-B rice, values under OF were 114.54%, 116.35%, 117.95%, 119.69% and 139.97% those under CF, respectively, suggesting that GS activity was greater under OF than CF in type-B rice. Type-C and D rice presented an opposite trend.

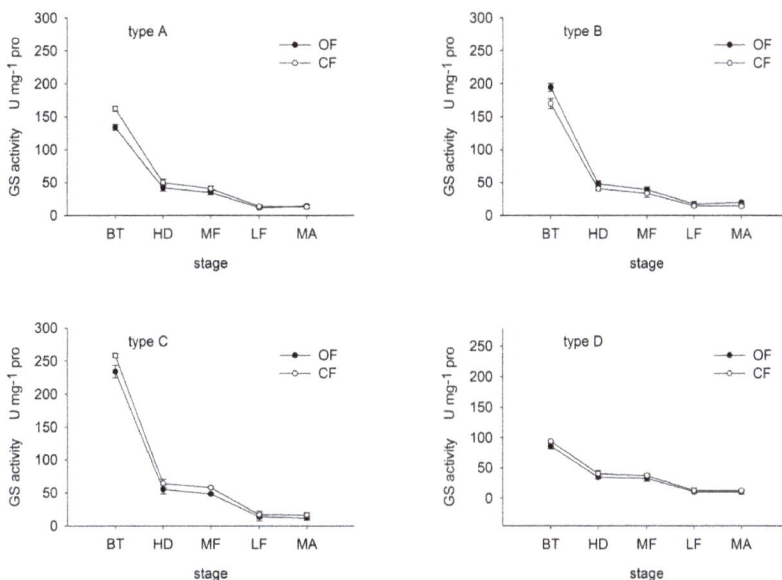

Figure 2. Effects of organic (OF) and conventional farming (CF) on GS activity in flag leaves of the four different types (**A**–**D**) of rice during key growth periods. BT, booting, HD, heading: MF, middle-filling; LF, late-filling, MA, maturity. Type-A, high REN under both OF and CF; type-B, high REN under OF whereas low REN under CF; type-C, low REN under OF whereas high REN under CF; type-D, low REN under both OF and CF. Values are means with standard errors shown by the vertical bars (*n* = 4).

These findings suggest that OF increased GS activity in flag leaves of type-B rice. Moreover, under OF, the decrease in GS in type-A and B rice was less than that under CF. In contrast, the decrease in type-C and D rice was greater under OF than CF. That is, OF slowed the decline in GS activity in N efficient varieties, thereby maintaining a high N assimilation rate in later growth stages.

3.4.2. GPT and GOT Activity

GPT and GOT are the most common forms of transaminase in plant. Leaf GPT and GOT activities were determined to gain a better understanding of the nitrogen assimilation response [27]. Under both OF and CF, GPT activity in all varieties decreased then increased before dropping once again at maturity (Figure 3). In type-C and D rice, GPT activity increased from the middle-filling stage, while in type-A and B the increase occurred earlier, at the heading stage. These findings suggest that GPT activity recovered more quickly in N efficient varieties, thereby maintaining the N transformation ability. GPT activity in type-B rice was particularly noteworthy, with values under OF 114.69%, 126.87%, 130.03%, 133.48% and 155.39%

of those under CF at booting, heading, middle-filling, late-filling and maturity, respectively. In contrast, the remaining three types showed lower GPT values under CF than OF at each corresponding stage.

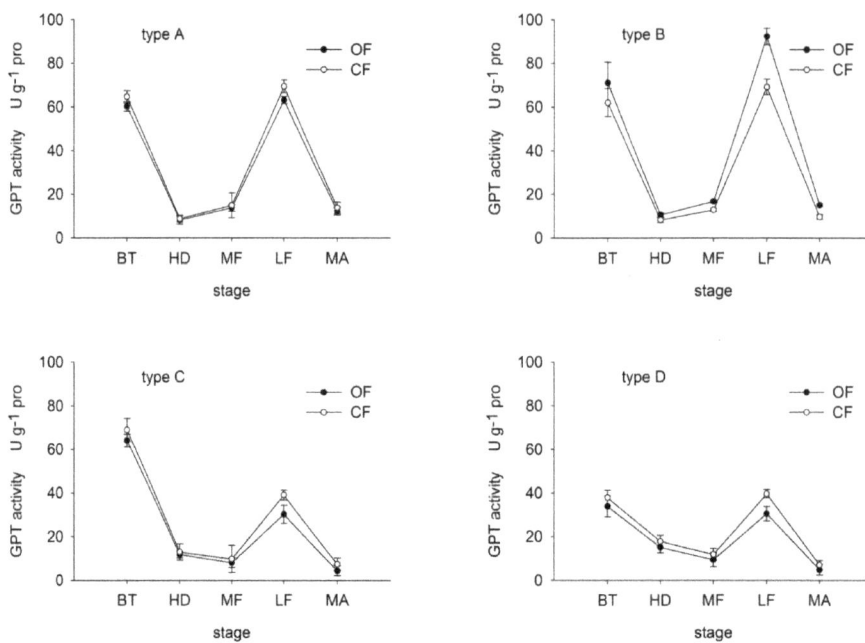

Figure 3. Effects of organic (OF) and conventional farming (CF) on GPT activity in the flag leaves of the four different types (**A–D**) during key growth periods. Abbreviations as in Figure 2.

In type-A and B rice, GOT activity first decreased then increased before dropping again at maturity, reaching a peak at the late-filling period (Figure 4). In type-C and D rice, GOT activity peaked at the heading and late-filling stages. The OF/CF ratio of GOT activity was similar to that of GPT activity. The ratio in type-A, C and D rice ranged from 91.15% and 92.86%, 61.26% and 85.90%, and 82.28% to 84.98%, respectively, throughout growth. In contrast, in type-B rice, the ratio ranged from 109.19% to 149.79%. These findings suggest that under OF, the decrease in GOT and GPT in type-A and B rice was lower than that in type-C and D. Moreover, under OF, organic N efficient type-A and B were able to maintain relatively high N transformation abilities.

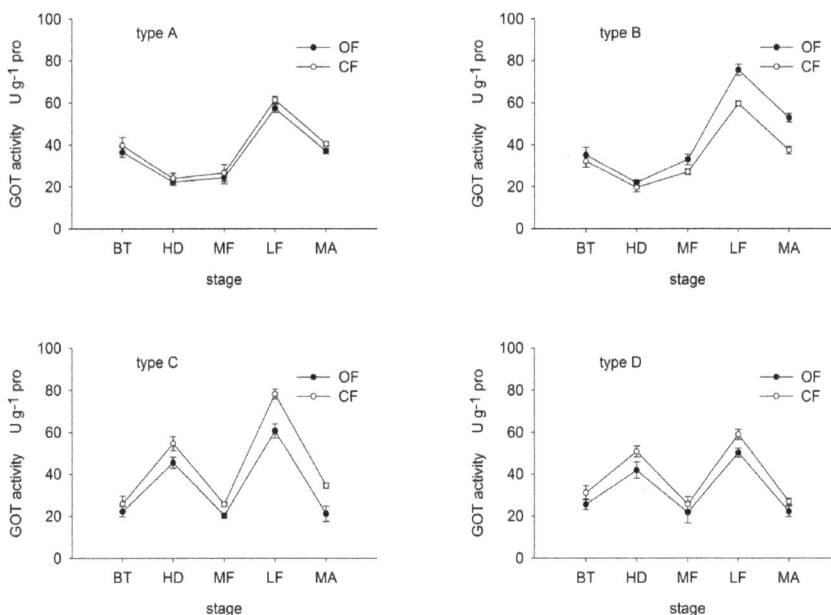

Figure 4. Effects of organic (OF) and conventional farming (CF) on GOT activity in the flag leaves of the four different types (**A–D**) during key growth periods. Abbreviations as in Figure 2.

4. Discussion

4.1. NUE Consistency and Conversion under CF and OF

In conventional cropping systems, different measurements of NUE are commonly applied; for example, REN, agronomic efficiency of nitrogen (AEN), physiological efficiency of nitrogen (PEN), and partial factor productivity of nitrogen (PFPN). Consequently, research results often vary according to these different indexes. REN is commonly used to evaluate the actual use efficiency and loss rate of N [26]. In a previous study, REN was found to range between 30% and 40% under CF in China [18], which is lower than in other developed countries, such as USA, Canada and so on. In the current study, REN revealed a synergistically high NUE type, high-low NUE transition type, low-high NUE transition type and synergistically low NUE type. PFPN, as an integrative index of the total economic output relative to utilization of all N sources, was suggested by Hasegawa et al. [30] as a useful criterion for organic cropping systems in which animal compost or organic fertilizer are applied. In future studies, to fully clarify the high grain yield and high NUE under OF, additional NUE indexes such as AEN and PEN will be used to comprehensively evaluate the N use mechanism.

The literature on NUE tends to focus on CF or inorganic/organic fertilization, rather than OF [26,31,32]. Application of organic-inorganic compound fertilizer was previously found to have a positive effect on soil organic carbon accumulation and crop productivity in rice fields, reducing chemical fertilizer use, optimizing the physical qualities of paddy soil, and improving long-term sustainability through increased N efficiency, possibly as a result of enhanced microbial activity [31–33]. However, OF differs from organic fertilizer experiments. In the district of Xinjiang, a high dose of organic fertilizer was found to result in lower REN and AEN values than under a low organic fertilizer dose, and far lower than under CF [16]. In this study, the average NUE of all varieties was higher (37.66%) under CF than under OF (22.61%), consistent with Sun et al. [16]. We also found that an increase in organic fertilizer resulted in an initial increase in NUE followed by a decrease, with the maximum value at a medium dose. In order to enhance NUE under OF, it is therefore important to control the dose of organic fertilizer applied.

We subsequently examined the possible reasons for the higher NUE under CF than OF. Liu et al. [25] suggested that under low soil fertility, chemical fertilizer provides nutrients quickly, thereby promoting rapid growth. In contrast, with organic fertilizer, the nutrient release rate is very late, and therefore there is insufficient available nitrogen for growth. Thus, based on the conversion of N, organic fertilizer acts as a slow-release N fertilizer, with nutrients released and absorbed in a step-by-step manner, lasting longer. This study focused on NUE, and therefore, requires further analysis of the long-acting release mechanism under OF.

4.2. Relationship between Yield and NUE under OF and CF

A technology gap is thought to exist between rice yield and environmental efficiency scores based on levels of pure N use [34,35]. As a result, farmers tend to increase the use of external nutrients such as N to compensate for potential yield losses during the initial OF conversion period [11,16]. In old alluvial soil of India, 40% nitrogen and 25% phosphate chemical fertilizer can safely be supplemented by low-cost, natural resource-based bio-fertilizer (*Azotobacter* sp.) at 12 kg·ha^{-1} and organic manure at 10.00 t·ha^{-1} to make rice cultivation more productive and profitable over a long period [3]. Conversely, in the case study in a farmer's fields in Japan, the highest grain yield was demonstrated to be obtained via internal nitrogen nutrient cycling of residues, such as rice straw, rice bran and weeds in lowland rice farming [36].

In our previous study, we found positive correlations between yield and dry matter accumulation at maturity, N uptake at maturity, REN and AEN in three high-quality rice varieties under OF [17]. In this study, we choose *Japonica* rice cultivars commonly grown in the middle and lower reaches of the Yangtze River, Jiangsu province. Under OF, spike number, grain number per panicle and spikelet number increased with an increasing dose of organic fertilizer. In contrast, the percentage of filled grains and 1000-grain weight showed an opposite trend, consistent with the conclusions of Ling et al. [23]. Similarly, this study demonstrated the effects of OF and CF on the various attributes of yield, revealing two synergy models: one showing consistently high yield, the other consistently low yield. Varieties showing high yield included the hybrid *Japonica* rice varieties Yongyou 8, Changyou 5 and Changyou 2, which often demonstrate high yield, achieving high output not only under CF but also at low soil fertility under OF. Those showing consistently low yield under both OF and CF included Suxiangjing 1 and Yingyu 2084. It is worth mentioning that the difference in yield between OF and CF was highly significant in some varieties; for example, Huaidao 5 and Wulingjing 1 showed high yield greater than 9 t·hm^{-2} under CF, but an average of only 5.98 t·hm^{-2} under OF.

Yield was always greater with CF than OF and the yield advantages were more prominent in the hybrid rice varieties than the conventional types under OF. At a late growth period under OF, the decrease in the range of leaf area and photosynthetic potential were found to be smaller, and moreover, the average population growth rate and percentage of dry matter accumulation were higher [29]. Therefore, to fully determine the optimal rice genotypes for OF, we need to screen, identify and evaluate various varieties under both OF and CF.

5. Conclusions

The results of this study suggest that when comparing NUE between CF and OF, some varieties obtain consistently high (type-A) and others consistently low NUE (type-D), whereas some transform from high to low (type-C) and others from low to high under OF compared to CF (type-B). From booting to maturity, higher values of SPAD and N content and less of a decline in GS, GPT and GOT activity in flag leaves were observed in type-A and B compared to type-C and D varieties under OF. Moreover, organic N-efficient type-A and B maintained relatively higher grain yield under OF. Accordingly, varieties with synergistically high NUE and high grain yield under OF were identified.

Acknowledgments: This study was supported by the National Natural Science Foundation of China (No. 31201154 and 31571596), Three New Agriculture Project of Jiangsu Province (SXGC[2016]212 and SXGC[2015]089), Jiangsu Key Research and Development Plan (BE2015340,BE2016351), Jiangsu Agriculture Science and Technology Innovation Fund (CX(16)1003), the Open Project Program of Key Laboratory of Crop Physiology (No. K12008) and the Priority Academic Program Development of Jiangsu Higher Education Institutions.

Author Contributions: Lifen Huang and Hengyang Zhuang conceived and designed the experiments; Jie Yang performed the experiments; Jie Yang and Lifen Huang analyzed the data; Jie Yang contributed reagents/materials/analysis tools; Lifen Huang wrote the paper; Xiaoyi Cui, Huozhong Yang and Shouhong Wang revised the paper. All authors read and approved the final manuscript.

Conflicts of Interest: The authors declare no conflict of interest.

Abbreviations

The following abbreviations are used in this manuscript:

NUE	nitrogen use efficiency
OF	organic farming
CF	conventional farming
CK	control check
LO	low level of organic fertilizer
MO	medium level of organic fertilizer
HO	high level of organic fertilizer
BT	booting
HD	Heading
MF	middle-filling
LF	late-filling
MA	maturity
type-A	high NUE under both OF and CF
type-B	high NUE under OF whereas low NUE under CF
type-C	low NUE under OF whereas high NUE under CF
type-D	low NUE under both OF and CF
GS	glutamine synthetase
GPT	glutamic-pyruvic transaminase
GOT	glutamic oxaloacetic transaminase
SPAD	soil and plant analyzer development
REN	recovery efficiency of nitrogen
PFPN	partial factor productivity of nitrogen
AEN	agronomic efficiency of nitrogen
PEN	physiological efficiency of nitrogen

References

1. Pimentel, D.; Hepperly, P.; Hanson, J.; Douds, D.; Seidel, R. Environmental, energetic, and economic comparisons of organic and conventional farming systems. *Bioscience* **2005**, *55*, 573–582.
2. Reganold, J.P.; Elliott, L.F.; Unger, Y.L. Long-term effects of organic and conventional farming on soil erosion. *Nature* **1987**, *330*, 370–372.
3. Mukhopadhyay, M.; Datta, J.K.; Garai, T.K. Steps toward alternative farming system in rice. *Eur. J. Agron.* **2013**, *51*, 18–24.

4.   Delmotte, S.; Barbier, J.M.; Mouret, J.C.; Le Page, C.; Wery, J.; Chauvelon, P.; Sandoz, A.; Lopez Ridaura, S. Participatory integrated assessment of scenarios for organic farming at different scales in Camargue, France. *Agric. Syst.* **2016**, *143*, 147–158.

5.   Kleemann, L.; Abdulai, A.; Buss, M. Certification and access to export markets: Adoption and return on investment of organic-certified pineapple farming in Ghana. *World Dev.* **2014**, *64*, 79–92.

6.   Hole, D.G.; Perkins, A.J.; Wilson, J.D.; Alexander, I.H.; Grice, P.V.; Evans, A.D. Does organic farming benefit biodiversity? *Biol. Conserv.* **2005**, *122*, 113–130.

7.   Garratt, M.P.D.; Wright, D.J.; Leather, S.R. The effects of farming system and fertilisers on pests and natural enemies: A synthesis of current research. *Agr. Ecosyst. Environ.* **2011**, *141*, 261–270.

8.   Jin, L.D. Production status and strategies of organic rice in China. *Chin. Rice* **2007**, *3*, 1–4. (In Chinese)

9.   Ma, S.M.; Joachim, S. Review of history and recent development of organic farming worldwide. *Agric. Sci. China* **2006**, *5*, 169–178. (In Chinese)

10.   Katayama, N.; Baba, Y.G.; Kusumoto, Y.; Tanaka, K. A review of post-war changes in rice farming and biodiversity in Japan. *Agric. Syst.* **2015**, *132*, 73–84.

11.   Hokazono, S.; Hayashi, K. Variability in environmental impacts during conversion from conventional to organic farming: A comparison among three rice production systems in Japan. *J. Clean Prod.* **2012**, *28*, 101–112.

12.   Nowak, B.; Nesme, T.; David, C.; Pellerin, S. Nutrient recycling in organic farming is related to diversity in farm types at the local level. *Agric. Ecosyst. Environ.* **2015**, *204*, 17–26.

13.   Rashad, F.M.; Saleh, W.D.; Moselhy, M.A. Bioconversion of rice straw and certain agro-industrial wastes to amendments for organic farming systems: 1. Composting, quality, stability and maturity indices. *Bioresour. Technol.* **2010**, *101*, 5952–5960.

14.   Moll, R.H.; Kamprath, E.J.; Jackson, W.A. Analysis and interpretation of factors which contribute to efficiency of nitrogen utilization. *Agron. J.* **1982**, *74*, 562–564.

15.   Hirel, B.; Gouis, J.L.; Ney, B.; Gallais, A. The challenge of improving nitrogen use efficiency in crop plants: Towards a more central role for genetic variability and quantitative genetics within integrated approaches. *J. Exp. Bot.* **2007**, *58*, 2369–2387.

16.   Sun, X.P.; Li, G.X.; Meng, F.Q.; Guo, Y.B.; Wu, W.L.; Yili, H.M.; Chen, Y. Nutrients balance and nitrogen pollution risk analysis for organic rice production in Yili reclamation area of Xinjiang. *Trans. Chin. Soc. Agric. Eng.* **2011**, *27*, 158–162. (In Chinese)

17.   Huang, L.F.; Yu, J.; Yang, J.; Zhang, R.; Bai, Y.C.; Sun, C.M.; Zhuang, H.Y. Relationships between yield, quality and nitrogen uptake and utilization of organically grown rice varieties. *Pedosphere* **2016**, *26*, 85–97.

18.   Peng, S.B.; Huang, J.L.; Zhong, X.H.; Yang, J.C.; Wang, H.G.; Zou, Y.B.; Zhang, F.S.; Zhu, Q.S.; Buresh, R.; Witt, C. Research strategy in improving fertilizer-nitrogen use efficiency of irrigated rice in China. *Sci. Agric. Sin.* **2002**, *35*, 1095–1103. (In Chinese)

19. Haefele, S.M.; Jabbar, S.M.A.; Siopongco, J.; Tirol-Padre, A.; Amarante, S.T.; Sta, C.P.; Cosico, W.C. Nitrogen use efficiency in selected rice (*Oryza sativa* L.) genotypes under different water regimes and nitrogen levels. *Field Crop. Res.* **2008**, *107*, 137–146.

20. Wilkinson, J.Q.; Crawford, N.M. Identification of the Arabidopsis CHL3 Gene as the Nitrate Reductase Structural Gene MA2. *Plant Cell* **1991**, *3*, 461–471.

21. Bao, A.; Zhao, Z.; Ding, G.; Shi, L.; Xu, F.; Cai, H. Accumulated expression level of cytosolic glutamine synthetase 1 Gene (*OsGS1;1* or *OsGS1;2*) alter plant development and the carbon-nitrogen metabolic status in rice. *PLoS ONE* **2014**, *9*, e95581.

22. Chen, S.; Zhang, X.G.; Zhao, X.; Wang, D.Y.; Xu, C.M.; Ji, C.L.; Zhang, X.F. Response of Rice Nitrogen Physiology to High Nighttime Temperature during Vegetative Stage. *Sci. World J.* **2013**.

23. Ling, Q.H.; Zhang, H.C.; Ju, Z.W.; Dai, Q.G.; Huo, Z.Y. Effect of different dosages of Sanan bio-organic fertilizer on organic rice production, quality and utilization of nitrogen absorption. *Chin. Rice.* **2010**, *1*, 17–22. (In Chinese)

24. Yoshida, S.; Forno, D.; Cock, J.; Gomez, K. *Laboratory Manual for Physiological Studies of Rice*; International Rice Research Institute: Metro Manila, Philippines, 1976.

25. Liu, L.J.; Chen, T.T.; Wang, Z.Q.; Zhang, H.; Yang, J.C.; Zhang, J.H. Combination of site-specific nitrogen management and alternate wetting and drying irrigation increases grain yield and nitrogen and water use efficiency in super rice. *Field Corp. Res.* **2013**, *154*, 226–235.

26. Zhang, Z.J.; Chu, G.; Liu, L.J.; Wang, Z.Q.; Wang, X.M.; Zhang, H.; Yang, J.C.; Zhang, J.H. Mid-season nitrogen application strategies for rice varieties differingin panicle size. *Field Crop. Res.* **2013**, *150*, 9–18.

27. Wu, L.H.; Jiang, S.H.; Tao, Q.N. Colorimetric method for plant transaminase (GOT and GPT activity). *Chin. J. Soil Sci.* **1998**, *29*, 136–138. (In Chinese)

28. Liu, S.L.; Huang, D.Y.; Chen, A.L.; Wei, W.X.; Brookes, P.C.; Li, Y.; Wu, J.S. Differential responses of crop yields and soil organic carbon stock to fertilization and rice straw incorporation in three cropping systems in the subtropics. *Agric. Ecosyst. Environ.* **2014**, *184*, 51–58.

29. Huang, L.F.; Zhang, R.; Yu, J.; Jiang, L.L.; Su, H.D.; Zhuang, H.Y. Effects of organic farming on yield and quality of hybrid *Japonica* Rice. *Acta Agron. Sin.* **2015**, *41*, 458–467. (In Chinese)

30. Hasegawa, H.; Furukawa, Y.; Kimura, S.D. On-farm assessment of organic amendments effects on nutrient status and nutrient use efficiency of organic rice fields in Northeastern Japan. Agric. *Ecosyst. Environ.* **2005**, *108*, 350–362.

31. Pan, G.X.; Zhou, P.; Li, Z.P.; Smith, P.; Li, L.Q.; Qiu, D.S.; Zhang, X.H.; Xu, X.B.; Shen, S.Y.; Chen, X.M. Combined inorganic/organic fertilization enhances N efficiency and increases rice productivity through organic carbon accumulation in a rice paddy from the Tai Lake region, China. *Agric. Ecosyst. Environ.* **2009**, *131*, 274–280.

32. Zhao, J.; Ni, T.; Li, J.; Lu, Q.; Fang, Z.; Huang, Q.; Zhang, R.; Li, R.; Shen, B.; Shen, Q. Effects of organic–inorganic compound fertilizer with reduced chemical fertilizer application on crop yields, soil biological activity and bacterial community structure in a rice–wheat cropping system. *Appl. Soil Ecol.* **2016**, *99*, 1–12.

33. Zhou, H.; Fang, H.; Mooney, S.J.; Peng, X.H. Effects of long-term inorganic and organic fertilizations on the soil micro and macro structures of rice paddies. *Geoderma* **2016**, *266*, 66–74.

34. Marchand, S.; Guo, H.X. The environmental efficiency of non-certified organic farming in China: A case study of paddy rice production. *China Econ. Rev.* **2014**, *31*, 201–216.

35. Ahmed, S.; Humphreys, E.; Salim, M.; Chauhan, B.S. Growth, yield and nitrogen use efficiency of dry-seeded rice as influenced by nitrogen and seed rates in Bangladesh. *Field Crop. Res.* **2016**, *186*, 18–31.

36. Tanaka, A.; Toriyama, K.; Kobayashi, K. Nitrogen supply via internal nutrient cycling of residues and weeds in lowland rice farming. *Field Crop. Res.* **2012**, *137*, 251–260.

Conventional, Partially Converted and Environmentally Friendly Farming in South Korea: Profitability and Factors Affecting Farmers' Choice

Saem Lee, Trung Thanh Nguyen, Patrick Poppenborg, Hio-Jung Shin and Thomas Koellner

Abstract: While organic farming is well established in Europe a nd USA, it is still catching up in Asian countries. The government of South Korea has implemented environmentally friendly farming that encompasses organic farming. Despite the promotion of environmentally friendly farming, it still has a low share in South Korea and partially converted farming has emerged in some districts of South Korea. However, the partially converted farming has not yet been investigated by the government. Thus, our study implemented a financial analysis to compare the annual costs and net returns of conventional, partially converted and environmentally friendly farming in Gangwon Province. The result showed that environmentally friendly farming was more profitable with respect to farm net returns. To find out the factors affecting the adoption of environmentally friendly farming, multinomial logistic regression was implemented. The findings revealed that education and subsidy positively and significantly influenced the probability of farmers' choice on partially converted and environmentally friendly farming. Farm size had a negative and significant relationship with only environmentally friendly farming. This study will contribute to future policy establishment for sustainable agriculture as recommended by improving the quality of fertilizers, suggesting the additional investigation associated with partially converted farmers.

Reprinted from *Sustainability*. Cite as: Lee, S.; Nguyen, T.T.; Poppenborg, P.; Shin, H.-J.; Koellner, T. Conventional, Partially Converted and Environmentally Friendly Farming in South Korea: Profitability and Factors Affecting Farmers' Choice. *Sustainability* **2016**, *8*, 704.

1. Introduction

Agriculture creates benefits for humans by providing fiber, food and fuel. However, intensively managed farms have increased various adverse effects including soil degradation, biodiversity loss, water pollution and agro-chemical pollution. Due to heavily managed intensive farming targeting yield maximization, environmental concerns over negative externality of agricultural production have been increasing. Therefore, sustainable agriculture has been developed as the alternative under conservation of environmental quality and the scarcity of natural resources. One of the alternatives can be several advanced farming management practices such as organic, environmentally friendly and partially converted and low-pesticides farming.

Organic farming is one of the most widespread farming techniques that balance social, economic and environmental sustainability. Although there are many definitions of organic farming [1], it is generally defined that it avoids the use of synthetic chemical fertilizers and pesticides, and regulates the application of agriculture practices [2,3]. Organic farming emphasizes ecological processes, human health and renewable resources adapted to the local agricultural system [4]. Despite the contentious issue on economic and environmental effects, organic farming has the potential to reduce environmental pollution [5–7], with higher farm household income and benefits to rural economies [8,9]. Moreover, in response to consumer demand for healthy food products, many farmers are converting their production method from conventional to organic farming [10].

In contrast with other developed countries like those in the European Union, which have adopted strict organic farming, the government of South Korea has adopted environmentally friendly farming since 1999. Due to the more flexible regulations than those supporting organic farming in the European Union, environmentally friendly farming in South Korea includes organic, no-pesticide and low-pesticide farming [11]. While the use of chemical fertilizers and pesticides of organic farming is forbidden like in other developed countries, the no-pesticide farming standard in South Korea allows the use of a certain level of chemical fertilizers. The low-pesticide farming allowed the use of both a certain level of chemical fertilizers and synthetic pesticides was abolished in 2015. The government of South Korea has implemented a long-term plan to promote environmentally friendly farming since 2000. The plan aimed to extend cultivated areas, to decrease the synthetic chemical inputs such as fertilizers and pesticides and to expand the organic products market [12]. This plan increased certified areas of environmentally friendly farming up to 172,674 ha cultivated by 160,628 farm households in 2011. These produced and supplied 1,819,228 tons of environmentally friendly agricultural products in 2011. The main cultivated products of environmentally friendly farming were vegetables (38.5%), fruits (23.8%), and cereal crops (22.3%). The area of environmentally friendly cultivation was approximately 10% in 2011 [12]. However, organic agricultural area was only about 1.1% in 2011 (Table S1), still accounting for a small proportion [11]. This is similar to the global organic agricultural land, accounting for approximately 1% [13]. Although North America, Africa, and Asia are lagging behind Europe and Latin America that are leading the growth of organic farming, the proportion of land cultivated using organic farming method is still low all over the world [13].

In the context of the low adoption rate of organic farming all over the world, considerable research attention has been paid to economic differences between conventional and organic farming [7,14]. The differences between net returns and costs analyzed by previous studies show that organic farming can be profitable [15,16]. Considering higher willingness to pay for organic products and the price premium paid by consumers [8,17], organic farming is more financially lucrative than conventional farming [18,19]. The majority of previous studies examined the driving forces leading to organic farming in conjunction with biophysical, institutional, socio-economic and political factors influencing farmers' choices [20–23]. In South Korea, a number of studies have contributed to the development of environmentally friendly farming including organic farming in the context of the production, consumption and distribution for environmentally friendly farming [24–26].

Furthermore, in South Korea, through only our field survey, it was observed that partially converted farmers existed. Partially converted organic farming has emerged in some countries, however, it is not allowed in some other countries which require compliance with rigorous regulations for organic farming in the various developmental paths [27]. The partially converted farming is defined so that farmers can decide to use only part of their land for organic production [28]. In other words, the partially converted farmers are using both conventional and environmentally friendly farming practices according to their own choices. They can become completely organic farmers in the near future, but are starting by implementing some organic practices now. Consequently, their farms are less than "half-organic". While previous studies shed light on the profitability of different types of farming, including partially converted farming in Europe and Canada [28,29], less is known for Asian countries. Only some studies in this important world region focus on environmentally friendly or organic farming [30,31] and the issue of partially converted farming is not yet covered. The missing differentiation between fully and partially converted organic farming is certainly a limitation of current empirical studies on organic farming [8].

Therefore, the first objective of our research was to identify the profitability among different farming techniques; i.e., conventional farming (CF), partially converted farming (PCF) and environmentally friendly farming (EFF), hereafter abbreviated with CF, PCF and EFF respectively. The second objective was to examine the key factors influencing the adoption of farming techniques in South Korea. This paper draws crucial conclusions based on a detailed discussion of the financial analysis with descriptive statistics and multinomial logistic regressions. The findings and policy recommendations can make valuable contributions to development of policies to promote organic farming in South Korea and other Asian countries.

2. Method

2.1. Study Area and Background

This study was conducted in the Soyang catchment of Yanggu-Gun (Nam-Myeon, Yanggu-Eup and Haean-Myeon), Inje-Gun (Girin-Myeon, Nam-Myeon, Buk-Myeon, Sangnam-Myeon, Seohwa-Myeon and Inje-Eup) and Hongcheon-Gun (Nae-Myeon) in Gangwon Province, South Korea (Figure 1b). The study site was selected based on consideration of the low adoption level of organic farming in South Korea, as well as the potential hazard of water pollution within Soyang watershed from intensively managed practices in the Gangwon Province of South Korea. The Gangwon Province in South Korea plays a key role in protecting the water quality of the upper Soyang watershed, which provides water supplies to downstream residents of several, densely populated cities of South Korea. Accordingly, EFF has been promoted in the Gangwon Province, to improve the water quality in Soyang watershed, which comprises environmentally sensitive area. Despite the promotion of environmentally friendly farming by the local and central government, water quality issues coming from intensive farming activities in the area have been continued. Therefore, based on the low adoption rate of EFF and the desired reduction of water pollution from CF, we selected the main environmentally sensitive area, the three districts in Gangwon Province of South Korea as our study area.

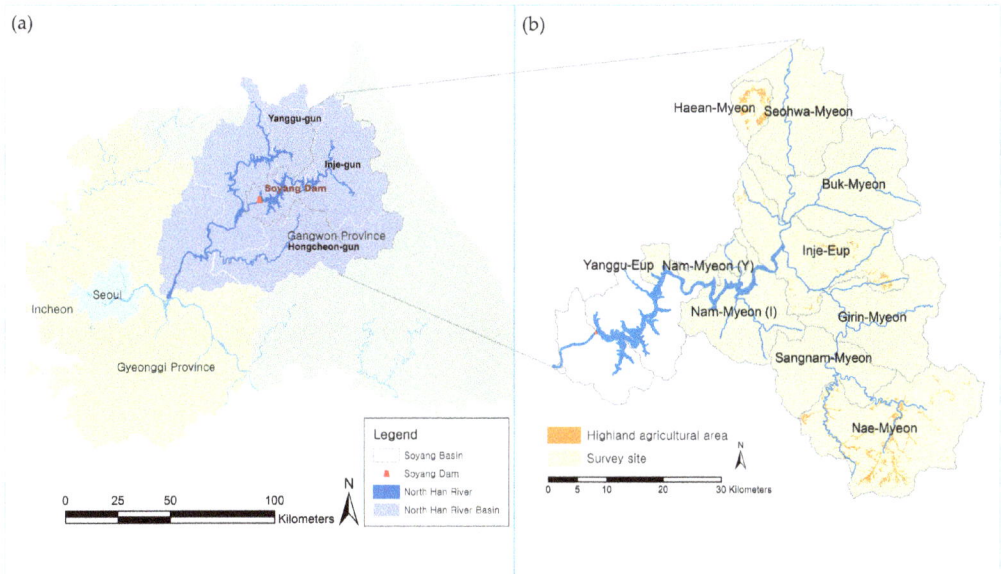

Figure 1. Location map of: (**a**) Soyang watershed in Gangwon Province; (**b**) three districts showing highland agricultural area.

The Gangwon Province of South Korea, which includes the catchment of the three districts, is located in the mountainous northeastern part of South Korea (latitude 37°02′N–38°37′N and longitudes 127°05′E–129°22′E). The Gangwon Province occupies around 20,569 km². The total agricultural area of the province was 109,496 ha which consists of rice paddies (41,086 ha) and field land (68,410 ha). The total population of farmers in the provin0ce was 191,922 in 2011 [32]. In 2011 the average farm size, from a total of 71,687 farm households in Gangwon Province, was 1.5 ha per farm, of which 0.57 ha was occupied by rice and 0.95 ha of field land, respectively. As of 2011, the total EFF cultivated farmland was 7962 ha with 5854 EFF farm households in this province; organic was 1976 ha and 1093 farm households, no-pesticide was 4899 ha and 3561 farm households, low-pesticide was 1088 ha and 1200 farm households [32]. The certified EFF area accounted for only about 4.6% of the whole of the certified areas in South Korea [26].

For this study, the three areas were selected within the watershed of Soyang Lake in Gangwon Province (Figure 1a). The Soyang watershed (2694.35 km²) is the largest reservoir and tributary located North of the Han River in South Korea. The watershed is important as one of the main drinking water sources of Seoul, capital of South Korea, and other metropolitan areas in South Korea. The residents in the downstream area of the watershed utilize the water resource overwhelmingly due to high population density of the capital area that shared 48.3% of the country's population in 2011. In other words, the water pollution in surrounding environmentally sensitive areas, especially in the selected area, is seriously affected by intensive farming, and can seriously damage fresh drinking water use of the citizens.

During the 2006 monsoon period, the water quality was seriously reduced brought by Typhoon "EWINER", resulting in high levels of turbidity (328NTU (Number of Transfer Units)), which was nearly four times the turbidity level observed the previous year. At that time, the sediment yields (865,062 ton/year) within the watershed were substantially higher from agricultural practices in the mountainous area. In order to protect the water quality of the Soyang watershed for the province, since 2006, selected areas have been designated as initial nonpoint pollution source management areas, with the aim of reducing sediment yields from agricultural practices in the mountainous areas of South Korea [33].

The three major regions causing the water quality problem from farming activities accounted for about 82.7% of the watershed in Gangwon Province [33]. The main areas of the Soyang watershed affecting from agricultural practices were Yanggu-Gun (146.44 km²), Inje-Gun (1678.48 km²) and Hongcheon-Gun (447.83 km²). The landscape of the catchment area is dominated in highland regions by upland fields. Out of the total highland farmland area (7313 ha) of South Korea, the majority of the highland upland areas were found in Yagngu-Gun (143.97 km²), Inje-Gun (1636.32 km²) and Hongcheon-Gun (447.51 km²) of the Province [33]. Regarding the water pollution associated with farming activities, the crucial problem identified was over-use of pesticides and fertilizers on steep slopes and at relatively high altitudes [34–36]. The main crops cultivated in the mountainous area were Chinese cabbage and radish, which rely heavily on chemical fertilizers and pesticides. In these areas, intensive agricultural practices with high concentrations of phosphorous and nitrogen, led to eutrophication of the reservoir [37,38]. This negatively affected the habitat of endangered species in the aquatic ecosystems of the watershed. Considering that the adverse effects could appear occasionally, although stable drinking water quality has been maintained in South Korea, there is potential for degradation of water quality from intensively managed farming activities still remaining in the districts during monsoon climate.

2.2. Sampling of Farm Households and Data Collection

Data were collected by face-to-face interviews. The survey period was between 19 March 2012 and 6 April 2012. The lists of residential farmers were received from local leaders and governmental staff after focus group meetings. During a pilot survey, we found that partially converted farmers existed between conventional and environmentally friendly farmers. Thus, stratified random sampling was selected from two farming techniques (CF and EFF) to three farming techniques (CF, PCF, and EFF). The sampling was applied to draw an estimated 7% sample size based on total population of farmers in three regions, due to time and budget constraints. Before the main survey of farm households, we contacted the farmers in the list by phone to check their production method and arrange the interviews from the contact lists. In addition, after the survey, in order to obtain more exact information on the survey, a gift was offered to the participants. Due to no responses and outliers in the key questions (Figures S1–S4), 218 farm households' interviews were analyzed from a total of 224 interviews. The data consisted finally of 85 conventional farmers, 65 partially converted farmers and 68 environmentally friendly farmers.

For the questionnaire, a pilot survey was carried out in order to check accuracy of the questionnaire and modify sentences to avoid misunderstanding. Through discussions with heads of the local farm households and governmental staff that were responsible for EFF, a semi-structured questionnaire was constructed. Based on feedback from the pilot survey with trained interviewers, a final questionnaire was completed. All data were investigated based on their farming activities in 2011, a year previous to the survey period. In order to compare more reliable financial profitability by farming techniques, data related to livestock were excluded from the survey. The questionnaire included farm size and number of cultivated crops in arable areas. The farmers were asked about their financial returns such as agricultural revenue and subsidies, as well as their cultivation costs, including expenditures for labor, seeds, installation and management of green houses, fertilizers, pesticides and agricultural machines. The final part of the survey collected socio-economic information of the farmers such as their age, education and farming experience.

2.3. Analytical Framework for Data Analysis

Descriptive statistics were used to give basic information on farm households. The descriptive indicators were average values with standard deviation and frequency, which applied as independent variables in dummy or mean values of multinomial logistic regression. In addition, financial analysis was carried out to compare costs and profits among the three farming methods. The calculation is specified by the following formula: The total benefit (E) = total revenue (B) − total costs (A) + total supported subsidy (D) (Tables 2 and 3). All costs included labor, land rent, mechanical operations, installation, management and maintenance in 2011. The farmers' net returns determined by the costs were calculated based on the revenue and subsidies obtained in 2011.

In our study, a multinomial logistic regression model was used to analyze the influence of socio-economic characteristics of farm households on different farming techniques. Multinomial logistic regression is an extended binary logistic regression model that has more than two categories of unordered outcome variables. The multinomial logistic model was estimated using normalization with one category, which is regarded as the "base category." In this study, the explanatory variable took different from one to three depending on their farming techniques. CF was used as the base category, which took one in the model. PCF took two in the explanatory variable and environmentally friendly farmers, which took three in the explanatory variable. There are several factors leading to choice decisions in the context of socio-economic background, and what we are interested in lies in the effect of each explanatory variable on individual outcomes. Therefore, we considered seven independent variables; age, education, labor of farm household, farm size, ownership, net return, and subsidy, which were simultaneously hypothesized as vital factors for the farmers' decisions.

Thus, the outcome variable can take on the variables, $j = 1, 2, 3, \cdots j$, with j, a positive integer. The model explains the probability of CF ($j = 1$) or PCF ($j = 2$), EFF ($j = 3$). The determinants associated with each category can be contrasted with the base category, which is CF in this study. In addition, this is to find out ceteris paribus changes in the elements of that affect the response probabilities,

$$p(y_i = k \mid x_i) = \frac{\exp(\beta_k x_i)}{\sum_{j=1}^{J} \exp(\beta_j x_i)}, \ j = 1, 2, 3, \ldots, J,$$ where k is one of the sub-groups and $P(y_i = k)$ is the probability that the farmer belongs to the subgroup and where x_i describes farmer characteristics. In order to identify this model, constraints for the assumptions must be applied. A common approach is to assume that $\beta_1 = 0$ [39].

This normalization makes it possible to identify the coefficients relative to the base outcome. Applying the constraint, the model can be written as:

$$p\left(y_i = k| x_i\right) = \frac{\exp(\beta_k x_i)}{1+\sum_{j=2}^{J} \exp(\beta_j x_i)}, \; for \; K > 1$$

$$p\left(y_i = k| x_i\right) = \frac{1}{1+\sum_{j=2}^{J} \exp(\beta_j x_i)}$$

(1)

The multinomial logit model utilizes maximum likelihood estimation to evaluate the probability of a categorical group using the following equation:

$$L\left(\beta_2, ..., \beta_j| y, X\right) = \prod_{k=1}^{j} \prod_{y_i=k} \frac{\exp(\beta_k x_i)}{\sum_{j=1}^{J} \exp(\beta_j x_i)}, \; \text{where} \; \prod y_i = k \; \text{is the product over}$$

all cases for which $y_i = k$ [40]. Coefficients are interpreted using the relative risk ratios, which is the relative probability of $y_i = k$, for k > 1. The relative risk ratio is calculated without reference to the remaining two groups, PCF and EFF. This shows the underlying assumption that the model has independence from irrelevant alternatives which is regarded as binary independence [40,41]. Although statistical tests are available to confirm this proposition, the use is not recommended due to unreliable test results [42,43]. Thus, based on the recommendation by Amemiya [44], a multinomial logistic model was selected among three types of farming techniques. Overall, the model helps to indicate significant differences between PCF and EFF in the study area, relative to CF. The utilized data were analyzed by IBM SPSS statistics. The parameter estimates for the vectors that maximize the log likelihood function can be achieved [45]. Relative risk ratios, meaning probabilities of choice, can be calculated from Equation (2):

$$\frac{\partial P_{iy}}{\partial x_i} = P_{ij}\left[\beta_j - \sum_{k=1}^{J} P_{ik}\beta_{k0}\right] \; for \; J = 1,2,....J$$

(2)

Applying Equation (2), we can observe changes in probabilities for their choice in farming techniques due to a small change in one of the farmers' characteristics, when all other independent variables are fixed [46]. The relative risk ratios for the multinomial logistic model were obtained by exponentiation of the coefficient. The exponent of the coefficients are commonly interpreted as odds ratios like logistic regression models and regarded as a marginal effect. The interpretation of the relative risk ratios is for a unit change in the predictor variable The relative risk ratio of base outcome relative to the reference group is expected to change by the factor of a respective parameter estimate, given the variables in the model are held constant.

Based on the findings of earlier studies, our study hypothesized that social and economic characteristics of farmers can be fundamental components in the adoption of farming practices. The age of farmers plays a significant role on the farmer's decision regarding conversion because younger farmers are expected to be more progressive and accepting of new farming techniques relative to older farmers [22,47]. The level of education is considered as an influencing factor. This is because well-educated farmers are more likely to utilize new advanced technologies efficiently and recognize the benefits for agricultural practices [48,49]. Farm size plays a crucial role in the conversion to EFF in terms of costs and benefits. Furthermore, higher costs of labor and time are inevitable during the conversion process [50,51]. Land ownership can be an advantage in terms of reducing the land rent cost [52]. As subsidies affect the profitability of EFF [53–55], farm net returns have also been identified as a key driver of the conversion to EFF [56,57].

Therefore, it is expected that the sign on the age variable will be negative because older farmers may set their sights on investments for farming activities over a short period of time. Education level is expected to have a positive influence on the adoption of EFF. The higher the education, the higher the probability that farmers may consider the benefits from EFF practices to recoup their costs and reap their future profits. The variable farm size was expected to have a negative sign due to the risk of income loss during the transition period and higher labor costs to convert farming techniques to EFF. The expected sign of the variable labor is negative. This is because labor is associated with additional costs and investments in the long term. Land ownership is expected to have a positive impact on the conversion to EFF in terms of fixed costs for farm management. It is clear that higher benefits were hypothesized to be positively associated with adoption of EFF. Obtaining a subsidy was perceived as a positive economic factor that affects farmers' choice on converting to EFF.

3. Results

The characteristics of the 218 farmers among three types of farming techniques are presented in Table 1. The general characteristics of the farmers are shown by descriptive statistics and the results of the one-way ANOVA. In regards to education, EFF farmers had the highest education level, with 17.6% university alumni and 25% high school graduates. The average farm size for CF was 3.4 ha. The average farm size for PCF was 4.0 ha, which included farmland area of 63.8% CF and 36.2% EFF. EFF occupied an average farm size of 2.3 ha, approximately half of the total PCF cultivated area. The age of farmers was homogenously distributed between the three groups. The group of CF was on average 55.7 years, whereas the group of partially converted farmers was on average 52.5 years old. The environmentally friendly farmers were on average 54.3 years of age. CF and PCF farmers had similar farming experience while EFF farmers had less farming experience. With respect to the EFF experience, environmentally friendly farmers had been doing EFF for nine years, about three years more experience compared to partially converted farmers.

The average number of cultivated crops for PCF farmers was 5.4 ha with a range of 2–9 crops in both farming techniques. CF and EFF had similar crop numbers (3.4 and 3.8 crops, respectively). The findings of the ANOVA analysis showed that the three farming techniques differ significantly in their farmland size ($F_{(218)} = 4.5$, $p < 0.10$)) and average number of cultivated crops ($F_{(218)} = 22.5$, $p < 0.01$)). The distribution of main crops among the three groups is shown in Table S1 of appendix.

Table 2 presents the results on differences for annual average costs and benefits per farm. PCF had the largest average costs per farm with most expenditure for farm management. CF had no big difference with PCF for land rental costs. EFF had the lowest land rental costs of 1.37 million KRW and fertilizer costs of 3 million KRW. Regarding the average cost of labor, PCF had the largest wage cost of 14.70 million KRW, compared to CF and EFF. PCF had the largest fertilizer expenditure, whereas EFF had the smallest fertilizer expenditure. In terms of cost of pesticides per farm, PCF had higher pesticide expenditures than that of EFF farms. PCF had the largest other costs compared to CF and EFF.

With respect to benefits per farm household, PCF had the largest annual revenues with 61.10 million KRW compared to CF and EFF. However, the EFF net income was the largest with 26.29 million in comparison to CF and PCF. The annual net income of a PCF farm household was the smallest which was similar to CF as the PCF farmers have the highest costs for their farming activities. Although EFF had the largest subsidies from the government or province, the amount of the annual subsidy among different farming techniques had no large difference. The total annual benefit (farm net income plus subsidies) per farm was the largest for EFF, about 1.5 times greater than the benefit of PCF and CF.

The results of annual average costs and benefits per ha are shown in Table 3. Compared to the costs per farm (Table 2), the results for costs per ha were somewhat different. The land rental cost per ha was almost the same for CF and EFF. There was no big difference in land rental costs per ha. PCF had the smallest costs for their farmland. Average labor costs per ha were the largest for EFF, which was the highest expenditure compared of all farming techniques. CF had the lowest expenditure for labor costs. Contrary to the result of fertilizer cost per farm, the costs of fertilizer were the largest for EFF. CF had the smallest fertilizer costs and PCF was the largest. In terms of cost of pesticides per ha, CF had the highest pesticide expenditures compared to that of PCF and EFF farms. Regarding other costs, PCF farmers spent the most on other costs, whereas CF farmers spent the least. Thus, total annual cost per ha of EFF was 12.85 million KRW. The EFF farmers had the largest annual costs compared to CF and PCF.

Table 1. Descriptive statistics of characteristics of farm types.

Characteristics	Description (Unit)	Conventional Farming CF (N: 85)	Partially Converted Farming PCF (N: 65)	Environmentally-Friendly Farming EFF (N: 68)	Total (N: 218)	F-value [a]
Education	Primary School (%)	38.8	23.1	26.5	30.3	
	Secondary School (%)	35.3	29.2	30.9	32.1	
	High School (%)	22.4	41.5	25.0	28.9	
	University (%)	3.5	6.2	17.6	8.7	
CF	Area under management (%)	100	63.8	NA	35.3	
EFF	Area under management (%)	NA	36.2	100	23.3	
		Mean (Std. Dev.)				F-value [a]
Farm size	(ha)	3.4 (3.8)	4.0 (4.2)	2.2 (1.8)	3.2 (3.5)	4.5 *
Age	(Years)	55.7 (10.2)	52.5 (7.9)	54.3 (9.4)	54.3 (9.3)	2.3
Farm experience	(Years)	29.7 (14.1)	29.0 (11.2)	25.9 (14.4)	28.3 (13.4)	1.6
EFF practices	(Years)	NA	6.1 (5.0)	9.1 (5.4)	7.6(5.4)	NA
Average number of crops	(N)	3.4 (1.4)	5.4 (1.8)	3.8 (2.2)	4.1 (2.0)	22.5 ***

[a] Generated from one way ANOVA; * Statistical significance at the 10% level; *** Statistical significance at the 1% level.

82

Table 2. Difference of the three different production modes in annual costs, revenues and total benefits per farm household per year. The numbers display the mean of all farms and in brackets the standard deviation.

	Conventional Farming CF (N: 85)	Partially Converted Farming PCF (N: 65)	Environmentally-Friendly Farming EFF (N: 68)	Total (N: 218)
Costs (10,000 KRW [a]/farm household/year)				
Land rent	230 (422)	233 (475)	137 (325)	202 (412)
Labor	942 (1884)	1470 (1898)	1084 (1904)	1144 (1899)
Fertilizer	463 (620)	649 (715)	300 (381)	467 (602)
Pesticides	503 (810)	545 (808)	182 (303)	416 (707)
Other costs [b]	874 (1500)	1390 (1979)	580 (685)	936 (1508)
Total cost (A)	3012 (4055)	4287 (4471)	2284 (2957)	3165 (3948)
Benefits (10,000 KRW [a]/farm household/year)				
Revenue (B)	4840 (6554)	6110 (5424)	4913 (6305)	5241 (6157)
Farm net income (C = B − A)	1828 (4711)	1823 (3566)	2629 (4156)	2076 (4221)
Subsidy (D)	109 (314)	119 (261)	131 (231)	119 (274)
Total benefit (E = C + D)	1936 (4744)	1942 (3609)	2760 (4220)	2195 (4266)

[a] Unit: 10,000 KRW = 7.56 euro; [b] Other costs mean extra costs for cultivating crops excepting the above mentioned costs, such as seeding, renting agricultural machinery, etc.

Table 3. The result for annual average costs, revenues and benefits standardized per hectare and year. All units are 10,000 KRW/ha/year. The numbers display the mean of all farms and in brackets the standard deviation.

	Conventional Farming CF (N: 85)	Partially Converted Farming PCF (N: 65)	Environmentally-Friendly Farming EFF (N: 68)	Total (N: 218)
	Costs (10,000 KRW [a]/farm household/year)			
Land rent	50 (68)	47 (59)	50 (100)	49 (77)
Labor	236 (389)	344 (389)	410 (610)	323 (473)
Fertilizer	157 (187)	195 (203)	251 (410)	197 (281)
Pesticides	155 (181)	144 (160)	109 (177)	137 (174)
Other costs [b]	303 (431)	540 (1442)	464 (680)	424 (916)
Total cost (A)	901 (862)	1270 (1564)	1285 (1322)	1131 (1258)
	Benefits (10,000 KRW [a]/farm household/year)			
Revenue (B)	1697 (1584)	2087 (2959)	2854 (2668)	2174 (2447)
Farm net income (C = B − A)	796 (1132)	817 (1764)	1570 (2183)	1044 (1735)
Subsidy (D)	57 (216)	49 (139)	89 (186)	65 (186)
Total benefit (E = C + D)	853 (1200)	866 (1805)	1658 (2151)	1108 (1756)

[a] Unit: 10,000 KRW = 7.56 euro; [b] Other costs mean extra costs for cultivating crops excepting the above mentioned costs, such as seeding, renting agricultural machinery, etc.

The farming technique with the largest annual revenue per ha was EFF, which made 28.54 million KRW. The annual revenues per ha of CF and PCF were 16.97 million KRW and 20.87 million KRW, respectively. EFF had the highest annual farm net income per ha with 15.70 million KRW. The net income of CF was 7.96 million KRW and the net income of PCF was 8.17 million KRW. In the case of their subsidy per ha, EFF had the largest subsidy, which was 0.89 million KRW. PCF had the lowest amount of subsidy in their farming activities at 0.49 million KRW. Therefore, total annual benefit per ha of EFF was the highest compared to CF and PCF. The difference of the benefits between EFF and other farming techniques was about double. The total benefit of CF and PCF was slightly different, as the total benefits of CF and PCF were 8.53 million KRW and 8.66 million KRW, respectively.

The result of multinomial logistic regression model is presented in Table 4. Based on the R^2 pseudo statistics and Chi-Square test, this multinomial logistic regression model shows that the estimated model is well fitted and statistically significant at the 1% level. It is important to note that likelihood ratio statistics indicated by X^2 statistics (52.57) are highly significant ($p = 0.0001$), suggesting that this makes the estimates obtained good enough for running this analysis. The Log likelihood value suggests that the model has adequately explained the farmers' choices on farming techniques. In all cases, the estimated coefficients are compared with the base category of conventional farming. Conventional farmers occupied 39.0% of our survey. The partially converted farmers accounted for 29.8%, whereas environmentally friendly farmers accounted for 31.2% of the sample.

The estimates for PCF and EFF relative to CF were observed differently with positive signs across the groups. The result showed that age, labor of farm household, land ownership, and farm net income were not statistically significant. However, education level, farm size and subsidy were significantly related to the farmers' choice on farming techniques. The coefficient for education level was statistically significant and positively correlated to the probability on PCF and EFF at 10% significance level, relative to CF. Farm size was found to be statistically significant at 5% significance level and positive correlation with the probability of adopting EFF, whereas farm size was not significantly related to the PCF. The coefficient for subsidy was highly significant for both farm groups relative to the base outcome at the 1% significance level. This indicates a strong positive relationship between the subsidy and the likelihood of farmers' adoption of PCF and EFF relative to CF. Therefore, these results show that as farmers' education level and subsidy increase, the likelihood of farmers' choice for PCF and EFF increases. Moreover, as the farm size decreases the probability of farmers' choice on EFF increases.

Table 4. Coefficient estimates and standard errors in parentheses for multinomial logistic regression model.

Variable	Partially Converted Farming PCF		Environmentally Friendly Farming EFF	
	Coeff.	Std. Error	Coeff.	Std. Error
Intercept	−1.212	1.293	−1.760	1.306
Age	−0.103	0.224	0.129	0.224
Education [1]	0.353 *	0.188	0.354 *	0.190
Farm size	0.015	0.050	−0.219 **	0.104
Labor of farm household [2]	0.247	0.437	0.361	0.436
Land ownership of land [3]	0.195	0.382	−0.586	0.391
Subsidy [4]	1.005 ***	0.356	1.649 ***	0.378
Farm net income	−0.035	0.049	0.047	0.054

Number of observations 218; Pseudo R^2: Cox and Snell 0.21; Nagelkerke 0.24; McFadden 0.11; LR chi^2(12) 52.57; Log likelihood −211.65.

Notes: * significant at 10%; ** significant at 5%; *** significant at 1%. [1] 0 = no education; 1 = primary education; 2 = secondary education; 3 = high school; 4 = college and university; [2] 0 = farmers who had no farm laborer, 1 = farmers who had own farm laborers; [3] 0= farmers who rented farmland, 1 = farmers who possess farmland; [4] 0 = farmers who did not receive subsidy, 1 = farmer who received subsidy.

The relative risk ratios of the multinomial logistic model are shown in Table 5. This result was obtained by the exponential of the coefficients, which provide estimates of the relative risks. The result showed that one unit change in education level had no significant differences between PCF and EFF, whereas relative risk ratios of the variable increased. It was expected that the relative risk of practicing PCF and EFF over CF (base category) increased by Exp. (0.35) = 1.42. If the farmers would increase their education level by one unit, the relative risk for PCF and EFF relative to CF would be expected to increase by the determinants of 1.42, given other variables in the model are held constant. With regard to farm size for their cultivated land, the relative risk ratio for EFF relative to CF would be expected to decrease by a factor of 0.80 given the other variables in the model are held constant. As farm size is negatively related to the EFF, an increase in farm size by one unit reduces the likelihood that a farmers' chose EFF by 80.3%. In addition, the relative risk ratios of the variable subsidy for PCF and EFF were 2.73 and 5.20, respectively. Given a one unit increase in subsidy, the relative risk of having adopted PCF and EFF would be 2.73 times and 5.20 times, respectively, more compared with the CF. This means farmers who received subsidies were more likely to choose PCF and EFF by a factor of 2.73 and 5.20, respectively, as partially converted and environmentally friendly farmers require subsides for the adoption.

Table 5. The results of marginal effects by multinomial logistic regression model.

Variable	Marginal Effect [1]	
	Partially Converted Farming PCF	Environmentally Friendly Farming EFF
Age	0.902	1.138
Education	1.423 *	1.425 *
Farm size	1.015	0.803 **
Labor of farm household	1.280	1.435
Land ownership	1.216	0.556
Subsidy	2.733 ***	5.200 ***
Farm net income	0.966	1.048

Note: * significant at 10%; ** significant at 5%; *** significant at 1%. [1] Marginal effect means exponentiation of the coefficients which is regarded as odds ratios for the predictors.

4. Discussion

Organic farming is one of several advanced farming techniques considered to provide environmental benefits and fit within the spectrum of sustainable economic development. In environmentally sensitive areas, organic farming supports water conservation as it reduces the rate of damaging runoff coming from insensitively managed farming. The national government of South Korea has adopted environmentally friendly farming (EFF) in order to move towards sustainable agriculture. The adoption rate of EFF in South Korea is, however, still low as it is in other developed and developing countries. Additionally, the selected area of our study is relative to other regions in South Korea more important with respect to farmers' decision on practices for watershed protection. Historically, during the monsoon period, in the selected area in Gangwon Province, the excess use of chemical fertilizers and pesticides has caused the permeation of these chemicals into surface waters, leading to negative effects on the water quality of the Soyang watershed, a main source of drinking water of South Korea. Thus, in order to identify which farming techniques are profitable and what factors influence farmers' choices, we compared the costs and benefits of various farming techniques and examined socio-economic factors affecting adoption of farming techniques, based on survey data. The findings of this study can contribute to the promotion and development of organic farming in South Korea. In addition, this study can be developed into similar studies in other Asia countries and in environmentally sensitive areas using multi-year data.

4.1. Environmentally Friendly Farming in South Korea

In South Korea, agriculture can be generally categorized into conventional farming (CF) and EFF. However, in this field survey, we found that partially converted farming (PCF) is emerging. Accordingly, the survey was conducted with the three types of farming techniques, namely CF, PCF and EFF. Moreover, the study site was a part of the nonpoint pollution sources management areas (Hongcheon-Gun, Inje-Gun, and Yanggu-Gun) within the catchment of the Soyang watershed in Gangwon Province, South Korea. The management area for nonpoint pollution sources was designated to prevent water quality degradation due to eroded soil from agricultural areas in this province. The Soyang catchment of this province has an important role in the supply of potable water for the metropolitan area Seoul. Despite the promotion of EFF by the local authorities and government of South Korea, the Gangwon Province contained a low certified area of EFF. Thus, with the importance of the study sites, this research aimed to identify which farming technique is more profitable by financial analysis and to examine which factors affect the adoption of farming techniques in South Korea using multinomial logistic regression.

4.2. Cost and Benefits of the Three Farming Techniques

The results of the financial analysis showed that the EFF labor costs per ha were higher than CF and PCF. This is in line with previous studies that have shown that organic farming has more labor requirements than CF [9,58]. In our study, fertilizer costs for EFF per ha were higher than for other farming techniques. This finding is inconsistent with the result of Sgroi et al. [59], who found that CF had higher fertilizer costs when compared to organic one. The reason for the higher fertilizer costs in this area might be caused by the use of low quality organic fertilizer, which led not only to less crop production but also caused higher costs. Due to a short history of EFF in South Korea, the adequate production, distribution and quality assurance of organic fertilizer are problematic and tend to increase their production costs [26]. This is in line with the studies of Bernal et al. [60] who mentioned that an increase in yields would require high compost quality and improved quality of organic fertilizer. Therefore, in order to promote the EFF, proper quality and quantity of fertilizers including different nutrients and ingredients should be investigated for the various crop choices reflected in different districts. An alternative way to reduce production costs substantially would be improved soil fertility, by promoting compost and nutrient management strategies. Considering water quality degradation of the catchment from soil erosion and nutrient run-off in this study area, the moderate application of fertilizers, dependent on the local geographical conditions, is required to protect the fresh water quality.

With regard to the benefits, financial net returns per farm and ha of EFF were higher compared to CF and PCF, when considering the total expenses, annual income and subsidies. This is coherent with the results of Kristiansen et al. [61], Delbridge et al. [16], Patil et al. [7] and Salvioni et al. [28] who showed the profitability of organic farming. In the benefit of EFF, the higher revenues per ha might be due to the price premium of the produce. This is consistent with findings of studies which indicated that the higher net returns can be attributed to the premium price of organic products [62,63]. In South Korea, with the certification system of EFF, a price premium incentivizes the farmers into the EFF products market like in other developed countries [64]. The price premium was about 1.2~2.0 times depending on different crop choices [26]. In the study area, we found with the personal interviews, that some farmers had contracts with a big market in the capital city as they guarantee relatively higher selling prices. Therefore, despite higher total costs per ha of EFF, compared to those of CF and PCF, the EFF was more financially attractive in this area with higher price premiums of the products. The results associated with profits in our study area were in contrast with the study by Kim et al. [26] that also surveyed in South Korea in terms of different crops in various provinces; they found that EFF cultivation of rice, vegetable and fruits had higher costs and lower benefits due to a transition period which caused low yields and hence income loss. Even though our work provides a number of interesting results, it should be extended in the future by interviewing more households in different areas of South Korea in different years so that the results can be generalized and are more robust. Thus, we suggest that future studies should survey more data in multiple years.

4.3. Factors Influencing the Adoption of Partially Converted Farming PCF and Environmentally Friendly Farming EFF

In our survey, most of the farmers that were interviewed as representatives of their farm households were male. With respect to the education level in our survey, EFF farmers were found to be better educated than the CF and PCF counterparts. Among the three agricultural groups, age differed only little, between one to three years on average. Among the farming techniques, the farming experience between CF and PCF was almost identical while the standard deviation for CF experience was slightly larger than the farming experience of PCF. Regarding the green farming experience, PCF farmers had less experience by about three years, compared with the EFF farmers. Farm size and number of crops were statistically significant as shown by ANOVA. The EFF had the smallest cultivated area, whereas PCF had largest farm size, which is in line with the results of the largest number of cultivated crops in PCF. PCF farms had a higher cultivated farm area per farm household than the South Korean average (1.23 ha in 2010).

To identify influencing factors determining the three farming techniques multinomial logistic regression (MNL) was used. Before implementing a variance inflation test was implemented to consider the risk of multicollinearity between selected explanatory variables. While the estimates of the parameter in MNL model gives the direction of the effect of predictors on the explanatory variable, the marginal effects in the model offer the actual magnitude of change in probability. Thus, in MNL, we showed the coefficients and marginal effects indicating relative risk ratios (Tables 4 and 5) are significant determinants that have an influence on the likelihood of the farmers' choice on farming techniques. The MNL model included important socio-economic variables such as age, education level, farm size, labor, land ownership, subsidies and net returns per farm household. Although we considered both subsidies and net income in this model simultaneously, the interpretation of the effects of these factors should be done with care, since they might be a causality problem due to an econometric simultaneity issue. The results showed that age, whether or not farmers have laborers and ownership over their farmland, and net farm income were not significantly related to any of the three farming techniques.

However, as expected, education level of farmers was positively correlated to PCF and EFF. This result is hardly surprising as more educated farmers would have acquired the knowledge and would adopt advanced techniques relatively easily. This implied that the higher the education of the farmers, the greater the likelihood that farmers choose to adopt PCF and EFF, by 1.42 times. This finding confirms that of Weir and Knight [48], and Lapar and Ehui [49] who argue that an increase in farmers' education level increases the likelihood of adopting advanced farming techniques.

Moreover, farm size had a negative and significant relationship with EFF. This implies that the farm size decreases the tendency of adopting EFF by 0.80. Our finding supports the previous study by Khaledi et al. [29] who found that farmers with smaller farmland can more easily manage their fields to certified regulations. In addition, relatively small farmlands could be easier to manage within the regulations and standards of organic farming. This is inconsistent with the results of Karki et al. [50], showing that larger farm size is likely to adopt organic farming. This means the larger farm size has the potential for higher costs in labor and inevitable larger income loss during their transition period after they adopt EFF. In addition, according to Padel [65], the conventional and partially converted farmers could adopt organic farming later. The result is in line with Läpple and Rensburg [5] suggesting that larger farms are less likely to adopt organic farming which causes more intensive labor and is associated with higher costs and relatively higher risks.

The variable indicating whether or not farmers receive subsidies had a highly positive influence on the probability of the farmers' adoption of PCF and EFF. As a result of marginal effects of subsidies for PCF and EFF, the relative risk ratio for PCF and EFF relative to CF would be expected to increase by a factor of 2.73 and 5.20, respectively. The result demonstrates that receiving subsidies is the most significant positive influence on farmers' decisions. Moreover, similar studies found a positive relationship between the conversion process as an institutional factor [66,67].

This revealed that the subsidy can be considered as a key factor to encourage farmers to convert to EFF and expand arable land area of EFF [68]. Considering the importance of the subsidy, it should be noted that the direct payment program for EFF in South Korea is important to stimulate the farmers to change their farming techniques to EFF. In order to extend the EFF, the improvement of direct payment program for EFF is required as an incentive for compensating the income loss of environmentally friendly farmers during their transition period. The improvement measure to enhance the program of direct payment could be the unit price adjustment, changes in the payment period and the compensation by crop types [69].

4.4. Partially Converted Farming PCF in Our Study

The results of the characteristics of PCF indicated that the partially converted farmers had the largest farm size and the highest number of crops. Although some PCF farmers went through the transition period in order to adopt EFF and the higher costs for implementing PCF, they continued to practice the PCF. This can play an important role in extending agricultural land of EFF. Therefore, viewed this way, the partially converted farmers in the districts might be considered as a bottleneck in promotion of EFF. Monitoring the developments of the agricultural sector among different types of farming techniques could be a key issue in the promotion policy of the local and national government.

Furthermore, throughout the interviews with farmers in the field survey conducted for this research, we found that partially converted farmers exist. The PCF is not officially recorded by the government as PCF farmers might be normally grouped in CF or in EFF under official data of the government. Therefore, extra studies related with PCF might be needed. Specifically, regarding the PCF, there is still little research on how PCF has developed, how they affect the market and how they influence the decision of other farmers. Accordingly, several questions occur: Can they be considered as a potential barrier to promote EFF, or are they in a transition period towards EFF? How high is the possibility that they return to CF or persevere with PCF? In this respect, PCF is especially important, as these farmers have the potential to compare both farming techniques and output of the sectors.

5. Conclusions and Policy Recommendations

The process of moving toward sustainability through organic farming has led to the emergence of partially converted farming in South Korea. These new partially converted farmers are not officially recorded and not investigated in South Korea. Partially converted farms could be a potential barrier for promotion of organic farming. Therefore, to extend organic agricultural land area, an up-to-date official database for partially converted farmers including production costs and revenues should be established in each district. In addition, while environmentally friendly farming is more profitable in our study area, the probability of higher costs is still remaining and could be one of the obstacles to extending organic agricultural land. Therefore, the government should provide more detailed support for reducing production costs. In particular, higher fertilizer costs are required in order to invest in improvement of the quality and investigation of the appropriate quality for organic fertilizers. Ultimately, in order to promote compliance with international standards of organic farming, improved measures for enhancing fertilizer management should be implemented by the government. Farmers' choice behavior can be driven by the utility perceived and net benefit from farming techniques. This is beyond the aim of the current study, which has focused on financial profitability and determinants affecting their decisions. Further research would be necessary to investigate farmers' perception and behavior reflecting different local conditions. Considering varying socio-economic characteristics and different factors affecting farming techniques in different regions, research projects on promotion of organic farming would be beneficial to design more targeted policy for sustainable agriculture.

Supplementary Materials: The following are available online at www.mdpi.com/2071-1050/ 8/7/704/s1. Figure S1: Distribution of costs per farm including outliers (N: 224), Figure S2: Distribution of benefits (red) per farm including outliers (N: 224), Figure S3: Distribution of costs per ha including outliers (N: 224), Figure S4: Distribution of benefits (yellow) per ha including outliers (N: 224), Table S1: Total and organic cultivated area and the consumed quantity per ha of chemical fertilizers and pesticide in South Korea, Table S2: Main crops in percentage of farmers cultivating it and its average farm size split by farming techniques (Conventional farming CF, Partially Converted farming PCF and Environment-Friendly Farming EFF).

Acknowledgments: This study was carried out as part of the International Research Training Group TERRECO (GRK 1565/1) funded by the Deutsche Forschungsgemeinschaft (DFG) at the University of Bayreuth, Germany and the National Research Foundation of Korea (NRF) at Kangwon National University, Chuncheon, South Korea. We would like to thank Joe Premier for providing English corrections. This publication was funded by the German Research Foundation (DFG) and the University of Bayreuth in the funding programme Open Access Publishing.

Author Contributions: All authors developed the research design and contributed to the writing of the paper. Saem Lee collected the data and did the statistical analysis. Trung Thanh Nguyen, Patrick Poppenborg, Hio-Jung Shin and Thomas Koellner contributed to the data analysis, and reviewed and edited the manuscript.

Conflicts of Interest: The authors declare no conflict of interest.

References

1. Stanhill, G. The comparative productivity of organic agriculture. *Agric. Ecosyst. Environ.* **1990**, *30*, 1–26.
2. Aldanondo-Ochoa, A.M.; Casasnovas-Oliva, V.L.; Arandia-Miura, A. Environmental efficiency and the impact of regulation in dryland organic vine production. *Land Use Policy* **2014**, *36*, 275–284.

3. Argyropoulos, C.; Tsiafouli, M.A.; Sgardelis, S.P.; Pantis, J.D. Organic farming without organic products. *Land Use Policy* **2013**, *32*, 324–328.
4. Stolze, M.; Lampkin, N. Policy for organic farming: Rationale and concepts. *Food Policy* **2009**, *34*, 237–244.
5. Läpple, D.; Rensburg, T.V. Adoption of organic farming: Are there differences between early and late adoption? *Ecol. Econ.* **2011**, *70*, 1406–1414.
6. Leifeld, J. How sustainable is organic farming? *Agric. Ecosyst. Environ.* **2012**, *150*, 121–122.
7. Patil, S.; Reidsma, P.; Shah, P.; Purushothaman, S.; Wolf, J. Comparing conventional and organic agriculture in Karnataka, India: Where and when can organic farming be sustainable? *Land Use Policy* **2014**, *37*, 40–51.
8. Uematsu, H.; Mishra, A.K. Organic farmers or conventional farmers: Where's the money? *Ecol. Econ.* **2012**, *78*, 55–62.
9. Lobley, M.; Butler, A.; Reed, M. The contribution of organic farming to rural development: An exploration of the socio-economic linkages of organic and non-organic farms in England. *Land Use Policy* **2009**, *26*, 723–735.
10. Scialabba, N.E.-H. *Organic Agriculture and Food Security*; FAO: Rome, Italy, 2007.
11. Jeong, H.-K.; Moon, D.-H. *Response Strategy to the Abolishment of Low-Pesticide Agricultural Product Certification*; Korea Rural Economic Institute (KREI) Report; Korea Rural Economic Institute: Seoul, Korea, 2013. (In Korean)
12. Ministry of Agriculture, Food and Rural Affairs (MAFRA). *The 3rd Environment-Friendly Agriculture Promotion 5-Year Plan*; Environment-Friendly Agriculture Division: Sejong ni, Korea, 2013. (In Korean)
13. The World of Organic Agriculture–Statistics and Emerging Trends 2016. Available online: https://shop.fibl.org/en/article/c/statistics/p/1698-organic-world-2016.html (accessed on 1 March 2016).
14. Rattanasuteerakul, K.; Thapa, G.B. Status and financial performance of organic vegetable farming in northeast Thailand. *Land Use Policy* **2012**, *29*, 456–463.
15. Acs, S.; Berentsen, P.; Huirne, R.; Van Asseldonk, M. Effect of yield and price risk on conversion from conventional to organic farming. *Aust. J. Agric. Resour. Econ.* **2009**, *53*, 393–411.
16. Delbridge, T.A.; Coulter, J.A.; King, R.P.; Sheaffer, C.C. A Profitability and risk analysis of organic and high-input cropping systems in Southwestern Minnesota. In Proceedings of the Agricultural and Applied Economics Association, Annual Meeting, Denver, CO, USA, 25–27 July 2010.
17. Offermann, F.; Nieberg, H. *Economic Performance of Organic Farms in Europe*; Organic Farming in Europe: Economics and Policy; University of Hohenheim: Stuttgart, Germany, 2000.
18. Mahoney, P.R.; Olson, K.D.; Porter, P.M.; Huggins, D.R.; Perillo, C.A.; Kent Crookston, R. Profitability of organic cropping systems In Southwestern Minnesota. *Renew. Agric. Food Syst.* **2004**, *19*, 35–46.

19. Oude Lansink, A.; Jensma, K. Analysing profits and economic behaviour of organic and conventional dutch arable farms. *Agric. Econ. Rev.* **2003**, *4*, 19–31.

20. Padel, S. Conversion to Organic Farming: A typical example of the diffusion of an innovation? *Sociol. Ruralis* **2001**, *41*, 40–61.

21. Sarker, M.A.; Itohara, Y.; Hoque, M. Determinants of adoption decisions: The case of organic farming in Bangladesh. *Ext. Farming Syst. J.* **2010**, *5*, 39–46.

22. Mabuza, M.L.; Sithole, M.M.; Wale, E.; Ortmann, G.F.; Darroch, M.A.G. Factors influencing the use of alternative land cultivation technologies in Swaziland: Implications for smallholder farming on customary Swazi Nation Land. *Land Use Policy* **2013**, *33*, 71–80.

23. Ullah, A.; Shah, S.N.M.; Ali, A.; Naz, R.; Mahar, A.; Kalhoro, S.A. Factors affecting the adoption of organic farming in Peshawar-Pakistan. *Agric. Sci.* **2015**, *6*, 587–593.

24. Hwang, J.-H. A Study on the marketing strategy of environment-friendly agricultural products. *Korean J. Org. Agric.* **2009**, *17*, 327–345.

25. Kim, H.-J. The Features of eco-agricultural producers, producers' organization and the potentialities of alternative agro-food system: A case study of the Pulmu Life Cooperative. *Korean Soc. Sci. Res. Rev.* **2008**, *24*, 185–212.

26. Kim, C.-G.; Jeong, H.-K.; Moon, D.-H. *Production and Consumption Status and Market Prospects for Environment-Friendly Agri-foods*; Korea Rural Economic Institute (KREI): Seoul, Korea, 2012. (In Korean)

27. Nieberg, H.; Offermann, F.; Zander, K. *Organic Farms in a Changing Policy Environment: Impacts of Support Payments*; EU-Enlargement and Luxembourg Reform; University of Hohenheim: Stuttgart, Germany, 2007.

28. Salvioni, C.; Aguglia, L.; Borsotto, P. The sustainability for an organic sector under transition: An empirical evaluation for Italy. In Proceedings of the 10th European IFSA Symposium, Aarhus, Denmark, 1–4 July 2012.

29. Khaledi, M.; Weseen, S.; Sawyer, E.; Ferguson, S.; Gray, R. Factors influencing partial and complete adoption of organic farming practices in Saskatchewan, Canada. *Can. J. Agric. Econ.* **2010**, *58*, 37–56.

30. Thapa, G.B.; Rattanasuteerakul, K. Adoption and extent of organic vegetable farming in Mahasarakham province, Thailand. *Appl. Geogr.* **2011**, *31*, 201–209.

31. Mondal, S.; Haitook, T.; Simaraks, S. Farmers' knowledge, attitude and practice toward organic vegetables cultivation in Northeast Thailand. *Kasetsart J. Soc. Sci.* **2014**, *35*, 158–166.

32. Province, Gangwon Statistical Yearbook, Gangwon Province, 2013. Available online: http://stat.gwd.go.kr/sub/sub03_10.asp# (accessed on 12 December 2015). (In Korean).

33. Jeon, M. *Device for Reducing Muddy Water in the Watershed of Soyang Dam*; Research Institute for Gangwon: Gangwon Province, Korea, 2008. (In Korean)

34. Jeon, M. *An Institutional Plan to Manage Areas in Gangwon Province that Are Vulnerable to Nonpoint Source Pollution*; Research Institute for Gangwon: Gangwon Province, Korea, 2015. (In Korean)

35. Poppenborg, P.; Koellner, T. Do attitudes toward ecosystem services determine agricultural land use practices? An analysis of farmers' decision-making in a South Korean watershed. *Land Use Policy* **2013**, *31*, 422–429.

36. Arnhold, S.; Lindner, S.; Lee, B.; Martin, E.; Kettering, J.; Nguyen, T.T.; Koellner, T.; Ok, Y.S.; Huwe, B. Conventional and organic farming: Soil erosion and conservation potential for row crop cultivation. *Geoderma* **2014**, *219–220*, 89–105.

37. Nguyen, T.T.; Ruidisch, M.; Koellner, T.; Tenhunen, J. Synergies and tradeoffs between nitrate leaching and net farm income: The case of nitrogen best management practices in South Korea. *Agric. Ecosyst. Environ.* **2014**, *186*, 160–169.

38. Hoang, V.-N.; Nguyen, T.T. Analysis of environmental efficiency variations: A nutrient balance approach. *Ecol. Econ.* **2013**, *86*, 37–46.

39. Long, J.S. *Regression Models for Categorical and Limited Dependent Variables*; Sage Publications: Thousand Oaks, CA, USA, 1997.

40. Tse, Y.K. A diagnostic test for the multinomial logit model. *J. Bus. Econ. Stat.* **1987**, *5*, 283–286.

41. Hausman, J.; McFadden, D. Specification tests for the multinomial logit model. *Econometrica* **1984**, *52*, 1219–1240.

42. Cheng, S.; Long, J.S. Testing for IIA in the multinomial logit model. *Sociol. Method. Res.* **2007**, *35*, 583–600.

43. Long, J.S.; Freese, J. *Regression Models for Categorical Dependent Variables Using Stata*, 2nd ed.; Stata Press: College Station, TX, USA, 2006.

44. Amemiya, T. Qualitative response models: A survey *J. Econ. Lit.* **1981**, *14*, 1483 1536.

45. Greene, W.H. *Econometric Analysis*, 6th ed.; Prentice-Hall: Upper Saddle River, NJ, USA, 2008.

46. Goktolga, Z.G.; Bal, S.G.; Karkacier, O. Factors effecting primary choice of consumers in food purchasing: The Turkey case. *Food Control* **2005**, *17*, 884–889.

47. Bullock, R.; Mithöfer, D.; Vihemäki, H. Sustainable agricultural intensification: The role of cardamom agroforestry in the East Usambaras, Tanzania. *Int. J. Agric. Sustain.* **2013**, *12*, 109–129.

48. Weir, S.; Knight, J. *Adoption and Diffusion of Agricultural Innovations in Ethiopia: The Role of Education*; Oxford University: Oxford, UK, 2000.

49. Lapar, M.L.A.; Ehui, S.K. Factors affecting adoption of dual-purpose forages in the Philippine uplands. *Agric. Syst.* **2004**, *81*, 95–114.

50. Karki, L.; Schleenbecker, R.; Hamm, U. Factors influencing a conversion to organic farming in Nepalese tea farms. *J. Agric. Rural Dev. Trop. Subtrop.* **2011**, *112*, 113–123.

51. Adesina, A.A.; Mbila, D.; Nkamleu, G.B.; Endamana, D. Econometric analysis of the determinants of adoption of alley farming by farmers in the forest zone of southwest Cameroon. *Agric. Ecosyst. Environ.* **2000**, *80*, 255–265.

52. Ayuya, O.A.; Waluse, S.K.; Gido, O.E. Multinomial logit analysis of small-scale farmers' choice of organic soil management practices in Bungoma county, Kenya. *Curr. Res. J. Soc. Sci.* **2012**, *4*, 314–322.

53. Flaten, O.; Lien, G.; Ebbesvik, M.; Koesling, M.; Valle, P.S. Do the new organic producers differ from the 'old guard'? Empirical results from Norwegian dairy farming. *Renew. Agric. Food Syst.* **2006**, *21*, 174–182.

54. Isin, F.; Cukur, T.; Armagan, G. Factors affecting the adoption of the organic dried fig agriculture system in Turkey. *J. Appl. Sci.* **2007**, *7*, 748–754.

55. Bowman, M.S.; Zilberman, D. Economic factors affecting diversified farming systems. *Ecol. Soc.* **2013**.

56. Koesling, M.; Flaten, O.; Lien, G. Factors influencing the conversion to organic farming in Norway. International Journal of Agricultural Resources. *Gov. Ecol.* **2008**, *7*, 78–95.

57. Pietola, K.; Lansink, A. Farmer response to policies promoting organic farming technologies in Finland. *Eur. Rev. Agric. Econ.* **2001**, *28*, 1–15.

58. Lampkin, N.H.; Padel, S. *Conversion to Organic Farming: An International Perspective*; Cab International: Wallingford, UK, 1994; pp. 295–313.

59. Sgroi, F.; Candela, M.; Trapani, A.; Foderà, M.; Squatrito, R.; Testa, R.; Tudisca, S. Economic and financial comparison between organic and conventional farming in sicilian lemon orchards. *Sustainability* **2015**, *7*, 947–961.

60. Bernal, M.P.; Alburquerque, J.A.; Moral, R. Composting of animal manures and chemical criteria for compost maturity assessment. A Review. *Bioresour. Technol.* **2009**, *100*, 5444–5453.

61. Kristiansen, P.; Merfield, C.H. Overview of organic agriculture. In *Organic Agriculture: A Global Perspective*; Kristiansen, P., Ed.; CSIRO: Clayton South, Australia, 2006; pp. 1–19.

62. Halpin, D.; Brueckner, M. The retail pricing, labelling and promotion of organic food in Australia. *The Australian organic industry: A profile*; Department of Agriculture, Fisheries, and Forestry: Canberra, Australia, 2004.

63. Characteristics of Conventional and Organic Apple Production in the United States. Available online: http://www.ers.usda.gov/publications/fts-fruit-and-tree-nuts-outlook/fts34701.aspx (accessed on 22 February 2016).

64. The World of Organic Agriculture-Statistics and Emerging Trends 2007. Available online: http://orgprints.org/10506/ (accessed on 12 March 2015).

65. Padel, S. Conversion to Organic Milk Production: The Change Process and Farmers Information Needs. Ph.D. Thesis, University of Wales, Aberystwyth, UK, 2001.

66. Tzouramani, I.; Liontakis, A.; Sintori, A.; Alexopoulos, G. Assessing organic cherry farmers' strategies under different policy options. *Mod. Econ.* **2014**, *5*, 313–323.

67. Lohr, L.; Salomonsson, L. Conversion subsidies for organic production: Results from Sweden and lessons for the United States. *Agric. Econ.* **2000**, *22*, 133–146.

68. Jánský, J.; Živělová, I. Subsidies for the organic agriculture. *Agric. Econ. Czech* **2007**, *53*, 393–402.

69. Kim, C.-G.; Jeong, H.-K.; Jang, J.-G.; Kwon, H.-M.; Moon, D.-H. *Improving Direct Payment Systems for Environment-Friendly Agriculture and Introducing Environmental Cross Compliance Program*; Korea Rural Economic Institute: Seoul, Korea, 2010. (In Korean)

Beef Cattle Farms' Conversion to the Organic System. Recommendations for Success in the Face of Future Changes in a Global Context

Alfredo J. Escribano

Abstract: Dehesa is a remarkable agroforestry system, which needs the implementation of sustainable production systems in order to reduce its deterioration. Moreover, its livestock farms need to adapt to a new global market context. As a response, the organic livestock sector has expanded not only globally but also in the region in search for increased overall sustainability. However, conversions to the organic system have been commonly carried out without analyzing farms' feasibility to do so. This analysis is necessary before implementing any new production system in order to reduce both the diversity of externalities that the variety of contexts leads to and the vulnerability of the DDehesa ecosystem to small management changes. Within this context and in the face of this gap in knowledge, the present paper analyzes the ease of such conversions and the farms' chances of success after conversion in the face of global changes (market and politics). Different aspects ("areas of action") were studied and integrated within the Global Conversion Index (GCI), and the legal requirement for European organic farming, organic principles, future challenges for ruminants' production systems, as well as the lines of action for the post-2013 CAP (Common Agricultural Policy) and their impacts on the beef cattle sector were taken into account. Results revealed that farms must introduce significant changes before initiating the conversion process, since they had very low scores on the GCI (42.74%), especially with regard to health and agro-ecosystem management (principle of Ecology). Regarding rearing and animal welfare (principle of justice/fairness), farms were close to the organic system. From the social point of view, active participation in manufacturing and marketing of products should be increased.

Reprinted from *Sustainability*. Cite as: Escribano, A.J. Beef Cattle Farms' Conversion to the Organic System. Recommendations for Success in the Face of Future Changes in a Global Context. *Sustainability* **2016**, *8*, 572.

1. Introduction

Organic livestock production has increased substantially in recent years in order to both increase farmers' income (trough agricultural subsidies and higher sale prices) and reduce farms' environmental impact. Moreover, the potential role of organic production in the socio-economic development of rural areas has been claimed by development agencies and has contributed to this trend.

In the region under study (Extremadura, SW Spain), the Dehesa ecosystem is a remarkable agroforestry system where small changes in farms' management can lead to important changes in both farm and agro-ecosystem sustainability. However, Dehesa farms' low economic performance has induced farmers to make adaptive changes, some of which (mainly intensification) have lead to reductions in the sustainability of both the Dehesa ecosystem and its livestock farms. One of the most recent adaptive changes has been the transition to the organic production system. The regional organic livestock sector is mainly composed of beef cattle, and was the fourth most important with regard to the number of this species and productive orientation (98 beef cattle farms, 5.63% of farms) [1] in 2014.

As part of the Mediterranean area, Dehesa benefits from the well-known ruminant pasture-based production systems. In this sense, the organic system apparently fits within this system's characteristics. In fact, Nardone *et al.* (2004) [2] predicted good feasibility of the conversion process towards the organic system in such an area.

In general terms, such transition has usually been carried out without plans of action in most cases. However, it is necessary to precisely assess such feasibility before implementing any new production systems (especially in sensitive ecosystems such as the Dehesa), since the diversity of contexts among countries and farms is too great to generalize assumptions and carry out business model changes based on them. The reason for this is that these transitions can be very variable, as they depend on several factors, *i.e.,* national regulations, certification bodies, the production system and the livestock species [3]. Thus, both positive and negative effects have been observed in cattle farms. For instance, productivity is often reduced and production costs increased, mainly during the first years of transition and due to the higher cost of organic feedstuff [4,5].

Thus, previous to the conversion, an in-depth study of the sector must be carried out, detecting its external and internal factors (SWOT: Strengths Weaknesses Opportunities and Threats Analysis), paying special attention to its future challenges on the basis of political, market and climatic changes. Thereby, it will be possible to predict the difficulties that farms will go through during the conversion process and establish tailor-made guidelines for each farm (or group of farms) in order to shape successful and sustainable business. Hence, it will finally be possible to design not only organic production systems but also sustainable ones in both local and global contexts.

In accordance, the present study aims to: (i) assess the feasibility of conversion of a sample of pasture-based beef cattle farms to a optimal production system designed not only on the basis of the organic regulation but also on its principles, taking into account future challenges for ruminants' production systems, as well as the lines of action for the post-2013 CAP (Common Agricultural Policy) and their impacts on the beef cattle sector; and (ii) establish specific measures to ease the conversion process and increase farms' chances of success after conversion is accomplished.

2. Materials and Methods

2.1. Study Area

The study area was the DDehesa ecosystem located in the Extremadura region (Spain). This region presents annual mean temperatures ranging between 16 and 17 °C. Summers are dry and hot (the mean temperature in July is over 26 °C, and the maximum is usually over 40 °C). Annual rainfalls are irregular during the year and also among years. The mean rainfall varies between 300 mm and 800 mm. Extremadura is located in SW Spain (between latitude 37°56′32″–40°29′15″ and longitude 4°38′52″–7°32′3″) and constitutes the core (geographically and in terms of hectares) of the Dehesa (Figure 1), grouping 2.2 million hectares from the 3 million hectare total area of Dehesa. This ecosystem is the most widely-used agroforestry system in Europe, and has been considered as a habitat to be protected under the European Habitats Directive, the cornerstone of Europe's nature conservation policy [6]. In it, trees, cereal-legume crops and extensive low-input farming systems based on grazing are integrated, where cork, firewood, hunting, and birdwatching, are also common and economically important. Soils are poor and, because of its continental Mediterranean climate, supply of grazing resources is scarce and irregular [7].

Livestock production systems have a great impact on overall sustainability in disfavored and/or sensitive areas (socially, economically and environmentally), which is even higher in traditional (extensive and mixed) production systems. In particular, Dehesa's traditional animal production systems were commonly diverse (mixed), where a mixture of agricultural uses (various livestock species—mainly beef, crops for animal feeding, hunting, and forestry) could be found.

From an economic point of view, the importance of the livestock sector in Extremadura is reflected by its contribution to the Agrarian Production in the region. In 2010, cattle, swine and small ruminants sectors generated (in terms of meat and livestock) a total of 629.52 million Euros at basic prices in Extremadura, which accounted for 32.23% of the regional Agrarian Production. If the products (milk and wool) are included, this value amounts to €396.46 M, reaching 36.10% [8].

From a social point of view, it is noteworthy to highlight that, in semiarid and rural regions such as Extremadura and Dehesa, extensive livestock production systems are often the main activity [9], and even the only source of livelihood [10]. In the case of Extremadura, the contribution of the livestock sector to the regional employment in 2010 was well above the national value (11.2% of the working population in Extremadura were related to the agricultural sector, while in Spain only 4.4% were in agricultural employment) [11]. Such dependence on the sector highlights the need to protect it and enhance it, as it contributes to create jobs, increase income and sustain the rural population, which is vital for the economy and rural development of these areas [12–15].

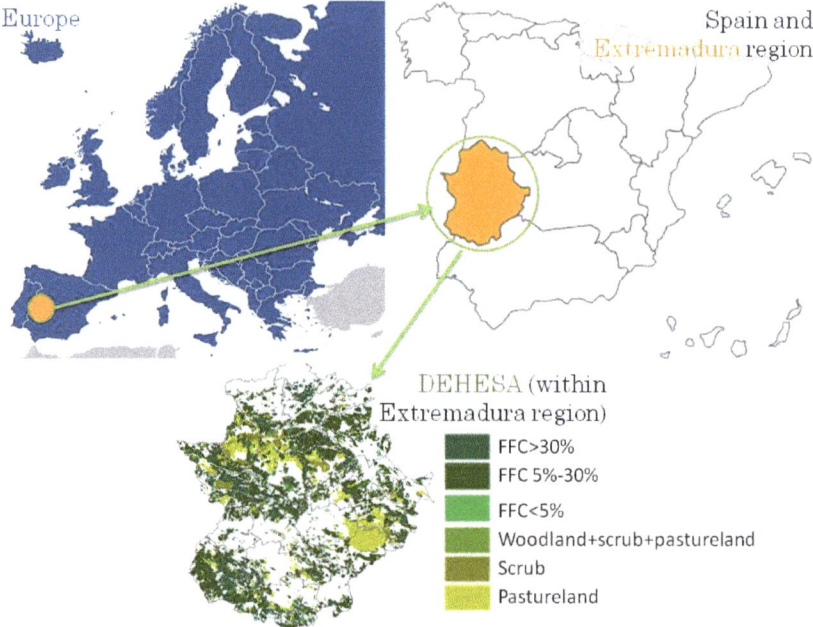

Figure 1. Dehesa location and different land cover characteristics. FFC (Forest Fractional Cover): Fraction of the land covered by the vertical projection of the tops of trees.

From a cultural and environmental perspective, traditional production systems are crucial for the conservation of cultural heritage and local identity, as well as landscape [16] and habitats of high ecological and aesthetic value [17,18]. This has its rationale in the fact that the Dehesa agro-ecosystem is an evolution of the original Mediterranean woodland by human activity in order to carry out agricultural activities. These production systems contribute to the improvement of soil and pastures, ensure biodiversity, and control coppice and woody scrub regrowth, thus reducing the risk of wildfire [3].

Different socioeconomic factors have led to abandonment and intensification processes that endanger the conservation of such traditional systems and the sustainability of the area in terms of three main pillars of sustainability (society, economy and environment), so that the evaluation and search for sustainable production systems are needed.

2.2. Farms Selection

Due to a lack of official statistics about figures and location of Dehesa farms in Extremadura, the sampling was nonprobabilistic by quotas. Forestry, livestock production, and farm size criteria were used to select the farms with the aim of obtaining a representative sample of the various subsystems of Dehesa, following the methodology used by previous studies [19–21]. The number of farms surveyed was 30, which is in accordance with other research studies on the topic (24 farms [22]; 31 farms [23,24]). More detailed information on these criteria can be found in the previous study of Escribano [3].

Sample Characterization: Conventional Farms

The sector located in the area under study is characterized by its scarce use of external feed resources (only for covering adults' maintenance nutritional requirements during summer). In most cases, calves are sold at weaning and fattened in feedlots, so that the value added to the farm is usually scarce. Moreover, cows' reproductive performance (weaned calves/cow/year) is low (0.81). All this reduces their income, bargaining power and competitiveness. To compensate for this situation, an important part of the farmer's activities is to carry out other economic activities that reduce economic risk and increase adaptation to sectorial changes [25].

Specifically, the farms analyzed had an average size of 275.80 ha of Utilized Agricultural Area (UAA), of which 64% was owned. Nearly 50% of it was covered by woody species, which has environmental, cultural and economic relevance. However, the presence of crops is almost non-existent, which increases farms' exposure to unstable and scarce on-farm feed resources. Average herd size reached 111.70 adult animals, of which 98% where cattle (adult cows and bulls). Total stocking rate was extremely high for the ecosystem under study (0.73 Livestock Units/ha), which suggests the necessity to lead Dehesa cattle production systems back.

Regarding management and herds' structure, it is notable that the use of reproductive techniques is typically scarce (estrus synchronization was only carried out in the 6.70% of farms; the same values as for implementation of artificial insemination). Genetically, farms were mainly composed of Purebred autochthonous cows (20.11% of total cows) and Purebred foreign bulls (86.98%).

2.3. Selection of Indicators

First, a literature review was carried out, based on the European legal requirements for organic beef cattle production [27] and subsequent amendments), as well as the principles of organic production set by IFOAM [26]. Requirements and principles were transformed into indicators.

Subsequently, other indicators that were considered important for the study were selected from the literature on the topic [21,22,25,28–30]. Furthermore, future political context and challenges for ruminants' production systems under extensive conditions in drylands were taken into account, as were the pillars and lines of action for the post-2013 CAP, and the impacts of the CAP on beef cattle sector [20,31]. In general terms, the 2014–2020 CAP has changed its philosophy, so that it is nowadays not only focused on ensuring the income of a certain segment of the population, but also on promoting the development of territories, promoting the efficient use of resources with an eye on a sustainable and diverse agricultural sector, paying even more attention on rural areas [32]. Thus, the challenges of "The CAP towards 2020" are the following: economics, food security, price volatility, and the agricultural sector's environmental impact (greenhouse gases, soil degradation, pollution, habitats and biodiversity). To do this, CAP has renewed instruments based on three main pillars (competitiveness, environmental sustainability, and rural development), where agri-food value chain (transparent and fair) and risk management are notable aspects and changes of the reform. Such instruments are as follows: market measures (reducing risks and improving risk management), agri-food value chain (fairness and transparency), research and knowledge transfer.

Although the boundaries of the system were the farm, some aspects of the food chain upstream (kilometers travelled by fodder) and downstream (products processing products by the producer, and direct sales) were also included, due to the importance of these aspects for the sustainability of these farms [33]. Finally, a list of 55 indicators (Table 1), complying with the criteria of relevance, practicability, and end user value recommended by [34] were selected. Then, aiming to increase the practical applicability of the results of the present study, indicators were grouped into "areas of action".

Table 1. Indicators selected.

Conversion Indicators and Areas of Action	Definitions and Units	Dummy	Optimal Value	Criteria [1]	Principle [2]	Weight (%)
Organic concentrate to fattening calves	Fattening calves were given organic concentrates.	Yes	1	Comp.		
Organic fodder to fattening calves	Fattening calves were given organic fodder/grass.	Yes	1	Comp.		
Organic concentrate to adults	Cows and bulls were given organic concentrates.	Yes	1	Comp.		
Organic fodder to adults	Cows and bulls were given organic fodder/grass.	Yes	1	Comp.		
Ration 60:40	At least 60% of daily ration was based on common forage and/or grass.	Yes	1	Comp.		
Feed management					1, 2	21.68
Number of veterinary medicines to calves	Calves (younger than one year) do not receive more than 1 treatment.	Yes	1	Comp.		
Number of veterinary medicines to adults	Adults neither receive more than 3 veterinary medicines nor 2 antiparasitics per year.	Yes	1	Comp.		
Preventive antiparasitics	The farmer did not use antiparasitics systematically as a preventive health management practice.	Yes	1	Comp.		
Preventive antibiotics	The farmer did not use antibiotics systematically as a preventive health management practice.	Yes	1	Comp.		
Healthy herd	The herd was healthy.	Yes	1	Comp.		

Table 1. *Cont.*

Conversion Indicators and Areas of Action	Definitions and Units	Dummy	Optimal Value	Criteria [1]	Principle [2]	Weight (%)
Cleaning products	Products used to clean the infrastructure were allowed by the Council Regulation.	Yes	1	Comp.		
Isolating for health reasons	Livestock was isolated when they were suffering infectious diseases easily contagious.	Yes	1	Comp.		
Quarantine	Livestock were isolated when coming from other farms.	Yes	1	Comp.		
Alternative medicine	The farmer used phytotherapy and/or homeopathy.	Yes	1	Comp.		
Water quality assessment	Water quality was assessed at least annually.	Yes	1	Comp.		
Vaccines	The farmer did not use of vaccines for non-endemic illnesses.	Yes	1	Comp.		
Health Management					1, 3, 2	20.83
Calves access to open spaces	Calves have permanent access to open spaces.	Yes	1	Comp.		
Adults access to open spaces	Adults have permanent access to open spaces.	Yes	1	Comp.		
Infrastructure (meters and facilities)	The infrastructure was adequate with regard to the meters and facilities needed for organic beef farms.	Yes	1	Comp.		
Calving period	The calving period was adequately organized in order to make better use of the available natural resources.	Yes	1	Comp.		

Table 1. *Cont.*

Conversion Indicators and Areas of Action	Definitions and Units	Dummy	Optimal Value	Criteria[1]	Principle[2]	Weight (%)
Infrastructure (cleanness)	The infrastructure was adequate with regard to cleanness.	Yes	1	Comp.		
Fattening period length	It was less than 3 months and less than 1/5 of their lifespan.	Yes	1	Comp.		
Protection against bad weather	The infrastructure has protection against sunlight, heat, raining, *etc.*	Yes	1	Comp.		
Weaning age	The lactation period lasted at least 3 months.	Yes	1	Comp.		
Isolating/Tying	Livestock was not isolated and/or tied up for reasons that were not related to health management.	Yes	1	Comp.		
Mutilations	The farmer does not carry out mutilations.	Yes	1	Comp.		
Autochthonous bovine breeds	75% or more of the cattle was autochthonous or considered as such in the region under study.	Yes	1	Comp.		
Animal welfare training	The farmer has attended to animal welfare courses.	Yes	1	Comp.		
Rearing and Animal Welfare					4, 1	9.17
Stocking rate	Livestock Units/ha.	No	0.33–0.5	Rec.		
Use of pesticides and/or herbicides	The farmer did not use pesticides and/or herbicides.	Yes	1	Comp.		
Use of mineral fertilizers	The farmer did not use mineral fertilizers.	Yes	1	Comp.		

Table 1. *Cont.*

Conversion Indicators and Areas of Action	Definitions and Units	Dummy	Optimal Value	Criteria [1]	Principle [2]	Weight (%)
Rotational grazing	Grazing was organized so that all plots were not grazed continuously.	Yes	1	Comp.		
% of wooded area	Wooded area/total area.	No	1.00	Max.		
Dung management	Qualitative scale (0–3). 0: There was no dung accumulation due to extensification; 1: No heaping, then spreading of immature dung; 2: There was some heaping (not enough), so that dung was not spread completely mature; 3: Proper compost elaboration, and spreading.	No	0 and 3	Rec.		
Legumes	Dichotomic. O: No. 1: Yes. Legumes were planted in isolation or associated with grains.	Yes	1	Max.		
Reforestation	Dichotomic. O: No. 1: Yes. The farmer plant autochthonous trees.	Yes	1	Max.		
Reduced tillage	Dichotomic. O: No. 1: Yes. Reduced or no-tillage agriculture was practiced.	Yes	1	Max.		
Crop rotation	Dichotomic. O: No. 1: Yes. The farmer practices fallow, crop rotation, *etc.*	Yes	1	Max.		
Crop association/Intercropping	Dichotomic. O: No. 1: Yes. The farmer practices crop associations/intercropping.	Yes	1	Max.		
Cover crops	Dichotomic. O: No. 1: Yes. The farmer used cover crops.	Yes	1	Max.		

Table 1. *Cont.*

Conversion Indicators and Areas of Action	Definitions and Units	Dummy	Optimal Value	Criteria [1]	Principle [2]	Weight (%)
Agro-Ecosystem Management					2, 1, 3	19.83
Self-sufficiency	% Mkcal of feed obtained from the = (1 − MKcal of external feedstuff)/nutritional requirements.	No	99.06	Max.		
Direct sales	Dichotomic. O: No. 1: Yes. Farmers sold products directly to consumers.	Yes	1	Max.		
Distance to the slaughterhouse	Km to the nearest slaughterhouses (organically certified ones in the case of organic farms).	No	12.48	Min.		
Elaboration of products	Dichotomic. O: No. 1: Yes. The farmer add value to the carcass/meat: packaging, *etc*.	Yes	1	Max.		
Km travelled by the fodder	Mean distance (km) to the market place.	No	0.00	Max.		
Self-Sufficiency and Agri-Food Chain Relationships					4, 2	15.33
Business diversification	Number of agricultural activities and products.	No	3	Max.		
Job creation potential	Total AWU [3] /100 ha UAA [4] (AWU/100 ha).	No	1.37	C3		
Workforce stability	(Fixed AWUs/Salaried AWUs) × 100 ha.	No	0.01	C3		
Satisfaction level	Qualitative scale (0–3). Increasing level of satisfaction level.	No	3	Max.		

107

Table 1. *Cont.*

Conversion Indicators and Areas of Action	Definitions and Units	Dummy	Optimal Value	Criteria [1]	Principle [2]	Weight (%)
Labor attractiveness of the farm	Number of External workers (AWU/100 ha).	No	0.67	C3	4	8.83
Human Well-Being and Rural World Opportunities						
Agroecology training	Farmer attended agroecology/organic production courses/seminars Dichotomic. O: No. 1: Yes. The farmer has been trained in agroecology.	Yes	1	Max.		
Level of studies	Qualitative scale (1–3). 1: No studies or basic educational level; 2: High school or and/or vocational education and training; 3: University degree or higher.	No	1	Max.		
Farmer' age	Years.	No	30	Rec.		
Data registering	Dichotomic. O: No. 1: Yes. Farmer records technical and economic data.	Yes	1	Max.		
Continuity/Future plans	Qualitative scale (1–4).1: Abandon; 2: Herd reduction or change of species; 3: Maintain specie(s) reared and herd size; 4: Increase the herd size.	No	3 and 4	Max.		
Human Capability Towards Implementing Sustainable Agricultural Practices					4	4.33

[1] Comp.: comply with organic regulations (1: comply with; 2: do not comply with); Max.: Maximum value of the sample; Min: Minimum value; Rec.: recommended value; [2] Principles of organic production defined by the International Federation of Organic Agriculture Movements (IFOAM, 2005) [26]: 1: Health (Healthy soil, plants, animals, humans = a healthy planet); 2: Ecology (Emulating and sustaining natural systems); 3: Care (For the generations to come); 4: Fairness (Equity, respect and justice for all living things); [3] AWU: Annual Work Units: work done by one employee during 228 days or 1826 h; [4] UAA: Utilized Agricultural Area (measured in ha—hectares).

2.4. Ordination of Indicators in Areas of Action, Allocation of Optimal Values and Relative Weights

After this process, optimum values were established for each indicator in accordance with stakeholders' and experts' opinions. Due to the fact that most of the indicators were created on the basis of organic regulations and principles, and/or on specific agro-ecological practices, most of these indicators were dummy or binomial (Table 1 provides detailed information in this regard), and their optimal value was 1, indicating "compliance". Hence, each variable acquired a mutually exclusive and unique value with a binomial Bernoulli distribution. In the case of quantitative indicators, the optimal values were established following the procedure of previous studies [8,35,36], where the author's previous experience and the characteristics of the sample were taken into account.

Finally, relative weights were attributed to each area of action, so that each Partial Conversion Index (PCI) had a different contribution to the GCI. The relative weighting of the areas of action is a recommended process as it allows aspects (areas) of greater impact on the ease of conversion and farms future success to be prioritized. Following the methodology of participatory research used by [25], numerical weights were established in a focus group meeting. This allowed the integration of local and scientific knowledge, taking into account participants' knowledge and values, following a collaborative procedure. Each participant was provided with a list of relative weights. Thus, the sum of the relative weights of each indicator of the same attribute will be equal to 100%. Their mean value corresponds to the final relative weight.

The selection of indicators and the establishment of the relative weights was based on the following criteria: (i) areas of action's importance with regard to compliance with the European regulations on organic production; (ii) principles of organic production and sustainability dimensions (mainly social and environmental); and (iii) farms' internal and external factors, as well as their future challenges in relation with predicted changes on agricultural policies and market. This allowed indicators to be derived, measured, and monitored as part of a systemic, participatory, interdisciplinary, and flexible process of evaluation.

The complete list of indicators of conversion, the principle they belong to, the areas of action in which they were included, and the relative weights of such areas are presented in Table 1.

2.5. Calculation of Conversion Index: Individual Conversion Scores, Partial Conversion Index (PCI) and Global Conversion Index (GCI)

There are previous studies that have analyzed the feasibility of conversion of farms to organic or agro-ecological production systems [22,28–30,37]. However, the above-mentioned methodology allows not only assess the feasibility of conversion

109

to the organic system set by European regulations and the principles of organic production, but also addresses important issues and challenges for the analyzed farms (SWOT and future challenges).

In order to apply this broader approach, methodological adaptations have been carried out. Such adaptations integrated the Organic Livestock Proximity Index developed by [29], the MESMIS Framework [38], and the AMOEBA approach [39]. MESMIS Framework (Framework for the Evaluation of Management Systems incorporating Sustainability Index) has been widely used through years to assess livestock systems' sustainability [40]. This methodology, along with the AMOEBA approach [39], allows the selection of indicators and their transformation on scores based on optimal values for each indicator, so that their initial values of the indicators are converted to percentage values (scores) according to their proximity to the optimum value. Thus, 100% is the optimal/desirable value. Thus, farms with several indicators with values of 100% in each area, would be more easily converted to organic. This analysis would generate the limit (maximum desirable value) for each area of action. To do this, three possible cases are faced:

(1)   When the indicators have an optimum value corresponding to either the maximum value of the sample or the value 1 for the case of qualitative variables, the scores were calculated as follows (Equation (1)):

$$\text{Individual conversion score} = (\text{initial value of the indicator}/\text{optimal value}) \times 100 \quad (1)$$

(2)   However, for indicators whose optimal value was the minimum value found in the sample (*i.e.*, distance traveled by feed), the scores were calculated as follows (Equation (2)):

$$\text{Individual conversion score} = (\text{optimal value}/\text{initial value of the indicator}) \times 100 \quad (2)$$

(3)   When optimum values were percentiles (range of values) or recommended values (such as farmer's age), the formula applied depended upon the magnitude of the indicator values and their optimal value. If the value of the indicator was lower than the optimum value, the Equation (1) was applied. In contrast, when the indicator value was higher than the optimal value, Equation (2) was applied. Moreover, for certain indicators, more details must be taken into account. Thus, in cases such as that of total stocking rate, either exceeding or not reaching the optimal value penalized the farm, since both high and low values lead to ecosystem degradation. In other variables (*i.e.*, self-sufficiency), the value of individual conversion scores remained 100% although the indicator value is greater than the optimum value. This allows for conversion rates for each indicator as the production systems are studied.

The next step is to calculate the conversion of each area of action (Partial Conversion Index: PCI). It was calculated as the average of the individual conversion scores grouped within the area of action.

$$PCI = \text{mean of individual conversion scores} \tag{3}$$

Finally, the GCI was the summation of multiplying each PCI by its relative weight.

$$GCI = \Sigma\,PCI \times \text{relative weight} \tag{4}$$

At this point, it is necessary to point out that any value was established as a threshold to decide whether farms could be converted to the organic system or not. The present study and methodological approach allowed increasing the understanding of the farms and make decisions regarding the conversion process based on the integrated study of all indicators and areas of action, taking into account the context of each of the farms (including the predicted global context changes mentioned in the introduction section). Moreover, conversion is justified if the recommendations and comments included in the discussion section are taken into account. After this, a reevaluation of the farms should be carried out and then the decision on conversion would be made again.

2.6. Statistical Analysis of Results

Once the PCI were obtained, farms were grouped based on them. For this purpose, a hierarchical cluster analysis (CA) using Ward's method and squared Euclidean distance was applied. Input variables for the CA were the PCI of the seven action areas. They were standardized by standard deviation. The CA allowed reducing the number of individual cases (farms) to a smaller (clusters or farms typologies), thus maximizing homogeneity within each typology and heterogeneity among them [41]. This facilitated the understanding of the farms involved in the study, since by means of an in-depth analysis, farm typologies' key features were identified. Then, it was possible to define measures for facilitating the conversion process and farms' success once issues were overcome.

Later, differences among farms' typologies were detected with regard to individual conversion scores and GCI. To do so, a single factor or one-way ANOVA was applied. Statistical analyses were carried out with the 2011 Statistical Package for Social Systems, version 20.0 [42]. The complete methodological procedure and steps can be observed schematically in Figure 2. It shows how different aspects (both the requirements for a ruminant production operation and other parameters of relevance for the sustainability of the sector under study) to be considered organic were converted into a checklist.

```
                    ┌─────────────────────────┐
                    │    FARMS SELECTION      │
                    └─────────────┬───────────┘
                                  │
                                  ▼
        ┌─────────────────────────────────────────────────┐
        │           SELECTION OF INDICATORS               │
        │   ┌─────────────────────────────────────────┐   │
        │   │         Bibliographic revision:         │   │
        │   │ Structure, technical, economic, land use,│   │
        │   │ sustainability and conversion indicators │   │
        │   │ (organic production legal requirements   │   │
        │   │ and principles).                         │   │
        │   │ Source: scientific literature, Council   │   │
        │   │ Regulation (EC) 834/2007 (and            │   │
        │   │ amendments), IFOAM (2005)                │   │
        │   └─────────────────────────────────────────┘   │
        └─────────────────────────┬───────────────────────┘
                                  │
                                  ▼
                    ┌─────────────────────────┐
                    │  Ordination of indicators│
                    │    into areas of action │
                    └─────────────────────────┘
```

Bibliographic revision:

Structure, technical, economic, land use, sustainability and conversion indicators (organic production legal requirements and principles).

Source: scientific literature, Council Regulation (EC) 834/2007 (and amendments), IFOAM (2005)

Ordination of indicators into areas of action

Questionnaire design

Establishment of:
(i) optimal values for the indicators
(ii) relative weights for areas of action

Data collection and data base elaboration

CALCULATION OF CONVERSION INDEX

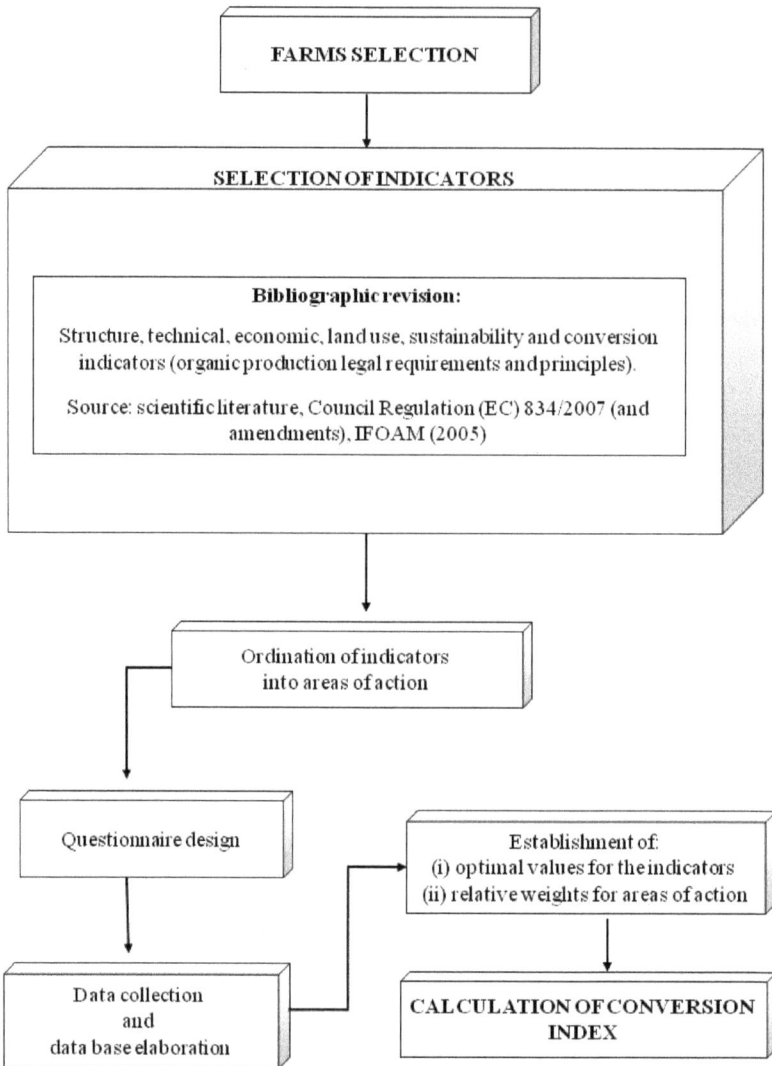

Figure 2. Process followed from the farms selection step to the calculation of conversion indexes.

3. Results

First, individual conversion scores for each indicator are shown (Tables 2–4).

Later, PCI and GCI scores are shown according to farms typologies (Table 5). The CA yielded the most significant results for a four-cluster solution (a better explanation thereof). The linkage distance used was in line with that of other studies on the topic [43], with a relatively short distance (< 20%). The resulting dendrogram is presented in Figure 3. The clusters (typologies) obtained from the CA were compared using an analysis of variance. For a better understanding of the characteristics of each typology, Table 5 is provided, showing the average values, standard error and significance level of the typologies regarding both the PCI and GCI. Figure 4 facilitates the comprehension regarding sample's partial scores (PCIs).

Table 2. Farms' scores with regard to cattle management. Areas of action: Feed Management, Health Management, and Rearing and Animal Welfare.

Areas of Action	Indicators	Sample Mean (%)	SE
Feed Management	Organic concentrate to fattening calves	00.00	00.00
	Organic fodder to fattening calves	00.00	00.00
	Organic concentrate to adults	00.00	00.00
	Organic fodder to adults	00.00	00.00
	Ration 60:40	83.33	6.92
Health Management	Number of veterinary medicines to calves	90.00	5.57
	Number of veterinary medicines to adults	93.33	4.63
	Preventive antiparasitics	3.33	3.33
	Preventive antibiotics	73.33	8.21
	Healthy herd	100.00	0.00
	Cleaning products	20.00	7.43
	Isolating for health reasons	90.00	5.57
	Quarantine	46.67	9.26
	Alternative medicine	0.00	0.00
	Water quality assessment	33.33	8.75
	Vaccines	100.00	0.00
Rearing and Animal Welfare	Calves access to open spaces	100.00	0.00
	Adults access to open spaces	100.00	0.00
	Infrastructure (meters and facilities)	0.00	0.00
	Calving period	33.33	8.75
	Infrastructure (cleanness)	100.00	0.00
	Fattening period length	83.33	6.92
	Protection against bad weather	43.33	9.20
	Weaning age	96.67	3.33
	Isolating/Tying	100.00	0.00
	Mutilations	100.00	0.00
	Autochthonous bovine breeds	13.33	6.31
	Animal welfare training	93.33	4.63

SE: Standard error.

Table 3. Farms' scores with regard to agro-ecosystem management. Area of action: Agro-ecosystem management.

Area of Action	Indicators	Sample Mean (%)	SE
Agro-Ecosystem Management	Stocking rate	61.47	4.60
	Use of pesticides and/or herbicides	73.33	8.21
	Use of mineral fertilizers	63.33	8.95
	Rotational grazing	4.67	0.93
	% of wooded area	45.55	7.92
	Dung management [1]	28.33	6.65
	Legumes	10.00	5.57
	Reforestation	0.00	0.00
	Reduced tillage	50.00	9.28
	Crop rotation	10.00	5.57
	Crop association/Intercropping	0.00	0.00
	Cover crops	6.67	4.63

SE: Standard error. [1] Number of measures/agricultural management practices implemented to reduce soil erosion and to improve soil quality. These include: cover crops, mulching, intercropping, crop rotation, plot rotation, fallow, and use of compost.

Table 4. Farms' scores with regard to social aspects. Areas of action: Self-sufficiency and agri-food chain relationships, Human well-being and rural world opportunities, and Human Capability towards Implementing Sustainable Agricultural Practices.

Area of Action	Indicators	Sample Mean (%)	SE
Self-Sufficiency and Agri-Foodchain Relationships	Self-sufficiency	65.18	4.80
	Direct sales	0.00	0.00
	Distance to the slaughterhouse	88.18	1.84
	Elaboration of products	0.00	0.00
	Km travelled by the fodder	95.36	1.43
Human Well-Being and Rural World Opportunities	Business diversification	38.89	2.81
	Job creation potential	43.64	3.83
	Workforce stability	10.00	5.57
	Satisfaction level	52.87	3.77
	Labor attractiveness of the farm	23.94	6.37
Human Capability Towards Implementing Sustainable Agricultural Practices	Agroecology training	26.67	8.21
	Level of studies	70.00	5.14
	Farmer' age	68.29	2.99
	Data registering	66.67	8.75
	Future plans	85.67	4.57

SE: Standard error.

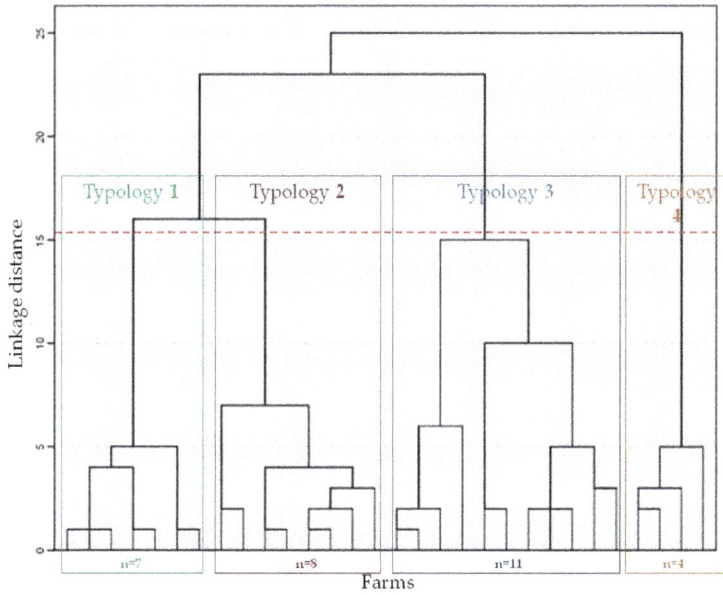

Figure 3. Dendrogram: Hierarchical cluster analysis of farms using Ward's method and the squared Euclidean distance.

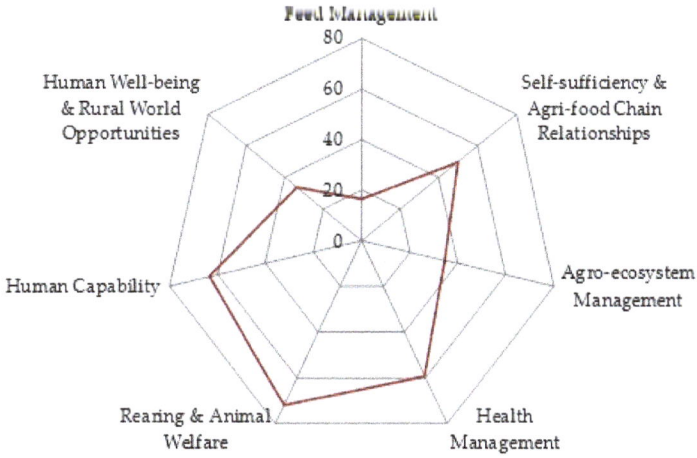

Figure 4. GCI scores of the farms.

Table 5. Partial Conversion Indices (PCIs) and Global Conversion Index (GCI). Mean scores and standard error for farm typologies and sample.

Conversion Index and Areas of Action	Mean (±SE)					F; sig.
	T1 (n = 7)	T2 (n = 8)	T3 (n = 11)	T4 (n = 4)	Sample of Conventional Farms (n = 30)	
Feed Management	20.00a (± 0.00)	20.00a (± 0.00)	18.18a (± 1.82)	0.00b (± 0.00)	16.67 (± 1.38)	31.06; 0.000
Health Management	57.15b (± 1.68)	54.55b (± 1.72)	66.12a (± 1.77)	52.28b (± 2.28)	59.09 (± 1.36)	11.55; 0.000
Rearing and Animal Welfare	76.19b (± 1.19)	75.00a (± 1.57)	68.94a (± 1.98)	66.67a (± 3.40)	71.94 (± 1.16)	4.65; 0.010
Agro-Ecosystem Management	18.59b (± 3.37)	46.76a (± 3.09)	34.73a (± 3.15)	37.67a (± 6.72)	34.56 (± 2.55)	9.87; 0.000
Self-Sufficiency and Agri-Food Chain Relationships	51.85ab (± 1.08)	53.14b (± 1.54)	47.27a (± 1.99)	46.06a (± 2.65)	49.74 (± 1.04)	3.04; 0.047
Human Well-Being and Rural World Opportunities	23.96b (± 2.18)	32.21a (± 2.97)	43.78a (± 5.29)	27.27ab (± 3.04)	33.87 (± 2.59)	4.38; 0.013
Human Capability Towards Implementing Sustainable Agricultural Practices	54.45ab (± 6.63)	78.53b (± 7.33)	64.74ab (± 4.91)	45.55a (± 7.84)	63.46 (± 3.70)	3.88; 0.020
Global Conversion Index	39.34b (± 1.10)	46.24a (± 1.08)	44.84a (± 0.85)	35.92b (± 1.20)	42.74 (± 0.85)	16.82; 0.000

Notes: SE (Standard Error); F: F-score; Sig.: level of significance.

Farms' Typologies: PCI and GCI Scores

Typologies 3 (36.67% of the sample) and 1 (23.33%) obtained intermediate (44.84% and 39.34%, respectively) GCI scores. Typology 1 stands out for two reasons: (i) health management was very close to that required in organic production; and (ii) it made a high contribution to human well-being and rural world opportunities. With regard to T1, it is worth mentioning their farming methods with regard to rearing and animal welfare. This typology also showed the highest scores (along with T2) with respect to Feed management, which was due to an adequate 60:40 (forage:concentrate) ratio.

T4, consisted of 13.33% of farms in the sample had the lowest scores in terms of the GCI (35.92%), and for almost all areas of action. T2 (26.67% of farms) scored the highest for most areas of action (feed management, agro-ecosystem management, self-sufficiency, agri-food chain relationships, and human capability towards implementing sustainable agricultural practices), and for the GCI (46.24%).

4. Discussion

The GCI shows the proximity and feasibility of converting conventional Dehesa beef cattle farms to the organic system (Table 5). Moreover, the study according to areas of action has allowed observing the areas in which farms would find less difficulty in carrying out the conversion. Therefore, the conversion process would require major changes in this regard. Regarding a practical application and the decision, it should be based on the areas of action and specific indicators of conversion, so that the farmers can reduce the weaknesses of each system, providing solutions and specific improvement measures.

4.1. Feed Management

As shown in Table 2, farms had low scores in terms of feed management because all farms analyzed were conventional ones so they did not provide organic feed.

4.2. Health Management

The fact that Dehesa beef cattle farms are extensive and that the local climate is hot and dry, means that veterinary actions are limited. In this sense, it could be expected that the conversion process is simple from the health management perspective. However, it has been observed that in the conventional farms analyzed, health management systematically (either there are clinical signs of illness or not) relies on the use of 1 or 2 antiparasitic products as a preventive measure. Moreover, calves entering the fattening period receive antibiotic treatments in order to prevent disorders typically related to this period (diarrheas and pneumonia). These findings are noteworthy because of their impacts in terms of public health (antibiotic resistance) and environmental pollution. Therefore, aiming to avoid such negative implications of the current health management, and in order to be consistent with the organic methods, farm health management should be based on preventive health management. As land is commonly not a limitation in Dehesa farms, this could be done by establishing grazing plan transitions: (i) prevent access to flooded areas (almost non-existent); (ii) reserve ungrazed plots for young animals; and (iii) integrate other non-host species of parasites, so that parasitic load is reduced. In addition, the level of stocking rates must be reduced in some farms.

Regarding the use of cleaning products, farms had low scores, so farmers should change the products used to disinfect and clean the facilities, which will not be complicated since [44] allows for the use of common commercial products used in many conventional farms.

The use of alternative medicine products was non-existent. Due to the lack of knowledge and commercially available products, health management must be based on grazing management, stocking rates and agro-ecosystem management practices oriented to increase microbiota competition (*i.e.*, intercropping, cropping diversity, habitat maintenance, *etc.*). Fortunately, Dehesa extensive production systems, its climate and the diversity within farms allow for this management, and additional health measures will not be necessary.

Therefore, in general it can be said that the Dehesa beef cattle farms could easily convert to the organic system in terms of livestock health management. However, it is necessary that these farms reduce the systematic (not necessary) use of preventive of antiparasitics (mainly in summer for ticks) and antibiotics at the start of the fattening period.

4.3. Rearing and Animal Welfare

Overall, farms had high scores (71.94%) for this area of action. Low scores were only observed regarding the degree of protection against bad weather and on the presence of autochthonous breeds. The low scores for proper infrastructure were due to the absence of facilities in cattle farms (more than to the lack of them), which is in turn due to the extensiveness on the system and the low number of farms fattening animals. In fat farms, fattening animals used to be done in plots (instead of feedlots). Infrastructures were clean enough, non-therapeutic mutilations were not common, and animals were isolated only for health purposes. Overall, animal welfare requirements are not apparently a major concern to convert the farms that were analyzed.

4.4. Agro-Ecosystem Management

As shown in Table 3, farms obtained poor scores (34.56%) on this area of action. This was mainly due to the use of pesticides/herbicides and of mineral fertilizers

fertilizers in many farms. These products are prohibited by Council Regulation (EC) No. 834/2007 ([27] and amendments), so that these farms should stop using them in order to be organic. This would not be difficult since crops are rainfed and their destination is livestock direct feeding (farmers are not looking to produce high amounts of highly priced crops).

However, regarding self-sufficiency, it must be mentioned that there is a remarkable decoupling between livestock and agriculture (previously observed in the Mediterranean area by Dantsis *et al.* [45]). This integration is important due to the interactions between livestock and plant-soil interface, and the low self-reliance of the farms (low purchasing power to buy external feed). Moreover, the use of protective-conservative agri-environmental practices is almost non-existent. Their implementation is important due to the fragility of the ecosystem. They would allow: (1) minimizing the disturbance; (2) maximizing soil surface; and (3) stimulating biological activity. This would allow better use of resources, increase ecosystem services, the landscape value and carbon sequestration, thus increasing long-term ecosystem's functionality, as well as its stability, economic performance and sustainability [46,47]. Although they are not mandatory, they are in accordance with IFOAM principles and agricultural sustainability, so that they should be implemented. Here, farmers would find great difficulties, since their application requires major changes such as training producers and designing production systems.

4.5. Self-Sufficiency and Agri-Food Chain Relationships

Despite the importance of self-reliance in pasture-based livestock systems [43], farms have shown low scores in this sense. This has been identified as one of the major weaknesses to convert to the organic system in the Mediterranean area [2,22]. As mentioned in the previous section, crop area should be increased and feed should be conserved for shortage periods. This is even more important if the higher price of organic feedstuff is taken into account. Agricultural business management out of the farm gate (to add value to the products and direct marketing) is essential for livestock production, even more in added value foods, so that farmers receive a higher price for their products. This has been identified as one of the main factors that determine the profitability of organic farms [4,33,48–50]. Moreover, these practices increase social interaction, opportunities in the rural world and the social and environmental impact of the food chain upstream. In the farms studied, direct sales by producers (a practice that is often linked to the organic sector), was nonexistent. This is still common and is due to the high resources (financial and human) required. This is even more remarkable in the livestock sector, since the perishable nature of fresh meat and hygiene regulations make it difficult and expensive due to cooling and logistics costs, even more if one takes into account that the main markets in terms of sales are mainly located in foreign countries (Figure 5); thus, the domestic market development is a challenge. To successfully overcome this barrier, producers must also focus on the European market, where consumption per capita is high (Figure 6), despite country sales being lower). Moreover, the low development of the organic industry as well as the low demand for organic meat products would hinder the transition to the organic system [3].

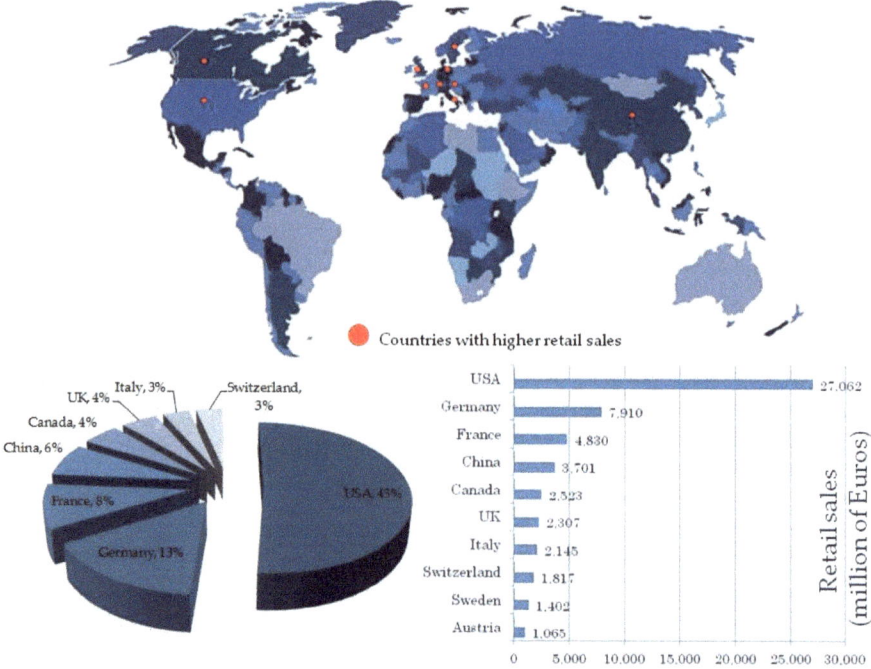

Figure 5. Countries with the largest markets for organic food. Retail sales: Million Euros. Upper map: ©TrueMiltra-FreeVectors.com. Charts: Own elaboration from FIBL (2016) [51].

4.6. Human Well-Being and Rural World Opportunities

This area of action is closely related to the principle of fairness, sustainable development and the food chain (above discussed). The study of this area is particularly important on farms and the ecosystem under study, due to the interdependent triad among the pasture, rural populations and livestock. In this regard, farms must improve their degree of business diversification, the number of jobs created and their stability (these are the main factors that both attract and retain people in rural areas). In general terms, conversion can be positive regarding job creation [52].

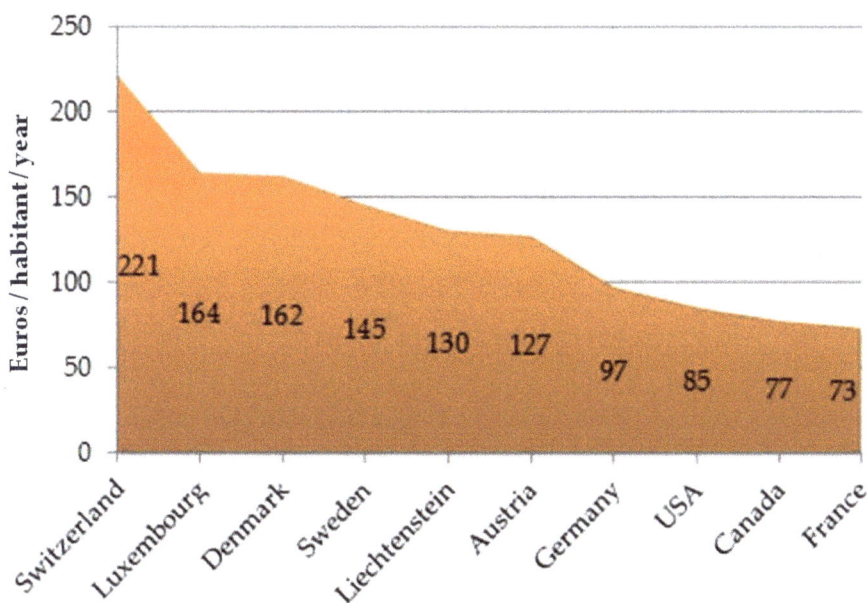

Figure 6. Countries with the highest per capita consumption (euros/habitant/year) in 2014. Own elaboration from FIBL (2016) [51].

4.7. Human Capability towards Implementing Sustainable Agricultural Practices

This area of action allowed understanding farmers' level of knowledge regarding organic production systems, their willingness to adopt them (farmers' age and future plans), and their adaptability to manage them. Farmers obtained low scores partially due to lack of interest (motivation) to find new sustainable business models in the face of future global changes (market and agricultural policies). Also, the lack of formal training on the field contributed negatively. As a recommendation, consulting and training initiatives through extension services, as well as trade unions and agricultural organizations must be enhanced [3].

Fortunately, the similarity between conventional and organic extensive pasture systems allows providing a process of simple conversion. However, the implementation of organic systems requires increased training, especially in business and agro-environmental terms.

4.8. Other Aspects Worthy of Discussion: Barriers, Perspectives and Solutions

It is worth mentioning the relationship between the results obtained and farms future sustainability. In this sense, the farms were classified with regard to different organic principles that are interrelated with sustainability dimensions. Farms'

121

classification based on cluster analysis has allowed having a deeper understanding of farms' group (typology) situations with regard to sustainability dimensions.

In this sense, T2 scored the highest in terms of Agro-ecosystem management (which is related to environmental sustainability). Regarding social sustainability, T2 showed higher results for self-sufficiency and agri-food chain relationships, while T3 had higher scores for human well-being and rural world opportunities. In terms of human resources/capital, T2 stood out from the rest. In general terms, it has been observed that most of the farmers did not focus on sustainability (social dimension: animal welfare, human health, creation of jobs; environmental dimension: environmental protection; economic dimension: local economy, short marketing channels). One of the main reasons for this has been the fact that many farms could easily comply with the organic regulations without carrying out environmentally-friendly management practices in their agro-ecosystems. Therefore, there is a real need for increasing managers' level of knowledge in the sustainability of agricultural practices. Special attention has not been paid to animal welfare, which could also be due to the fact that it is commonly assumed that animals under extensive production systems have a higher welfare status.

In order to implement such aspects, income plays a more important role than farmers' motivations, thus environmental quality and welfare status should be awarded and or supported (either via price premium and consumer's awareness, or agricultural subsidies). The review of Escribano [3] paid due attention to the market side because it was identified as key. However, marketing of organic animal products is not simple, since national demand is low, which requires export (and consequently a higher level of knowledge and costs) (Figures 5 and 6).

Regarding livestock management, there is a need to design feeding strategies that provide adequate nutrition, which is important to ensure a high level of health status based on prevention. From the economic viewpoint, feed management must be more focused on local resources in order to avoid the high costs of external organic feedstuff. This also has a consequence on the environmental side (nutrient cycling). Moreover, regulations should both unify criteria and facilitate the production of feed additives by companies because the consequences of it could be really important and positive for the organic livestock sector and for the sustainability of the food system.

Animal health in organic farming constitutes a challenge in many areas. Fortunately, due to the climate in the area under study (dry) and soils with a scarcity of organic matter, the prevalence of infectious/parasitic diseases is reduced, thus facilitating the conversion to the organic system. However, this fact should not make veterinarians feel too confident. On the contrary, the knowledge of the veterinarians with regard to animal health management must be improved. Furthermore, health care protocols based on preventive medicine must be developed, and epidemiological data should be part of the veterinary arsenal.

The perspectives for the organic beef cattle sector in the area seem not to be very promising. In fact, the regional census of organic beef cattle farms has decreased rather than increased in recent years [1]. During the interviews, organic farms were also analyzed, and many of their managers conveyed their intention to turn back to the conventional system, due to the difficulty of marketing their product as organic, despite the efforts carried out (transition period, bureaucracy load, *etc.*).

However, among the organic farms analyzed as a part of the research project (INIA-RTA2009-00122-C03-03 of the Spanish Ministry Economy and Competitiveness), success stories were also found and published by Escribano *et al.* [25] in their comparative sustainability assessment of extensive beef cattle farms in Dehesas (both conventional and organic ones). In this study, two subgroups of organic farms were identified: (i) a major group of farms that were just certified as organic but did not fatten their calves nor sell them as organic; and (ii) a second group constituted of very well organized full-cycle farms selling organic fattened calves (characterized by belonging to the organic farmers' association, having organic crops and mill, fattening animals, having trucks to transport them, and signing contracts with supermarkets).

Therefore, the advantage for organic beef cattle farmers in the area not belonging to the second group of organic farms (this is a closed group not allowing more farmers to join) is the benefit obtained from greening. However, their low productivity and competitiveness do not allow them to be sustainable, since their unique product was selling recently weaned calves (5–6 months and around 220 kg live weight) to be fattened either in other farms or in feedlots under the conventional system.

Thus, the domestic market development remains a challenge. In order to improve the contribution of the region to the organic market, structural changes in marketing channels must be made, but this will not be possible if consumers do not increase their demand of organic beef, which is still low due to low purchasing power in the area, the current national financial crisis, and a low level of knowledge and awareness regarding organic products [53]. The fact that regional citizens are used to extensive production systems could also be playing a role, as differences between conventional and organic products are not so clearly observed by the local population, which is the first step in generating demand.

5. Conclusions

The GCI data allowed assessing the feasibility of conversion of Dehesa beef cattle farms to the organic system. The integrative approach of the present study allowed taking into account not only the European legal requirements concerning organic farming and its principles, but also the particularities and future challenges of pasture-based beef cattle farms located in semi-arid regions.

The present study has revealed that the farms analyzed must carry out adaptations in all areas of activity that allow them overcome the conversion process successfully, especially with regard to health management and the agro-ecosystem (environmental, ecology principle). Rearing, animal welfare and management issues seem not to be of major concern. However, from the social point of view (principle of fairness), active participation in adding value to the products and on direct sales must be enhanced. In addition, the farms' self-reliance is a key issue in these farms that must be increased, as it would improve the economic results and ecological soundness (nutrients cycling, agro-biodiversity, *etc.*) of farms. In response to the environmental dimension and the principle of ecology, they should implement more environmentally-friendly farming practices (including reduction of total stocking rates, increase of crop area). Finally, transversal support measures are necessary, for example, training consumers' level of awareness regarding organic food and their willingness to pay premium prices.

Acknowledgments: Research funded by project INIA-RTA2009-00122-C03-03 of the Spanish Ministry Economy and Competitiveness. The author thanks the farmers, practitioners and experts that contributed to this research. The author also acknowledges the pre-doctoral financial support of the Fundación Fernando Valhondo Calaff. Special thanks to my Doctoral Thesis supervisors for their guidance (Paula Gaspar, Francisco J. Mesias, Miguel Escribano).

Conflicts of Interest: The author declares no conflict of interest.

References

1. Magrama 2015. Ministerio de Agricultura, Alimentación y Medio Ambiente. Agricultura Ecológica. Estadísticas 2014. Available online: http://www.magrama.gob.es/es/alimentacion/temas/la-agricultura-ecologica/estadisticas_ae_2014_definitivopdf_tcm7-405122.pdf (accessed on 14 March 2016).
2. Nardone, A.; Zervas, G.; Ronchi, B. Sustainability of small ruminant organic systems of production. *Livest. Prod. Sci.* **2004**, *90*, 27–39.
3. Escribano, A.J. Organic livestock farming: Challenges, perspectives, and strategies to increase its contribution to the agrifood system's sustainability. In *Organic Farming—A Promising Way of Food Production*, 1st ed.; Konvalina, P., Ed.; InTech: Rijeka, Croatia, 2016; pp. 229–260.
4. Benoit, M.; Veysset, P. Conversion of cattle and sheep suckler farming to organic farming: Adaptation of the farming system and its economic consequences. *Livest. Prod. Sci.* **2003**, *80*, 141–152.
5. Blanco-Penedo, I.; López-Alonso, M.; Shore, R.F.; Miranda, M.; Castillo, C.; Hernández, J.; Benedito, J.L. Evaluation of organic, conventional and intensive beef farm systems: Health, management and animal production. *Animal* **2012**, *6*, 1503–1511.
6. European Commission. EEC Council Directive 92/43/EEC of 21 May 1992 on the Conservation of Natural Habitats and of Wild Fauna and Flora. 1992. Available online: http://ec.europa.eu/environment/nature/legislation/habitatsdirective/index_en.htm (accessed on 14 March 2016).

7. Moreno, G.; Pulido, F.J. The functioning, management and persistence of Dehesas. In *Agroforestry in Europe: Current Status and Future Prospects*, 1st ed.; Rigueiro-Rodríguez, A., McAdam, J., Mosquera-Losado, M., Eds.; Springer Science + Business Media B.V.: Dordrecht, The Netherlands, 2009; pp. 127–160.

8. Sánchez, J. Las macromagnitudes agrarias. In *La Agricultura y la Ganadería Extremeñas. Informe 2012*; Facultad de Ciencias Económicas y Empresariales/Escuela de Ingenierías Agrarias, Universidad de Extremadura, Caja de Badajoz: Badajoz, Spain, 2013; pp. 37–52.

9. Easdale, M.H.; Aguiar, M.R. Regional forage production assessment in arid and semi-arid rangelands—A step towards social-ecological analysis. *J. Arid Environ.* **2012**, *83*, 35–44.

10. Peacock, C.; Sherman, D.M. Sustainable goat production—Some global perspectives. *Small Rumin. Res.* **2010**, *89*, 70–80.

11. Prudencio, C. El mercado de trabajo. In *La Agricultura y Ganadería Extremeña. Informe 2012*; Facultad de Ciencias Económicas y Empresariales/Escuela de Ingenierías Agrarias, Universidad de Extremadura, Caja de Badajoz: Badajoz, Spain, 2013; pp. 266–270.

12. Marsden, T.; Banks, J.; Bristow, G. The social management of rural nature: Understanding agrarian-based rural development. *Environ. Plan. A* **2002**, *34*, 809–825.

13. Pretty, J. *Agriculture: Reconnecting People, Land and Nature*; Earthscan: London, UK, 2002; p. 261.

14. Boyazoglu, J.; Hatziminaoglou, I.; Morand-Fehr, P. The role of the goat in society: Past, present and perspectives for the future. *Small Rumin. Res.* **2005**, *60*, 13–23.

15. De Rancourt, M.; Fois, N.; Lavín, M.P.; Tchakérian, E.; Vallerand, F. Mediterranean sheep and goats production: An uncertain future. *Small Rumin. Res.* **2006**, *62*, 167–179.

16. Cocca, G.; Sturaro, E.; Gallo, L.; Ramanzin, M. Is the abandonment of traditional livestock farming systems the main driver of mountain landscape change in Alpine areas? *Land Use Pol.* **2012**, *29*, 878–886.

17. MacDonald, D.; Crabtree, J.R.; Wiesinger, G.; Dax, T.; Stamou, N.; Fleury, P.; Lazpita, J.G.; Gibon, A. Agricultural abandonment in mountain areas of Europe: Environmental consequences and policy response. *J. Environ. Manag.* **2000**, *59*, 47–69.

18. Gellrich, M.; Baur, P.; Koch, B.; Zimmermann, N.E. Agricultural land abandonment and natural forest re-growth in the Swiss mountain: A spatially explicit economic analysis. *Agric. Ecosyst. Environ.* **2007**, *118*, 93–108.

19. Gaspar, P.; Mesías, F.J.; Escribano, M.; Pulido, F. Sustainability in Spanish extensive farms (Dehesas): An economic and management indicator-based evaluation. *Rangel. Ecol. Manag.* **2009**, *62*, 153–162.

20. Franco, J.A.; Gaspar, P.; Mesías, F.J. Economic analysis of scenarios for the sustainability of extensive livestock farming in Spain under the CAP. *Ecol. Econ.* **2012**, *74*, 120–129.

21. Escribano, A.J.; Gaspar, P.; Mesias, F.J.; Pulido, A.F.; Escribano, M. A sustainability assessment of organic and conventional beef cattle farms in agroforestry systems: The case of the Dehesa rangelands. *ITEA* **2014**, *110*, 343–367.

22. Mena, Y.; Nahed, J.; Ruiz, F.A.; Sánchez-Muñoz, J.B.; Ruiz-Rojas, J.L.; Castel, J.M. Evaluating mountain goat dairy systems for conversion to the organic model, using a multicriteria method. *Animal* **2012**, *6*, 693–703.

23. Toro-Mújica, P.; García, A.; Gómez-Castro, A.G.; Acero, R.; Perea, J.; Rodríguez-Estévez, V.; Aguilar, C.; Vera, R. Technical efficiency and viability of organic dairy sheep farming systems in a traditional area for sheep production in Spain. *Small Rumin. Res.* **2011**, *100*, 89–95.

24. Toro-Mújica, P.; García, A.; Gómez-Castro, A.; Perea, J.; Rodríguez-Estévez, V.; Angón, E.; Barba, C. Organic dairy sheep farms in south-central Spain: Typologies according to livestock management and economic variables. *Small Rumin. Res.* **2012**, *104*, 28–36.

25. Escribano, A.J.; Gaspar, P.; Mesías, F.J.; Escribano, M.; Pulido, F. Comparative sustainability assessment of extensive beef cattle farms in a high nature value agroforestry system. In *Rangeland Ecology, Management and Conservation Benefits*, 1st ed.; Squires, V.R., Ed.; Nova Publishers: New York, NY, USA, 2015; pp. 65–85.

26. IFOAM. Principles of Organic Agriculture. International Federation of Organic Agriculture Movements. 2005. Available online: http://www.ifoam.org/en/organic-landmarks/principles-organic-agriculture (accessed on 22 January 2016).

27. The Council of the European Union. Council Regulation (EC) No. 834/2007 of 28 June 2007 on organic production and labelling of organic products and repealing Regulation (ECC) No. 2092/91. *Off. J. Eur. Union* **2007**, *50*, 1–23.

28. Mena, Y.; Nahed, J.; Ruiz, F.A.; Castel, J.M.; Ligero, M. Proximity to the organic model of dairy goat systems in the Andalusian Mountains (Spain). *Trop. Subtrop. Agroecosyst.* **2009**, *11*, 69–73.

29. Nahed-Toral, J.; Sanchez-Munoz, B.; Mena, Y.; Ruiz-Rojas, J.; Aguilar-Jimenez, R.; Castel, J.M.; de Asis Ruiz, F.; Orantes-Zebadua, M.; Manzur-Cruz, A.; Cruz-Lopez, J.; *et al.* Potential for conversion of agrosilvopastoral systems of dairy cattle to the organic production model in southern eastern Mexico. *J. Anim. Vet. Adv.* **2012**, *11*, 3081–3093.

30. Nahed-Toral, J.; Sánchez-Muñoz, B.; Mena, Y.; Ruiz-Rojas, J.; Aguilar-Jimenez, R.; Castel, J.M.; de Asis Ruiz, F.; Orantes-Zebadua, M.; Manzur-Cruz, A.; Cruz-López, J.; *et al.* Feasibility of converting agrosilvopastoral systems of dairy cattle to the organic production model in south eastern Mexico. *J. Clean. Prod.* **2013**, *43*, 136–145.

31. Boatto, V.; Trestini, S. The role of post-2013 common agricultural policy on the sustainability of Italian beef production. *Agric. Conspec. Sci.* **2013**, *78*, 137–141.

32. The European Parliament and the Council of the European Union. Regulation (EU) No 1306/2013 of the European Parliament and of the Council of 17 December 2013 on the Financing, Management and Monitoring of the Common Agricultural Policy and Repealing Council Regulations (EEC) No 352/78, (EC) No 165/94, (EC) No 2799/98, (EC) No 814/2000, (EC) No 1290/2005 and (EC) No 485/2008. Available online: http://eur-lex.europa.eu/legal-content/EN/ALL/?uri=celex:32013R1306 (accessed on 16 June 2016).

33. Wittman, H.; Becky, M.; Hergesheimer, C. Linking local food systems and the social economy? Future roles for farmers' markets in Alberta and British Columbia. *Rural Soc.* **2012**, *77*, 36–61.

34. Leach, T.; Baret, P.V.; Stilmant, D. Sustainability indicators for livestock farming. A review. *Agron. Sustain. Dev.* **2013**, *33*, 311–327.

35. Nahed, J.; Castel, J.M.; Mena, Y.; Caravaca, F. Appraisal of the sustainability of dairy goat systems in Southern Spain according to their degree of intensification. *Livest. Sci.* **2006**, *101*, 10–23.

36. Ripoll-Bosch, R.; Díez-Unquera, B.; Ruiz, R.; Villalba, D.; Molina, E.; Joy, M.; Olaizola, A.; Bernués, A. An integrated sustainability assessment of Mediterranean sheep farms with different degrees of intensification. *Agric. Syst.* **2012**, *105*, 46–56.

37. Kerselaers, E.; Cock, L.D.; Lauwers, L.; Huylenbroeck, G.V. Modelling farm-level economic potential for conversion to organic farming. *Agric. Syst.* **2007**, *94*, 671–682.

38. Masera, O.; Astier, S.; López-Ridaura, S. Sustentabilidad y manejo de los recursos naturales. In *El Marco de Evaluación MESMIS*, 1st ed.; Masera, O., Astier, S., López-Ridaura, S., Eds.; Mundi-Prensa, IE-UNAM: Mexico, 1999; p. 160.

39. Ten Brink, B.J.E.; Hosper, S.H.; Colin, F. A quantitative method for description and assessment of ecosystems: The AMOEBA approach. *Mar. Pollut. Bull.* **1991**, *23*, 265–270.

40. Astier, M.; García-Barrios, L.; Galván-Miyoshi, Y.; González-Esquivel, C.E.; Masera, O.R. Assessing the Sustainability of Small Farmer Natural Resource Management Systems. A Critical Analysis of the MESMIS Program (1995–2010). *Ecol. Soc.* **2012**, *17*, 25.

41. Hair, J.F.; Anderson, R.F.; Tathakm, R.L.; Black, W.C. *Análisis Multivariante*, 5th ed.; Hair, J.F., Anderson, R.E., Tatham, R.L., Eds.; Prentice Hall Iberia: Madrid, Spain, 1999; p. 832.

42. SPSS (Statistical Package for the Social Sciences). *Manual del Usuario del Sistema Básico de IBM SPSS Statistics 21*; IBM Corporation: Armonk, NY, USA, 2012; p. 464.

43. Ripoll-Bosch, R.; Joy, M.; Bernués, A. Role of self-sufficiency, productivity and diversification on the economic sustainability of farming systems with autochthonous sheep breeds in less favoured areas. *Animal* **2013**, *4*, 1–9.

44. The Commission of the European Communities. Commission Regulation (EC) No. 889/2008 Laying down Detailed Rules for the Implementation of Commission Regulation (EC) No. 834/2007 on Organic Production and Labelling of Organic Products with Regard to Organic Production, Labelling and Control. Available online: http://eur-lex.europa.eu/LexUriServ/LexUriServ.do?uri=OJ:L:2008:250:0001:0084:en:PDF (accessed on 16 June 2016).

45. Dantsis, T.; Loumou, A.; Giourga, C. Organic agriculture's approach towards sustainability; its relationships with the agro-industrial complex, a case study in Central Macedonia, Greece. *J. Agric. Environ. Ethics* **2009**, *22*, 197–216.

46. Andriarimalala, J.H.; Rakotozandriny, J.N.; Andriamandroso, A.L.H.; Penot, E.; Naudin, K.; Dugué, P.; Tillard, E.; Decruyenaere, V.; Salgado, P. Creating synergies between conservation agriculture and cattle production in crop-livestock farms: A study case in the lake Alaotra region of Madagascar. *Exp. Agric.* **2013**, *49*, 352–365.

47. Sanderson, M.A.; Archer, D.; Hendrickson, J.; Kronberg, S.; Liebig, M.; Nichols, K.; Schmer, M.; Tanaka, D.; Aguilar, J. Diversification and ecosystem services for conservation agriculture: Outcomes from pastures and integrated crop-livestock systems. *Renew. Agric. Food Syst.* **2013**, *28*, 129–144.

48. Hrabalová, A.; Zander, K. Organic beef farming in the Czech Republic: Structure, development and economic performance. *Agric. Econ. Czech* **2006**, *52*, 89–100.

49. Tzouramani, I.; Sintori, A.; Liontakis, A.; Karanikolas, P.; Alexopoulos, G. An assessment of the economic performance of organic dairy sheep farming in Greece. *Livest. Sci.* **2011**, *141*, 136–142.

50. Argyropoulos, C.; Tsiafouli, M.A.; Sgardelis, S.P.; Pantis, J.D. Organic farming without organic products. *Land Use Pol.* **2013**, *32*, 324–328.

51. The World of Organic Agriculture. *Statistics & Emerging Trends*; FIBL & IFOAM: Rheinbreitbach, Switzerland, 2016; p. 333. Available online: https://shop.fibl.org/fileadmin/documents/shop/1698-organic-world-2016.pdf (accessed on 16 June 2016).

52. Lobley, M.; Butler, A.; Reed, M. The contribution of organic farming to rural development: An exploration of the socio-economic linkages of organic and non-organic farms in England. *Land Use Pol.* **2009**, *26*, 723–735.

53. García-Torres, S.; López-Gajardo, A.; Mesías, F.J. Intensive *vs.* free-range organic beef. A preference study through consumer liking and conjoint analysis. *Meat Sci.* **2016**, *114*, 114–120.

Section 2:
Farming to Food Systems

Can Organic Farming Reduce Vulnerabilities and Enhance the Resilience of the European Food System? A Critical Assessment Using System Dynamics Structural Thinking Tools

Natalia Brzezina, Birgit Kopainsky and Erik Mathijs

Abstract: In a world of growing complexity and uncertainty, food systems must be resilient, i.e., able to deliver sustainable and equitable food and nutrition security in the face of multiple shocks and stresses. The resilience of the European food system that relies mostly on conventional agriculture is a matter of genuine concern and a new approach is called for. Does then organic farming have the potential to reduce vulnerabilities and improve the resilience of the European food system to shocks and stresses? In this paper, we use system dynamics structural thinking tools to identify the vulnerabilities of the conventional food system that result from both its internal structure as well as its exposure to external disturbances. Further, we evaluate whether organic farming can reduce the vulnerabilities. We argue here that organic farming has some potential to bring resilience to the European food system, but it has to be carefully designed and implemented to overcome the contradictions between the dominant socio-economic organization of food production and the ability to enact all organic farming's principles—health, ecology, fairness and care—on a broader scale.

Reprinted from *Sustainability*. Cite as: Brzezina, N.; Kopainsky, B.; Mathijs, E. Can Organic Farming Reduce Vulnerabilities and Enhance the Resilience of the European Food System? A Critical Assessment Using System Dynamics Structural Thinking Tools. *Sustainability* **2016**, *8*, 971.

1. Introduction

Food is of key relevance to human health and survival. Europeans take their food and nutrition security (FNS) for granted and rely on a food system in which most of the food is produced by conventional farmers subsidized from the Common Agriculture Policy (CAP) [1]. Over the last decades this system, hugely depending on public support, has achieved tremendous improvements in productivity [2]. As a result, nowadays more food is supplied than demanded at historically low prices. This allows European consumers to spend only a small percentage of their household disposable income on food [1,3].

These FNS achievements in Europe are, however, far from ideal and looking ahead Europeans may not be as food secure as they perceive themselves to be. Most of the European consumers rely on a complex system, in which conventional farmers, driven by profit maximization, are continuously intensifying, specializing, standardizing, expanding their operations and becoming even more dependent on the application of off-farm sourced modern tools such as chemicals to manage fertility and pests, diesel-powered machines, biotechnology and proprietary seeds [2]. These processes and practices, in turn, feed back to the environment and to society with numerous unintended consequences, *inter alia*, soil degradation, nutrient runoff, greenhouse gas (GHG) emissions, biodiversity loss, pesticide-born health damage and socio-economic decline in rural communities. These consequences pose risks to FNS and well-being of future generations [4]. Moreover, much of the productivity advances and associated trends in the European food system were realized in times of relatively stable climate, when natural resources seemed to be infinite, and the human population was considerably smaller [5,6]. In the face of already observed changing climate, deteriorating natural resources, growing population largely driven by migration as well as many other emerging challenges and uncertainties, there are growing concerns that the European food system is vulnerable and thus unable to withstand disturbances without undesirable outcomes [1,5,7–14].

In order to cope with the challenges and uncertainties, we need a new approach to agriculture in the food system [7,8,15]. Such an approach must change both the farming practices as well as the socio-economic organization of food production to increase the food system's resilience and its ability to deliver sustainable and equitable FNS today and in the future [1,5,7–9]. One of the potential candidates is organic farming [5,7,9,16], which from all the alternate approaches is the only one that has been regulated and supported at EU level by a vast array of legal, financial and knowledge-based policy instruments for several decades [17,18]. Accordingly, the number of organic farms, the extent of organically farmed land, funding devoted to organic farming and the market size for organic foods have steadily increased across Europe [18].

Given this development, an important question that arises is whether organic farming can reduce the vulnerabilities and enhance the resilience of the European food system and hence deliver sustainable and equitable FNS? Organic farming seems to be a promising approach as it is built on four systemic principles formulated by the International Federation of Organic Agriculture Movements (IFOAM): "health", "ecology", "fairness" and "care". Organic farming thus aims to produce wholesome food in an environmentally-friendly way, as well as to contribute to economic sustainability and social justice [19–21]. In research and public debate, however, organic farming has a history of being contentious [21]. At the same time, understanding and operationalization of the concepts of the food system's vulnerability and resilience themselves is limited [22]. On the one hand, many studies provide evidence for organic farming's ability to balance the multiple sustainability goals [19,21] and build resilience to disturbances, especially at farm level [23–27]. On the other hand, critics consider organic farming as an inefficient approach to FNS, one that will become irrelevant in the future, because of too many shortcomings and poor solutions to agriculture problems [4,19–21]. Furthermore, some argue that organic farming undergoes 'conventionalization' and is a mere substitution of inputs rather than a redesign of farming operations [28]. Consequently, organic farming may violate many of the ecologically, socially and economically progressive principles originally valued [20,21,28], further exacerbating vulnerabilities and undermining resilience of the European food system [5].

With regard to the nature of the assessments on which the debate draws, the majority is based on comparisons of outcomes delivered by organic versus conventional farming system (e.g., crop yields, profitability, environmental impacts, etc.) (e.g., [21,29–33]) as well as individual causal connections (e.g., the effect of organic farming practices on biodiversity, food quality or crop yield, etc.) ([34–36]), at a given point in time. A system's perspective over time is, however, missing. Food systems, no matter whether they are based on conventional, organic or any other food production approach, are dynamic and complex social-ecological systems (SES) [37]. Their structures are formed by many internal and external variables which interact with each other often across multiple, hierarchically linked subsystems [38] and through feedback mechanisms to generate outcomes [39,40]. These feedback mechanisms are largely masked to farmers, consumers and policymakers [11]. They also involve nonlinearities, time delays and accumulations, which complicate information and material flows in the food system and hence lead to counterintuitive system behavior [15,41]. Inherent to these features of food systems such as SES are the synergies and trade-offs between outcomes that they produce [37,40,41]. Given the dynamic complexity inherent in food systems, it is not immediately apparent where and how the vulnerabilities to disturbances occur in the system and how resilience is generated. Therefore, it is challenging for decision makers to design and implement effective strategies to farming and other aspects related to food systems that would reduce its vulnerabilities and enhance its resilience [15].

Conceptualizing and modelling of SES has the potential to assist decision makers in managing complex human-environment relationships that form the basis of food systems [42,43]. The development of SES models is, however, challenging as it requires *inter alia* integration of knowledge scattered across many disciplines on variables and their relationships from both the social and the ecological domains as well as explicit modelling of feedbacks between the social and ecological systems along with their cross-scale and cross-level interactions [38,42–44]. There are various approaches to the interdisciplinary modelling of SES with differing underlying assumptions and anchored in different scientific perspectives, so there is always the likelihood that another model of a particular food system might give diverse outcomes [43,44].

In this paper, we adopt a system dynamics approach [45] to understand the European food system's vulnerabilities and to assess the potential of organic farming to reduce them and enhance its resilience [15] through sustainability lenses. System dynamics is a computer-aided modelling approach to policy analysis and design that takes an explicit feedback perspective and enables capturing the dynamic complexity of SES, such as the food systems [40–42]. This approach is based on the underlying assumption that the internal structure and the feedback processes in a system determine its dynamic behavior over time and how it responds to disturbances [15,45]. By adopting this approach we do not provide new data, introduce new variables or measure the strengths of a particular causal-effect link. Rather, our main contribution is the reorganization of existing knowledge and the promotion of structural insights from variables already established in the literature. More specifically, we combine an in-depth literature review and secondary data analysis with system dynamics diagramming tools to fulfill three objectives. The first objective is to understand the different sources of vulnerabilities in the European food system based on conventional agriculture by analyzing its internal structure and feedback processes, where the entry points for disturbances are, and the mechanisms by which the disturbances are transmitted throughout the system. The second objective is to assess whether organic farming is a viable strategy to reduce the vulnerabilities and enhance the resilience of the European food system. The third objective is to illustrate throughout the analyses how the system dynamics approach can address some of the current challenges posed by SES modelling. Ultimately, we provide decision makers—e.g., policymakers, NGOs, farm associations, etc.—at EU level with a framework that could support them in developing more effective strategies for the European food system.

The remainder of the paper is organized as follows: after a brief overview of the conceptual background and methodology, we articulate the dynamic problem (i.e., select the system's boundary) and conceptualize the European food system from a feedback perspective. Next, we qualitatively analyze the food system's vulnerabilities by focusing on the interplay between internal structure and feedback processes of the system and external disturbances. Finally, we discuss organic farming as an alternative approach and close the paper with conclusions.

2. Methodology: System Dynamics Structural Thinking Tools for Food System and Vulnerability Analysis

Food systems are coupled SES formed by many internal and external variables that are interconnected through feedback processes at various scales and levels and that determine FNS along with other environmental and socio-economic outcomes [38–40,46,47]. When exposed to various and unforeseen disturbances, the emergence of undesirable outcomes indicates that somewhere in the food system a critical capacity is failing and that the structure and processes driving the functioning of the system make it vulnerable [47]. We thus define vulnerability as a system's inability to respond to disturbances without generating undesirable outcomes. In vulnerable food systems, even small disturbances may cause detrimental changes from which it is difficult to recover [15,39,47]. Resilience, on the other hand, is the capacity of a food system to withstand disturbances and continue providing the same or possibly even improved desirable outcomes [47]. Vulnerability and resilience are dynamic and normative concepts in the sense that the value judgement of what is desirable and what constitutes improvement or damage over what period of time depends on the observer [47,48]. Hence, to assess whether a food system is resilient or vulnerable we have to define: (1) the boundaries of the system (vulnerability/resilience of what); (2) relevant disturbances (vulnerability/resilience to what) and (3) what constitutes desirable change over what time frame and to whom [47,48]. We address these questions by adopting a system dynamics approach.

System dynamics is an approach designed to examine and manage complex systems that change over time. It is applicable to any dynamic systems of which integral features are interdependence, mutual interaction and feedback loops [45,49,50]. System dynamics modelling is an iterative process that begins with defining dynamic problem, proceeds through developing dynamic hypothesis and modeling stages, to building confidence in the model and analyzing policy implications [45,49,50].

Conceptually, the central principle of this modelling approach is that the endogenous structure of a system determines its dynamic behavior over time and how it responds to disturbances and policy changes [16,46]. Thus, in system dynamics the emphasis is given to a continuous view (i.e., 'the large picture'), shifting the attention from events to behavior to structure. The endogenous point of view implies that the causes are contained within the internal structure of the system itself, while exogenous disturbances are seen at most as triggers of system behavior. Feedback loops are central for conceptualizing the internal structure of complex systems. These closed loops of causal links involve delays and nonlinearities as well as processes of accumulation and draining. The internal structure of a system is a combination of such feedback loops, which by interacting with each other can generate all kinds of dynamic patterns of behavior. However, the concept of underlying feedback loops is not exhaustive for explaining the dynamic behavior of a system. The explanatory power of feedback understanding lies in the shifting interplay between loops, implying that different parts of a system become dominant over the others at different times [50–52].

The system dynamics methodology provides structural thinking tools—closed boundary, feedback loops, stocks, flows, etc.—used to communicate the boundary of the system and to represent its causal structure in a structural diagram. The goal of a system dynamics modeler is to assemble such a structural diagram that can endogenously, by itself, explain the dynamic problem. The closed boundary refers to the effort to view a system as causally closed as opposed to the open and closed systems in the general system sense. In turn, causality refers to causes as pressures which produce aggregated patterns of behavior rather than events, actions, individual stimuli and decisions [52]. This implies that feedback processes between levels (such as agriculture production and consumption) can be captured providing that the individual levels are modelled in an aggregated way [41]. The causally closed system boundary identifies the endogenous perspective as the feedback view pressed to an extreme. A causally closed structural diagram provides important qualitative insights into the system's behavior [15,53–55] and can facilitate the identification of leverage points for intervention in the system [15]. Based on structural diagrams computer simulation models can be created to experiment on how the system behaves under unanticipated disturbances or policy interventions [15,55,56].

The theoretical assumptions of system dynamics have been addressed in several studies (e.g., [51,52,54–63]), but usually system dynamists take them for granted. System dynamics appears to be ontologically a realist approach, as models are presented as abstract representations of the real physical and information flows in a system, with feedbacks implying that, *"decisions are not entirely 'free will' but are strongly conditioned by the environment"* [51]. However, this objective stance of system dynamics models mixes with subjectivity, as the purpose of system dynamics is also to engage with 'mental models'. These mental models range from hard, quantitative information to more subjective, or even judgmental aspects of a given situation [57–59]. In consequence, a model should be focused around a particular issue (dynamic problem). The focus on trying to understand the real-world phenomena reflects the practical engineering origin of system dynamics [58]. From social theoretic perspective, however, divergent practice within this field makes it difficult to place it in one paradigm. Superficially, system dynamics can be positioned within the functionalist sociology paradigm, its ideas seeming to be a version of social systems theory [58,62,63]. However, the practice of system dynamics, and hence its theory in use, has many features of more interactionist paradigm and also some links to interpretativism [58,62,63]. The uncertainty related to positioning system dynamics within a social theoretic perspective leads to the conclusion that this approach appears to be best locatable within those theories that try to integrate the agency and structure views of the social realm (for detailed analysis see [52,58,62,63]).

In this paper, we adapt the approach taken by Stave and Kopainsky [15]. They used system dynamics to promote qualitative structural insights on mechanisms and pathways of food supply vulnerability, arguing that *"any examination of food supply vulnerability to disturbances, or ability to withstand disturbances that could lead to food supply disruption, should start by examining the food system's components, causal connections, and feedback mechanisms and describing system interactions in terms of material and information flows that pass changes in one component on to other components"* [15]. The approach taken in this study consists of three iterative steps inspired by the system dynamics modelling process: 1. problem articulation; 2. system conceptualization as well as 3. vulnerability and policy analysis [45,49]. The implementation of these steps addresses the abovementioned three prerequisites for vulnerability/resilience assessment and hence leads to qualitative structural insights into the food system' vulnerability/resilience as presented in Figure 1. Quantitative analysis of system behavior when exposed to disturbances would require a fully specified computer simulation model and is beyond the scope of this paper.

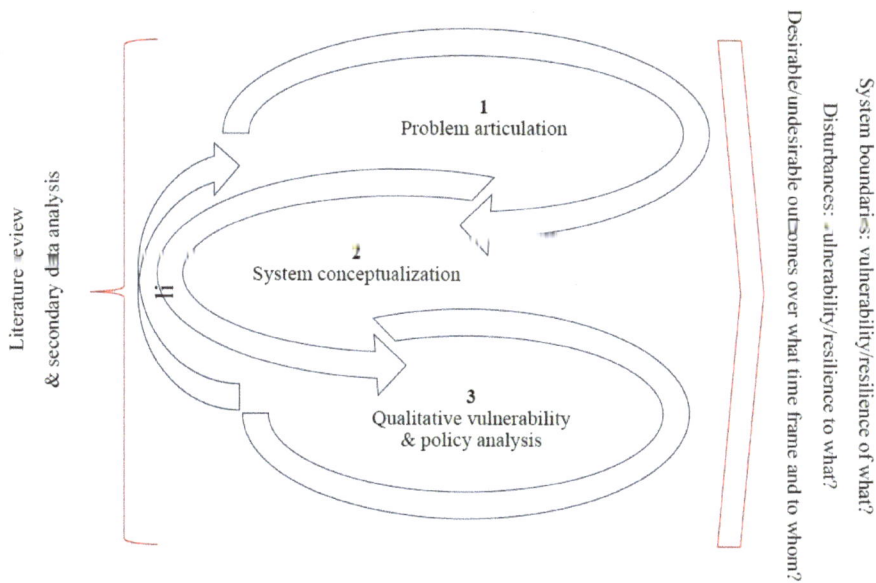

Figure 1. Three iterative methodological steps inspired by system dynamics approach.

The starting point of a system dynamics analysis is the identification of the dynamic problem at stake, that is, the pattern of behavior of the system's outcome of interest, unfolding over time, which shows how the problem arose and how it might evolve in the future [15,45,49]. The initial articulation of the dynamic problem predetermines the system's boundary and the scope of the iterative modeling effort.

To define the dynamic problem in our study and accordingly select the boundary of the modelled food system we analyzed relevant literature and time series of secondary data. Prior to an in-depth search in electronic databases, a general Google Scholar search was run to gather key documents. These papers were used to collect terms and phrases pertaining to the performance of conventional and organic farming in relation to their contribution to sustainable development as well as drivers of change influencing the food system in general, and of agricultural production in particular. Based on the terms and phrases we conducted an in-depth search from November 2015 to February 2016 without any restrictions on publication dates to ensure that the broadest set of data could be captured, yet with imposed limitation to English language publications only. The search strategy was applied to four databases: Web of Science (Thomson Reuters, New York, NY, USA), Scopus (Elsevier, Amsterdam, The Netherlands), ScienceDirect (Elsevier, Amsterdam, The Netherlands) and Organic E-prints (International Centre for Research in Organic Food Systems, Tjele, Denmark). In addition, we searched relevant organizational websites (e.g., European Commission, International Federation of Organic Agriculture Movement EU, Food and Agriculture Organization) in order to capture the grey literature. Reference lists of included publications were also hand-searched for additional relevant studies. The content of the pertinent papers was then manually reviewed with support of automatic word frequency and text search queries in NVivo11® (QSR International, Melbourne, Australia) (a software for qualitative data analysis) to elicit a list of key indicative outcomes of the European food systems based on conventional and/or organic agriculture along with related internal and external variables that are relevant for the subsequent analytical steps. The insights from literature and additional analyses of time series data obtained from EUROSTAT, FAOSTAT and FADN, allowed us to articulate the dynamic problem by specifying the several reference modes of historically observed trends in selected indicative outcomes of the European food system as well as of their desirable and undesirable developments in the face of disturbances. The analysis was conducted on a selection of outcomes being a simplified representation of the European food system's performance from different stakeholders' perspectives (e.g., price for consumer, profits for producers, etc.). We focused on the selective list of indicative outcomes to demonstrate the way in which the system dynamics approach can be used to study synergies and trade-offs in outcomes relevant for different stakeholders. For a comprehensive analysis many more outcomes delivered by the European food system and valued by various stakeholders would have to be further diversified.

Once the dynamic problem has been articulated over an appropriate time horizon, system dynamics modelers specify the model boundary by conceptualizing the system. The boundary of a system is defined in a causal rather than in a geographical way. It implies that system dynamists look for processes that explain observed or anticipated problematic behavior (the dynamic problem), irrespective of where these processes unfold. In system dynamics language, the modelers formulate a theory, called a dynamic hypothesis, which provides an endogenous explanation of the dynamics characterizing the problem at stake in terms of the underlying causal structure of the system. It is a hypothesis as it is always an interim, working theory, subject to reconsideration or abandonment as the knowledge base about the real world develops [45,49]. The concentration on endogenous explanations does not mean that exogenous variables are excluded from the model. They are included in models, but each of the candidate for an exogenous variable is carefully examined, to determine whether there is any relevant feedback from the endogenous variables to the candidate. If so, the boundary of the model is extended and the candidate exogenous variable is modelled endogenously [45].

To communicate the system conceptualization a variety of tools can be used. These range from qualitative structural thinking tools (e.g., causal loop diagrams, stock and flow maps), which visually represent different types of variables and their interconnectedness, to operational tools, which express relationships between variables in terms of mathematical equations [15].

In our study, the development of dynamic hypothesis started with insights from the Sustainability Institute [64]. Further, the dynamic hypothesis was enriched with internal and external variables and the associated causal connections elicited from the in-depth literature review, analyzed time series data, theory and general knowledge. Guided by the dynamic problem, we conceptualized the European food system in the form of causal loop diagram drawn in Vensim DSS®(Ventana Systems Inc., Harvard, United States) (i.e., software for system dynamics modelling), in which we marked important feedback processes forming the endogenous explanation. Specifically, we built the system's internal causal structure by tracing from the previously selected indicative food system outcomes (i.e., of which dynamic behavior was considered problematic) outward along the chains of cause and effect, variable-by-variable, rather than from system boundary inward. In developing our dynamic hypothesis we did not aim at explaining all possible dimensions of the food system outcomes. Instead, we focused on the key dimensions, represented by the selected indicative outcomes of the European food system, to exemplify how system dynamics structural thinking tools can be used to study complex food system issues.

Arrows represent the causal links between variables, which indicate both the direction of causality and whether the variables change in the same—a positive link (+)—or in the opposite—a negative link (−)—direction (Figure 2). For example, if price is a cause and supply is an effect, a positive link indicates that, *ceteris paribus*, an increase in price leads to an increase in supply. If, on the other hand, supply is a cause and price is an effect, a negative link means that, all else equal, an increase in supply causes a decrease in price or vice versa a decrease in supply causes an increase in price.

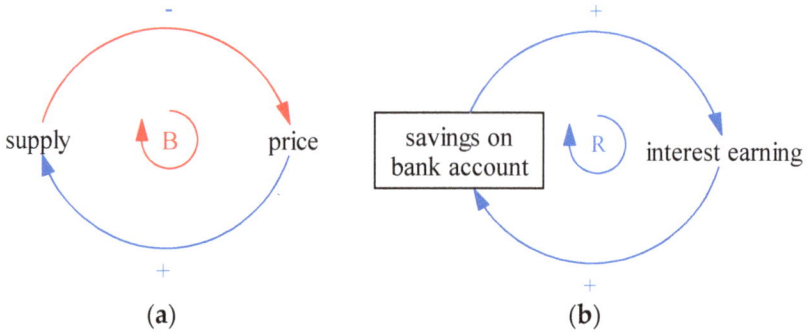

Figure 2. Indication of causal links, feedback loops and their nature: (**a**) balancing loop (B); (**b**) reinforcing loop (R) with signified stock (rectangle) and delay (crossing the causal link arrow).

When a feedback loop arises around two or more variables, we classify it either as a balancing (B; stabilizing, negative; Figure 2a) or a reinforcing (R; amplifying, positive; Figure 2b) feedback loop. To determine the polarity of the loops we trace the effect of change in one of the variables as it propagates around the loop. The classification rule is that if the feedback loop effect reinforces or amplifies the original change, it is a reinforcing loop (e.g., the more savings we have on a bank account, the more interest we earn and in turn the more savings we accumulate); if it counteracts or opposes the original change, it is a balancing loop (e.g., the higher the supply, the lower the price and in turn the lower the supply) [45,49]. Reinforcing are sources of growth, explosion, erosion, and collapse in systems. Balancing loops are self-correcting. For clearer and more insightful analysis, we also indicated in the causal loop diagrams important stocks in rectangles (Figure 2b). Stocks are accumulations, which characterize the state of the system and generate the information upon which decisions and actions are based. They create also inertia in systems that could either be source of disequilibrium dynamics (i.e., instability and oscillations) or filter out unwanted variability [45]. Other delays and flows are also inherent in the structural diagrams, but for readability purposes we did not signify them in any special form.

Once internal structure and feedback processes in the European food system that determine its outcomes were formulated, the resulting causal loop diagram guided the identification of entry points that expose the system to external drivers of change. Finally, we examined the systemic impacts of the internal processes and external unanticipated disturbances on the outcomes of the food system to assess qualitatively both (1) vulnerabilities of the European food systems and (2) the potential of organic farming to reduce the vulnerabilities and enhance resilience of the system. We assessed the direction of the change in the food system's outcomes that internal processes and unexpected disturbances cause. Specifically, we analyzed how the disturbances could be either intensified or reduced throughout the system internal structure and change its outcomes.

By formulation of the internal causal structure and the identification of the external disturbances we did not aim to capture the complete, real, yet only vaguely understood European food system as a SES. Alternatively, we illustrate how system dynamics structural thinking tools can be used to study where complex food systems might be vulnerable to external disturbance and how these disturbances are transmitted throughout the internal feedback structure; more generally what kind of insights can result from taking such an approach and how it addresses some of the challenges involved in SES modelling.

3. Problem Articulation: Boundary Selection

Given the definition of vulnerability above, we frame the dynamic problem as the concern that the European food system when subjected to disturbances of different nature and origin would be unable to withstand them and hence cause its outcomes to considerably or permanently diverge from their desired level. Ericksen [37,39] distinguishes three groups of outcomes that can indicate vulnerability of the food system, namely failure to provide FNS as well as collapse of environmental and socio-economic welfare. The prevailing European food system, which is based on conventional agriculture, continuously faces a trifold challenge of reconciling FNS, viability of rural societies (socio-economic welfare) and sustainable management of the EU's natural land-based resources (environmental welfare) [4,8,65].

In the following subsections we analyze the trifold challenge for policymaking in terms of historical trends in selected indicative outcomes that the European food system delivers. Table 1 summarizes our findings and outlines the desirable and undesirable trends in the outcomes that could result from an exposure of the system to shocks and stresses. These trends serve as reference modes to which we refer back throughout the following vulnerability analysis.

141

Table 1. Summary of historically observed trends * in indicative outcomes of the European food system along with their desired/undesired trends in the face of disturbances.

Indicative Outcome	Observed Trend [1]	Desirable Trend [1]	Undesirable Trend [1]	V/R [2]
Food and nutrition security				
Food production		supply = demand		+
Yield				−
Price of food [3]		stable	volatile	−
Socio-economic welfare				
Profits [4,5]			volatile	−
Environmental welfare				
Natural resource condition				−

* Time range of the historically observed trends are indicated in the text of the Sections 3.1–3.3; [1] arrow indicates direction of trend in the particular outcome over time; [2] qualitative assessment of vulnerability (V)/resilience (R) to the current impacts of driving forces, where (−) signifies vulnerability, (+) signifies resilience; [3] consumer perspective; [4] producer perspective; [5] profits are expressed at farm level and due to data availability proxied by farm income defined by the European Commission as the farm net value added (FNVA) per annual work unit (AWU) calculated as the sum of total production value plus direct payments minus intermediate consumption and depreciation.

3.1. Food and Nutrition Security

In the 1950s–1960s European food producers were primarily concerned with the quantity of foods they needed to supply to overcome the post-war shortages in food availability [65–67]. As a result, food production has experienced a leap forward, which has been attributed mainly to yield improvements rather than expansion of agricultural land. The story of English wheat is emblematic for the European context. It took nearly 1000 years for wheat yields to increase from 0.5 to 2 t/ha, but only 40 years to climb from 2 to 6 t/ha [2]. Simultaneously, despite the inherent tendency of agri-food markets to be volatile, the agricultural commodity prices and related food prices have exhibited a rather steady pattern of decline until about a decade ago. Accordingly, from the perspective of European consumers the food system has been uninterruptedly delivering desirable FNS outcomes. Food per each European has been available in surplus—from around 3000 kcal/day in the 1960s to over 3400 kcal/day in the 21st century in comparison with the needed 2000–2500 kcal/capita/day—and accessible at relatively low prices [1–3,68–70].

Yet within the new millennium several undesirable trends in crop yields and prices have emerged. The crop yields in some European regions (e.g., wheat in Northwest Europe or maize in South Europe) have reached or moved close to their plateaus. This implies that the yields have not increased for long periods of time following an earlier period of desired steady linear increase and thus raises concerns over future food availability [71,72]. As regards the prices of agricultural commodities and food, their volatility has increased in the last decade. More specifically, sharp increases in food prices in 2007–2008 and 2010–2011 were followed by recurring periods of often severe price depressions. The high volatility in prices has created an uncertain environment with many undesirable consequences for consumers' access to food. The price hikes caused a rapid increase in consumer food prices, which reduced average EU household purchasing power by around one percent. Low income households (especially the 16% of EU citizens who live below the poverty line) were hit even harder [73–75].

Furthermore, despite increasing food availability Europe has not managed to guarantee FNS for all citizens. About 10% of the European households have been persistently unable to access meat or a vegetarian equivalent every second day—an amount generally recommended in European dietary guidelines [75]. At the same time, the proportion of overweight or obese people has continuously increased to reach over 50% in 2010 [76]. Although both of these undesirable trends are more political and distributional problems rather than agricultural issues per se and hence their in-depth analysis remains beyond the scope of our study, they indicate important failures in the socio-economic organization of food production and downstream food system activities.

3.2. Socio Economic Welfare

FNS and consumers are only one side of the food system. On the other side are the food producers, in a broader sense rural communities, and their viability. While the increase in yields has brought benefits to both consumers and producers, the decline in prices of agriculture commodities has been undesirable for the latter. Accounting for inflation, from 1960s to 2005 European farmers experienced almost incessant (i.e., as one price peak in particular stands out—the so-called world food crisis of the 1970s) real price declines in output and input prices, but with the former decreasing faster. Since then the trend in input prices has reverted and they started to increase, further widening the gap between input and output prices [77,78]. This cost-price squeeze has caused an undesirable decline in the realized profits from farm operations and threatened the farm's viability in the long term.

The widening gap between output and input prices has been counterbalanced by significant gains in labor productivity achieved due to structural changes in the EU agricultural sector over the last several decades. The adjustments in structure have been manifested by, *inter alia*, reduction in farm labor, decrease in the number of farms and increase in the average farm size. To illustrate these trends, from 2002 to 2010 the agricultural labor input in the EU decreased by as much as 32% (a drop of 4.8 million full-time equivalent jobs), while between the 2005 and 2013 the annual average rate of decline in the number of agriculture holdings stood at -3.7% and the average size of each farm in EU-27 rose in terms of hectares from 11.9 to 16.1 as well as in terms of the economic size expressed in European Size Units (ESU) [78–81].

Although during the last several decades the increasing labor productivity have offset the undesirable trend in input and output prices, taking into account the total costs for own and other factors of production (land, labor, capital) still many of the European farms have remained unprofitable with market revenues alone [80–83]. To this end, since the early 1960s subsidies in different forms (i.e., within the years there was a gradual shift from price support to direct payments), have played an increasing role in farm profits [78,80–83]. As a result, the average dependence of farm profits on subsidies in the EU is now as high as 58% [83]. Moreover, in recent years the gains in labor productivity have been increasingly insufficient to compensate for the growing cost-price squeeze and the farm profits have become volatile and hence created a high level of uncertainty among food producers [78,84–86].

3.3. Environmental Welfare

Farmers represent only around 5% of the European Union's (EU's) working population, yet they manage over 40% of the EU's land area, and generate important impacts on the environment [87]. Hence in addition to FNS and other socio-economic welfare, environmental welfare is of great importance as both a condition for and an outcome of applied agriculture practices.

Over the past decades, the loss of traditional farming to intensive agriculture has contributed to the transgression of a number of critical planetary boundaries [88,89]. Inappropriate agricultural practices and land use have been responsible for adverse impacts on natural resources condition such as pollution of soil, water and air, fragmentation of habitats and loss of biodiversity. The reforms of the CAP in the 1990s, 2003 and 2008 have increasingly integrated environmental protection measures, including obligatory crop rotation, grassland maintenance, and more specific agri-environment measures, aimed at climate change mitigation and biodiversity conservation. In the latest CAP reform in 2010, even 30% of direct payments to farmers ("Pillar 1") were to become conditional on compliance with "greening measures" [90]. However, during the negotiations the new environmental prescriptions were so diluted, that most farmers are exempted from implementing them and they concern merely 50% of the EU farmland [91]. Effectively, the agro-environmental measures have brought about some improvements such as decreasing GHG emissions and pesticide use [91,92]. However, according to many academics and stakeholders these improvements have not been sufficient [91–95]. European agriculture still depends highly on external inputs, intensifies and specializes or abandons semi-natural grassland in less productive or accessible regions [91]. Consequently undesirable environmental outcomes like exceedance of nutrients, diffuse pollution to water and dramatic loss of biodiversity persist, further diminishing ecosystems' resilience [91]. More efforts are called for to balance food production and the environment [91,94,95].

3.4. Signs of Vulnerabilities and Resilience

European food production—one of the most important FNS outcomes—has been remarkably resilient to the impacts of distinct drivers of change over the last decades (Table 1). However, much of the food had been produced during a period of successful regional cooperation and supportive political environment, relatively stable climate, when farms were predominantly small-scale and diverse, natural resources appeared abundant and the human population was considerably smaller. Besides, despite the abundance of food production, apparently too much of the wrong kind of food at the wrong price has been provided, as the double burden of malnutrition (i.e., undernutrition and overweight) has continued in the EU.

A comparison of the observed trends in the remaining indicative outcomes—i.e., agriculture yield, price of food, profits, natural resources condition—with their desired levels, reveals emerging signs of the European food system's vulnerabilities to disturbances that have been at play so far (Table 1). The productivity of the current food system has come at the expenses of our natural and human resources. This poses severe risks to its continuity in delivering the fundamental FNS outcomes.

To conclude the analysis of indicative food system outcomes over time, it seems that the improvement of some FNS outcomes in the last decades have come at the expense of other food system outcomes and that the European food system is gradually becoming more vulnerable to a wide range of disturbances. If the undesirable developments continue, the existing vulnerabilities of the food system might be further exacerbated or give rise to new vulnerabilities endangering the food production.

4. System Conceptualization: Internal Processes and External Disturbances

Many processes underlie the trends described in Section 3. In this section, we adopt a feedback perspective and describe the underlying causal structure of the European food system likely to be generating the problematic trends. The structure is composed of several reinforcing feedback processes—*mechanization* (R1a), *intensification* (R1b) as well as *efficiency maximization* (R5)—that explain why food production grows regardless the direction of change in profits realized by food producers. When profits rise, food producers (re)invest in machinery and external inputs to increase food production, whereas when profits fall, food producers feel pressure to reduce production costs by maximizing efficiency and hence again increase food production using equal or even less inputs. Further, the central processes of *mechanization*, *intensification* and *efficiency maximization* are linked to other feedback loops of reinforcing (i.e., *labor reduction (R1c), compensation for degraded natural resources with external inputs (R2), organization of food production (R3), substitution of tacit with standardized knowledge (R4))* as well as balancing (i.e., *degradation of natural resources (B2), regeneration of natural resources (B2), loss of tacit knowledge (B3), supply (B4) demand (B5), trade (B6), market expansion (B7), cost minimization (B8))* nature. The interconnected feedback structure relates food production to other FNS, socio-economic and environmental outcomes. Based on this integrated feedback structure we explain how the ever rising food production emerges from within the same dynamics as the mounting pressures on human and natural resources that make the food production possible in the first place. Subsequently, we also identify entry points for external disturbances to which the food system might be exposed.

4.1. Internal Causal Structure Driving the European Food System

*Under conditions of high or rising profits, mechanization and intensification lead to growth in food production (*Figure 3*).* The structure of causes and effects linked together in a set of reinforcing feedback loops (Figure 3)—*mechanization (R1a), intensification (R1b)* and *labor reduction (R1c)*—operate in every capitalist market system. Food producers, having profit maximization as a goal, (re)invest in food producing inputs—land, labor (R1c, Figure 3), machinery (R1a, Figure 3) and external inputs (R1b, Figure 3) like fertilizers, plant protection products, seeds, feed, antibiotics, hormones, etc. The (re)investment is encouraged also by political and financial commitment of the EU to the agri-food industry (e.g., subsidies in different forms: direct payments, investment grants, intervention buying, private storage aid or export refunds, etc.). Explicitly, with the subsidies going into agriculture, food producers have both the security and the financial resources to (re)invest in production inputs.

The more inputs are used, *ceteris paribus*, the more output per hectare (or per animal), i.e., yield, can be achieved. In turn, multiplying the crop (or animal) yield by the limited amount of land area (or the number of animals) determines the food production that flows into the stock of food available for consumption. Food production, if sold on market, brings the producers profits. A share of the profits is reinvested in new production inputs, which are then used to increase the amount of food produced for sale. As long as profits are above breakeven point, implying that food producers are able to cover incurred production costs by received revenues (including subsidies) earning an income comparable to the rest of the economy, the reinforcing feedback loops—R1a, R1b, R1c (Figure 3)—function in the food system and lead to a boost in food production.

Yet having a limited budget and a goal of maximizing profits, the investment decision on 'what' and 'how' to produce involves relevant trade-offs and thus is not straightforward. As regards 'what' to produce, shifts between crop and animal production (not shown in Figure 3 for clarity reasons) result from changes in relative production profitability and consumption patterns of the population [41]. For instance, a growing demand for animal-based food products increases the attractiveness of animal production. Hence, food producers allocate more land and other production inputs to animal production at the expense of crop production [41]. Similar tradeoffs occur when considering agricultural production for food and for other uses than food like biofuels, textiles, etc.

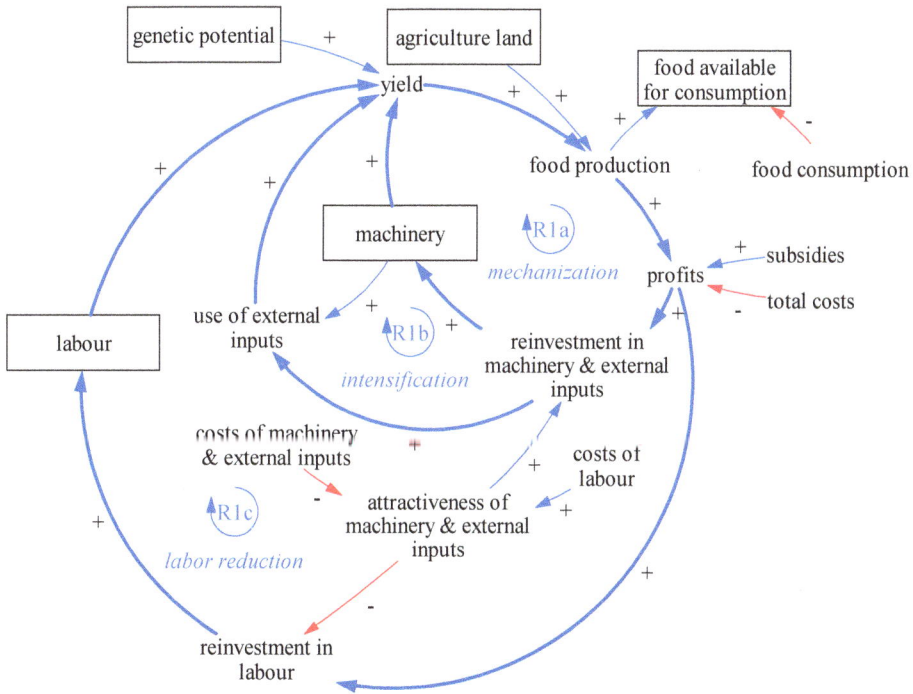

Figure 3. Causal loop diagram representing *mechanization* and *intensification* reinforcing feedback loops (respectively R1a, R1b) driving food production growth under conditions of rising profits; some links are omitted for visual clarity.

When deciding "how" to produce, no matter whether this concerns crop or animal production (or other uses), to a certain extent labor can be substituted with machinery and external inputs. The feedback mechanism in Figure 3 shows that when fossil fuel and other external inputs are available and inexpensive, there is a strong incentive to invest and use diesel-powered machinery and off-farm sourced inputs instead of labor to increase yields [2,9,10,96,97]. In other words, higher costs of labor increase the attractiveness of investing in and using machinery and external inputs instead. The success of machinery and external inputs in delivering higher yields, translating into higher food production and accordingly higher profits, strengthens itself leading to further *mechanization* (R1a, Figure 3) and *intensification* (R1b, Figure 3) of farm practices. Simultaneously, because of decreasing reinvestment in labor and hence its replacement with machinery and external inputs, the stock of labor is forced into a reinforcing downward spiral that gradually leads to *labor reduction* (R1c, Figure 3) [79,96,97].

Food production is embedded in ecosystems, implying that it is based on the condition of natural resources such as soil, water, air, biodiversity, nutrients and fossil fuels (Figure 4). As the natural resource base is limited, food production cannot grow infinitely. The worse the conditions of natural resources, the lower yield can be achieved and/or the less agricultural land is available for food production. The flows—degradation (outflow) and regeneration (inflow)—that influence the stock of natural resources are determined, among other things, by the implemented management of agroecosystems (i.e., the 'what' and 'how' to produce). Intensive food production practices that depend on use of external inputs tend to degrade the productive natural resources by their overexploitation (e.g., phosphate rock [98–100], fossil fuels [101,102], etc.) and pollution (e.g., nutrient leaching [88], GHG emissions [103], etc.) [7,8,104–107]. For instance, the stronger the reinforcing feedback loops driving use of diesel-powered machinery (R1a, Figure 3) and synthetic nitrogen fertilizers (R1b, Figure 3), the more of the non-renewable fossil fuels [108] are exploited and the more GHG are emitted to the atmosphere [109]. Likewise, the more pesticides are used to combat pests and diseases, the lower is the biodiversity and biological control potential on farmland [110,111]. These practices increase the rate of degradation and translate thus into a more degraded natural resource base. The degradation rate increases with increasing animal production, as animal-based food products are particularly resource-intensive [112,113]. At the same time, in intensive food production systems practices that treat natural resources in a more regenerative way are minimal or even none. As the outflow (degradation) of natural resources is higher than the inflow (regeneration) of natural resources, then the condition of natural resources worsens, jeopardizing the food production.

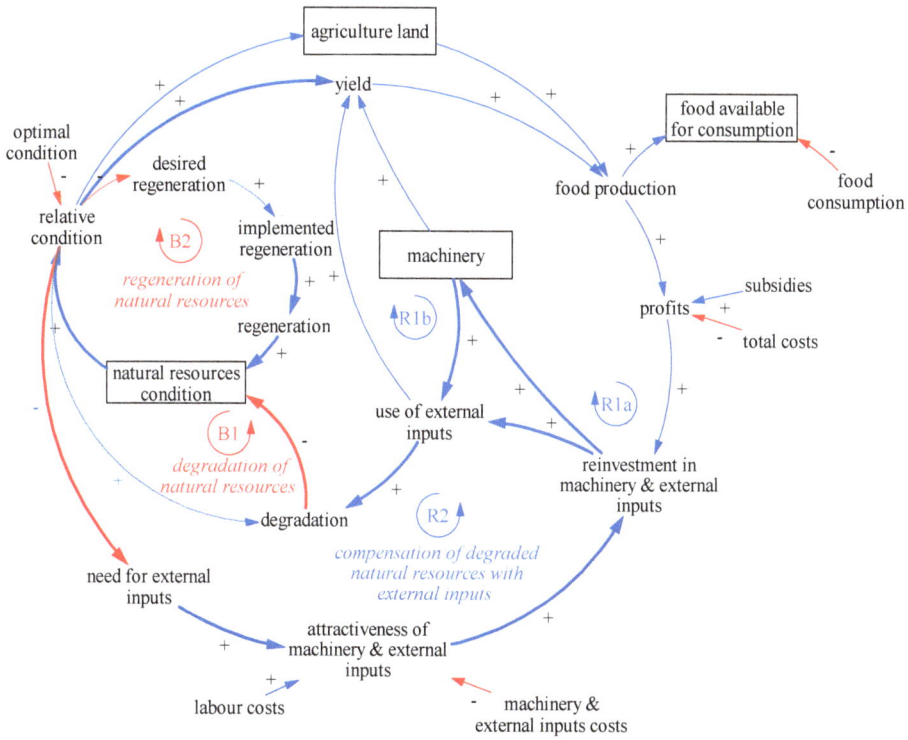

Figure 4. Causal loop diagram representing the relationship between food production and natural resources condition (B1, B2, R2); some links are omitted for visual clarity.

There are two balancing feedback loops that regulate *degradation* (B1, Figure 4) and *regeneration* (B2, Figure 4) *of natural resources*. The goal of the two balancing feedback loops is to maintain the condition of natural resources in a stable state. The balancing feedback loop B1 (Figure 4) sets limits to overuse or pollution (degradation) of natural resources as their condition worsens. The limit is signaled to food producers through, for instance, declining yield or rising costs of acquiring non-renewable natural resources (e.g., phosphate rock, fossil fuels) when they become scarce. However, the signal is often either missing or too weak and too delayed for food producers to notice it and implement on time more environmentally benign practices that decrease degradation (e.g., by reduced use of external inputs) and/or increase regeneration (signified with dashed lines in Figure 4) [11,15,64]. The longer the food producers do not recognize the worsening condition of natural resources and do not desire and effectively implement regenerative practices, the lower is the actual regeneration. With insufficient regeneration, all else equal, the conditions of natural resources move farther away from an optimal level, which should translate into increased need for regenerating natural resources (desired regeneration). However, because of the distorted flow of information about the relative condition of the natural resources, the desired and accordingly implemented regeneration is limited. That is, the desired regeneration is underestimated and impedes making informed decisions on implementation of appropriate food production practices.

Furthermore, external inputs can imitate some functions of the food producing natural resources (at least in the short-term). This feature allows food producers to substitute natural resources with external inputs in food production, when the condition of the former deteriorates [15,114]. As a result, food producers fall into a reinforcing spiral of *compensating for the degraded natural resources with the application of external inputs* (R2, Figure 4) rather than implementation of regenerative practices, which, in turn, further worsens the condition of natural resources. The reinforcing feedback loop driving substitution of natural resources with external inputs to produce food is a vicious circle that locks farmers into dependence on the use of external inputs.

Food producers require knowledge to know how to best organize food production (Figure 5) (i.e., to combine food production inputs with ecosystems to achieve the highest potential yield holding the costs constant). According to theorists, knowledge is perhaps the most relevant economic resource and learning the most important process [115]. In principle, the more one has of production inputs and knowledge, the more can be produced. Hence, the growth in food production is driven by accumulating inputs (e.g., land, labor, machinery, external inputs, etc.) (R1a, R1b, R1c, Figure 3) as well as the technical, agronomic knowledge that drives *organization of food production* processes (R3, Figure 5) [116].

Food producers gather knowledge while performing their activities and because of new learnings. Knowledge of food producers is a combination of tacit (or local) knowledge with standardized (or codified) knowledge [117]. The more knowledge of tacit and/or standardized nature food producers have, the stock of total knowledge increases. As a result food producers (at least theoretically) are able to better *organize food production* and realize higher yield (R3, Figure 5).

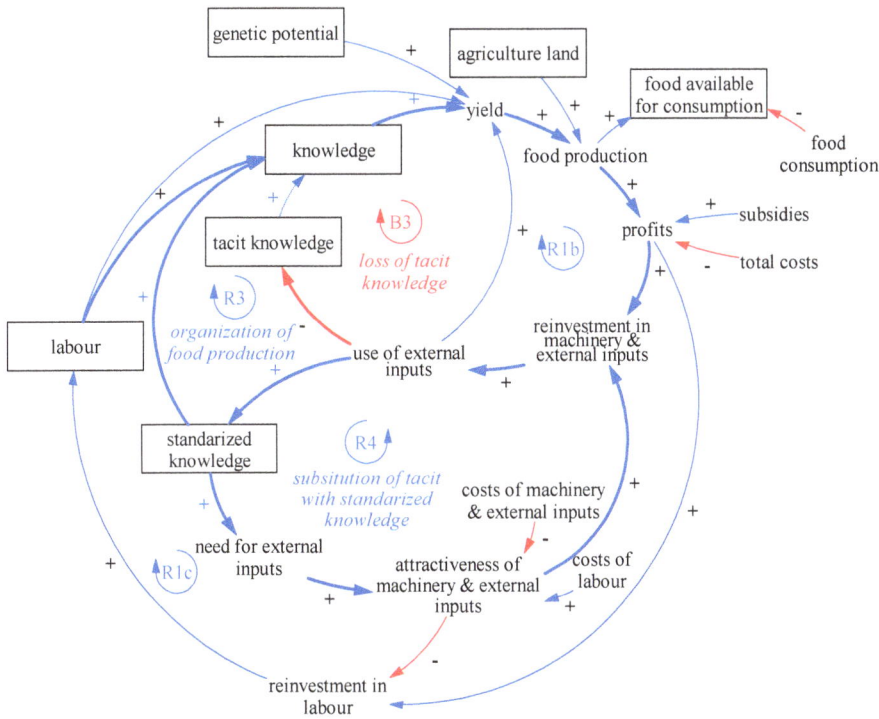

Figure 5. Causal loop diagram representing the relationship between food production and knowledge (B3, R4); some links are omitted for visual clarity.

In contrast to standardized knowledge, tacit knowledge of food producers implies an intimate knowledge of their landholding, its composition, fertility and so on, acquired through food producing practices (e.g., rotation, ploughing, etc.). The tacit knowledge is localized as it is closely tied to local ecosystem in which food production takes place. For instance, while the same principles of growing crops are widespread, tacit knowledge allows food producers to apply these principles differently in different local conditions and hence produce better results. With the widespread application of external inputs (*intensification*, R1b, Figure 3), which need not to be adapted to local circumstances as simple standardized instructions on their use provided usually by input industry are sufficient for food producers to achieve desired yield, the relationship between food producers and local ecosystems is disrupted. Accordingly, the stock of tacit knowledge required to manage the local ecosystems fades away, whereas uniform and spatially standardized knowledge accompanying use of external inputs builds up and replaces the former type of knowledge [117]. The function of the balancing feedback loop B3 (Figure 5) is to signalize the *loss of tacit knowledge* through decreasing yield. Yet the warning sign

151

is hugely disregarded by food producers or masked by the large and powerful institutions which lie upstream (and downstream) of the farm [11,117].

The longer the importance of accumulating tacit knowledge for achieving better yield in the long-term remains unnoticed by food producers, the *substitution of tacit knowledge with standardized knowledge* (R4, Figure 5) progresses. This development locks food producers into a vicious circle (R4, Figure 5) of increasing reliance on the use of external inputs [117].

Produced food is supplied on an agri-food market, which is a medium that allows consumers to access food (Figure 6). On a competitive agri-food market, price balances food production with food consumption. The functioning of such a market is characterized by the interplay between two balancing feedback loops of *supply* (B4, Figure 6) and *demand* (B5, Figure 6), both of which in a globalized setting are influenced by *trade* arrangements (B6, Figure 6).

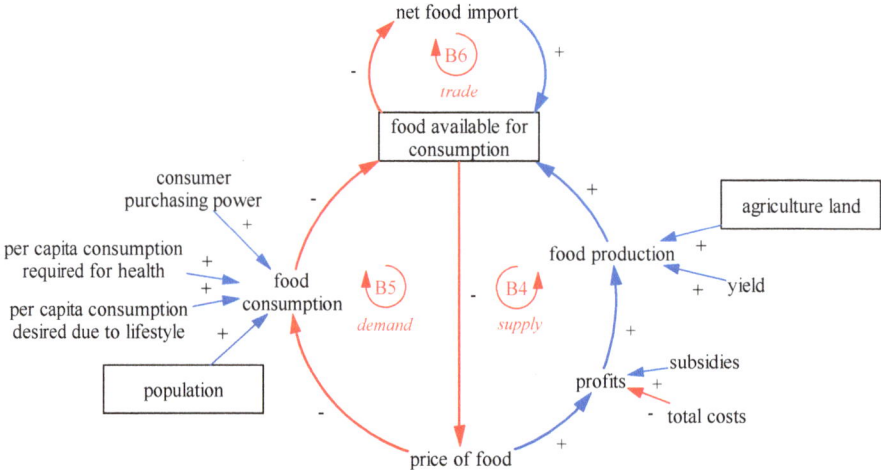

Figure 6. Causal loop diagram representing competitive market structure characterized by interplay between two balancing feedback loops of *supply* (B4) and *demand* (B5), both of which in a globalized setting are influenced by *trade* (B6); some links are omitted for visual clarity.

On the *supply* side (B4, Figure 6), a large number of food producers compete with each other. Specifically, producers reinvest (R1a, R1b, R1c, Figure 3) and produce food, increasing the amount of food available for consumption. Profits are realized when the amount of revenues gained from producing food exceeds the incurred production costs. As revenue is the product of the volume of produced food being sold and the price of the food, the higher the production and/or the higher price, *ceteris paribus*, the more profits food producers realize. Rising profits encourage existing food producers to reinvest and increase output (food production) as well as attract new entrants to the market. However, greater food production increases the stock of food available for consumption, which in turn, bids the price of food down. Declining price of food, all else equal, diminishes profits and hence discourages food producers from investing in increasing food production (B4, Figure 6).

On the *demand* side (B5, Figure 6), the population consumes (and wastes) the supplied food according to its purchasing power, dietary requirements for health and desires due to its lifestyle. The lower is the price of food, the more access to food people have and thus the more food is demanded. Higher food consumption diminishes the amount of food available for consumption, which translates into, all else equal, higher price of food (B5, Figure 6).

The state of the stock of food available for consumption indicates the balance between food production (as proxy for supply) and food consumption (as proxy for demand). The *supply* (B4, Figure 6) and *demand* (B5, Figure 6) balancing feedback loops cause the price of food to adjust until, in the absence of market imperfections and external disturbances, the market reaches an equilibrium characterized by a clearing price at which food production equals food consumption (i.e., the stock of food available for consumption is stable).

In a globalized world, however, in which markets are committed to open *trade*, there is an additional balancing loop B6 (Figure 6). Food producers export surplus of food production or are confronted with competitive imports, if the domestic food production is insufficient to meet the desired consumption. The imports add to the stock of food available for consumption, putting an additional downward pressure on price, and vice versa in case of exports. Hence, protective measures for reasons of FNS or employment are a natural response of governments.

In case of oversupply, food producers and governments put efforts to expand the market (Figure 7). Most agri-food markets are competitive, but not always perfectly. Market imperfections (caused by, for example, subsidies) add to the tendency to oversupply food relative to demand (i.e., the stock of food available for consumption increases). An extreme example of this phenomenon are the European Union's 'butter mountains and milk lakes' that occurred in the 1990s. As mentioned earlier, the oversupply pushes the price of food downward, reducing revenues and, *ceteris paribus*, consecutively profits. When the stock of food available for consumption increases, price of food declines and thus profits drop, food producers face pressure to expand the market in order to distribute the surpluses of food. Hence, usually along with governments they try to *expand markets* by, for instance, storing, exporting (B6, Figure 6), upgrading (e.g., ready-made meals instead of raw food products), advertising, or creating new uses of agricultural products (e.g., bioenergy from food crops). Accordingly, the consumption of food products goes up, reducing the stock of food available for consumption and pushing up the price and profits (B7, Figure 7). In fact, although there is no general consensus on the relative importance of the underlying causes for the 2007–2008 food price increase, the new, stimulated by policies, demand for biofuel feedstocks from grains and oilseeds has been widely cited as one of the major factors explaining the food price boom [118–124].

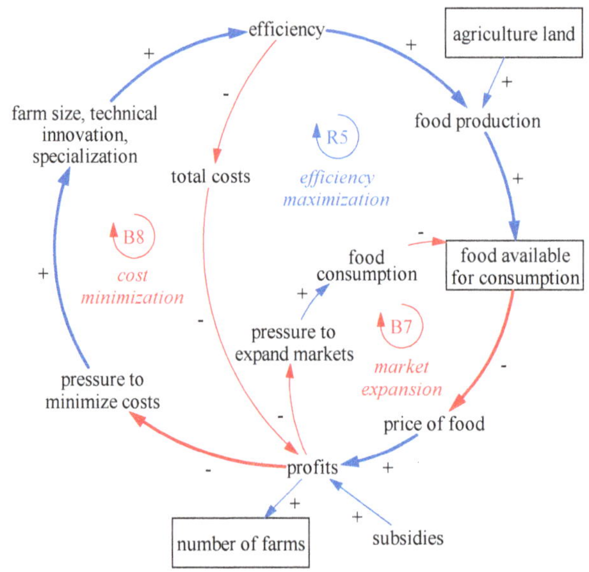

Figure 7. Causal loop diagram representing *efficiency maximization* reinforcing feedback loop (R5) driving food production growth under conditions of falling profits; some links are omitted for visual clarity.

However, when there is a general impression that the price of food is likely to rise from additional demand, existing food producers tend to speculate. Hence, they (re)invest in inputs to increase food production (R1a and R1b, Figure 3) in order to maximize profits from the expected price rise when the investment in new production is realized. If the price increase is considerable and progressive, also new producers are attracted to enter the agri-food market. As long as the price has not begun to fall, food producers extrapolate the price trend and are willing to believe that it will continue rising. After a time, depending on the delays involved in expanding production, the overproduction begins to be perceived and the price begins to decline again through the *supply* balancing feedback loop (B4; Figure 6). Food producers, if possible, rush with their products into agri-food market to avoid greater loss, dramatically worsening the imbalance between demand and supply. Consequently the price of food and producers' profits fall even harder and the agri-food market enters again into crisis.

Under conditions of low or falling profits, efficiency maximization leads to growth in food production (Figure 7). In addition to the amount of food sold, its price and subsidies, profits depend also on total costs incurred during food production. The higher are the total costs of production, the lower are the profits realized by food producers, all else being equal. If both trends—decreasing or stagnating price and growing costs of production—occur concurrently, food producers face a cost-price squeeze that causes profits to drop, farm debt to grow and a general loss of producer power. Food producers usually try to alleviate the undesirable downward pressure on their profits via a number of balancing processes aimed at *cost minimization* (B8, Figure 7) [70,125–131].

When the profits are negative, many food producers, particularly the small- and medium-scale ones, abandon the industry altogether (Figure 7). Only those food producers remain in the agri-food business that are most efficient and/or have the most optimistic expectations on the future price and costs [45,70,129–135]. This is evident in the declining number of farms. Meanwhile, however, the total number of cultivated hectares of land remains more or less constant. Hence, farm size increases, meaning that overall there are fewer but larger farms. In fact, scale economies along with technical innovations and specialization reinforce each other and are the most common routes to compensate for the falling profits by *minimizing costs* of food production through improved efficiency (B8, Figure 7) [129,130].

Although profits improve when food producers keep on *minimizing costs* through achieving higher efficiency (B8, Figure 7), a reinforcing mechanism resulting from *efficiency maximization* (R5, Figure 7) impedes their efforts. To produce food more efficiently means to produce more food with the same or less production inputs. The usual net result of *minimizing costs* (B8, Figure 7) and *maximizing efficiency* (R5, Figure 7) is that globally food production goes up, prices go down, and the realization of profits is again no longer possible even with the lower production costs. Food producers are locked into a vicious circle (R5, Figure 7), in which lower prices of food put a continuous pressure on food producers to minimize costs that forces them to become even more efficient if they are to survive at all. The farmers who lag behind and do not become efficient enough are lost in the price (or even cost-price) squeeze and leave room for the more successful food producers to expand [136].

4.2. Entry Points for External Disturbances in the European Food System

In addition to the internal causal structure, the functioning of the European food system is driven by multiple adverse and favorable external disturbances of various origin (e.g., socio-economic, ecological, technological, political etc.) [1,2,10,47]. Food system disturbances range from rapid and dramatic shocks (e.g., pest outbreaks, economic and political crises, weather events such as droughts, floods, or storms, fuel shortages, disease pandemics) to slow and moderate stresses (e.g., climate change, urbanization, population growth, changing consumption patterns), which do not function in isolation from one another, but co-occur and interact in many different ways [39,137,138].

In this section, we combine the individual causal loop diagrams from previous Section 4.1 into an integrated causal structure of the European food system to identify entry points that expose the system to external disturbances (Figure 8).

In principle most of the variables constituting the internal causal structure of the food system could be affected by a range of different shocks and stresses form outside the system. Examples of such entry points for external disturbances include (Figure 8):

- Food production can be affected by unfavorable weather conditions (e.g., severe droughts led to reoccurring famines in Russia) and pest outbreaks (e.g., potato disease caused crop failure that led to Great Irish Famine) that reduce crop yield, livestock diseases (e.g., avian flu or bovine spongiform encephalopathy) that lead to removal of large numbers of animals from the system as well as geopolitical dynamics causing disruptions to supply of external inputs used to maximize crop yield (e.g., phosphorus fertilizer, fossil fuels).
- Profits of food producers can be affected by an economic crisis that causes price of external inputs to increase considerably or become volatile as well as unfavorable political environment that leads to sudden or gradual removal of financial support (e.g., subsidies) for agriculture.
- Natural resource conditions can be affected by urbanization and population pressure that cause loss of agricultural land to other purposes, competition for resources (e.g., water, fossil fuels) from other industries that reduces the amount of natural resources available for food production or climate change impacts that disrupt provision of ecosystem services needed for food production.
- Labor employed in food producing activities can be affected by widespread disease outbreaks that reduce productivity of labor force or even the number of people able to produce food.
- Food consumption can be affected by population growth and ageing, changes in household incomes, changes in dietary patterns as well as routine and habits (e.g., food waste), values and ethical stances of consumers.

The integrated structure presented in Figure 8 allows us not only to identify what shocks and stresses the food system at stake, but also to systemically explore how the disturbances may be conveyed throughout key feedback processes in the food system and generate vulnerabilities. Examples of such pathways are provided in the following Section 5.

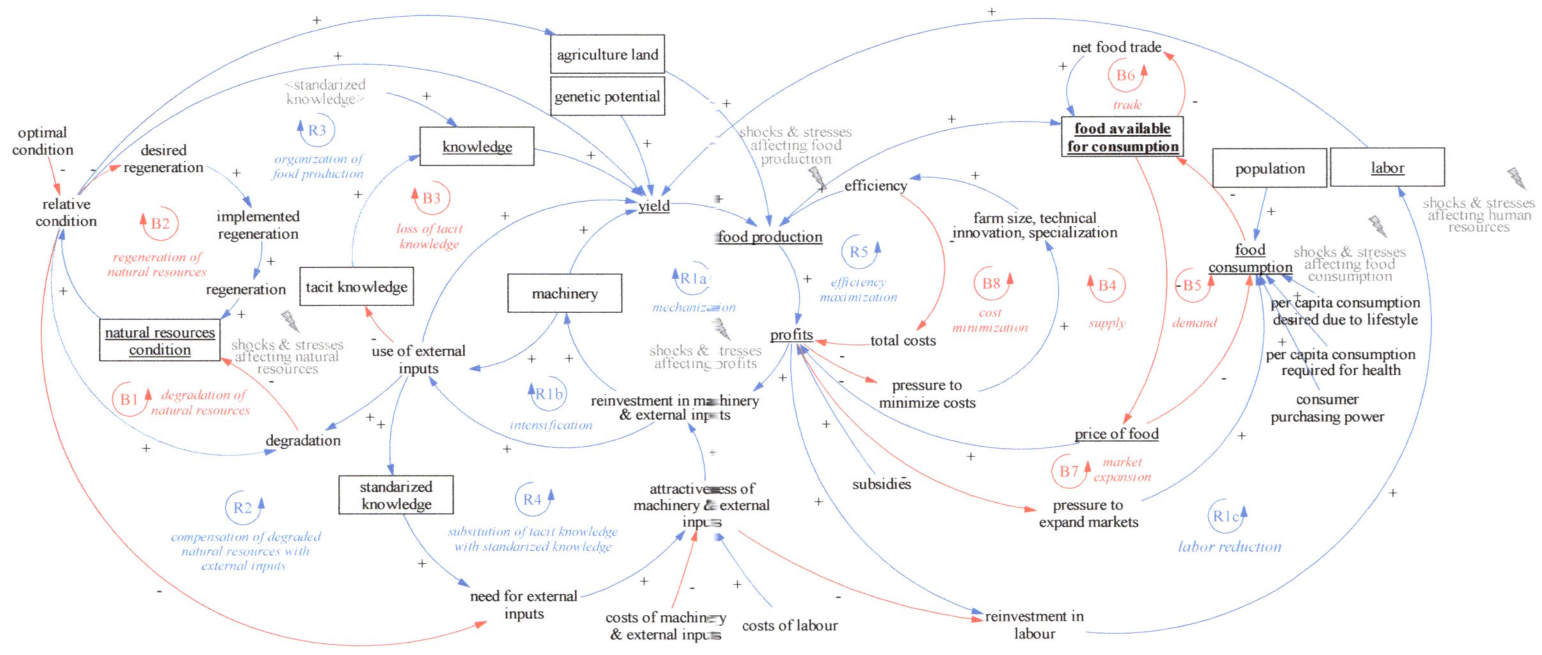

Figure 8. Integrated causal loop diagram of the conventional European food system with indicated exemplar entry points for external drivers of change; underline—key food system outcomes, grey & lightning ( )—examples of entry points for external drivers of change; some links are omitted for visual clarity.

5. Vulnerability Analysis: Interplay between Internal Causal Structure and External Disturbances

The analysis of internal causal structure in Section 4 shows how the conventional model of the European food system has been able to deliver ever greater quantities of cheap food. Specifically, increase in food production, despite the direction of change in profits, has been possible due to several strong reinforcing feedback loops that underlie reinvestment in machinery (*mechanization*, R1a, Figure 3) and external inputs (*intensification*, R1b, Figure 3) as well as *efficiency maximization* realized through scale economies, specialization and technological innovations (R5, Figure 7). From the perspective of many consumers these reinforcing feedback loops have been virtuous circles that have allowed them to access food at affordable prices. However, the benefits to consumers have come at a cost. Food producers have experienced the same reinforcing processes as vicious circles that have eroded their profits and involved them into a 'treadmill' where individual food producers must produce ever more just to stay in the agri-food industry [64,70,136]. Furthermore, the same strong reinforcing feedback loops have forced the European food system also into a counter-productive behavior and made it vulnerable to external disturbances. First, the reinforcing processes have been accompanied by several balancing feedback loops of which role is potentially to signal and minimize the undesirable socio-economic and environmental outcomes of the food system. However, in most cases the balancing processes have been either too weak or too delayed to do so. As a result, the strong reinforcing feedback loops underlying food production growth have generated numerous unintended negative impacts on human (e.g., reduction of rural employment, loss of knowledge) and natural resources (e.g., loss of biodiversity, soil degradation, pollution of water and air) which themselves are preconditions for food production. Second, they have given rise to two additional strong reinforcing processes—i.e., *compensation for the degraded natural resources with the application of external inputs* (R2, Figure 4) and *substitution of tacit knowledge with standardized knowledge* (R3, Figure 5)—that have made food producers increasingly reliant on the use of external inputs. Below, we outline five key vulnerabilities that stand out from our analysis of the internal causal structure.

Vulnerability I: Degrading natural resources

The stock of natural resources is a proxy for diversity and thus a kind of a buffer that absorbs ecological disturbances such as weather shocks or plague of pests in the system. Natural resources determine also how many options for adaptation and alternative solutions food producers have. Assuming relatively stable climate and abundance of natural resources, the reinforcing loops driving food production and at the same time degrading natural resources through *mechanization* (R1a, Figure 3), *intensification* (R1b, Figure 3) and *efficiency maximization* (R5, Figure 7) (all of which facilitate monoculture) could be strong, while degradation and regeneration of natural resources (B1, B2, Figure 4) weak and delayed (as they have been so far). However, we now know that the natural resources are finite (or take a lot of time to regenerate) and once they are depleted, food production is not possible anymore, while there will be few (or even no) options for adaptation to this and other disturbances. Especially because implementation of alternative solutions such as low external input practices usually requires a good condition of natural resources to achieve the desired yield without external inputs. Moreover, given that the literature on climate change predicts weather disturbances and pest infestations to become more extreme and more regular, by deteriorating the condition of natural resources food producers reduce ecological resilience of their agroecosystems to these shocks and ultimately endanger food production.

Vulnerability II: Trading tacit with standardized knowledge

Tacit knowledge is necessary for food producers to be able to continuously adapt their practices to changing conditions and thus keep producing enough food. In the conventional European food system strong reinforcing feedback loops that drive *intensification* (R1b; Figure 3) and *efficiency maximization* (R5, Figure 7) erode the traditional, local, ecosystem-sensitive (tacit) forms of knowledge and replace it with standardized, codified forms of knowledge. At the same time, the balancing loop (B3, Figure 5), of which role is to minimize the *loss of tacit knowledge*, is weak as many food producers disregard it. Consequently, the feedback processes reduce the capacity of food producers and the scope of options available for them that are necessary to make autonomous decisions regarding what they produce, how they produce it, and why. Food producers could afford to trade tacit with standardized knowledge for increased output when the conventional model of food system is not exposed to disturbances and works smoothly—producing ever more food at ever lower price with few unnoticeable side effects. However, as the food system moves into a crisis caused by, for instance, visible environmental impacts of conventional food production or consumers' concerns about food quality, the vulnerability of food producers becomes apparent. Food producers would need a lot of time (e.g., in case of crops, several growing seasons) to rebuild the stock of tacit knowledge indispensable for effective implementation of alternative practices and thus achievement of the same or better level of food production.

Vulnerability III: Dependence on external inputs and governmental subsidies

The strong reinforcing feedback loops that drive food production through *intensification* (R1b, Figure 3) and *mechanization* (R1a, Figure 3) and concurrently degrade natural resources and erode tacit knowledge, give rise also to two additional strong reinforcing processes through which *degraded natural resources are compensated for with external inputs* (R2, Figure 4) and *tacit knowledge is substituted by standardized knowledge* (R4, Figure 5). Both of the latter reinforcing feedback loops are examples of unintended processes that increasingly lock food producers into dependence on external inputs, the companies that provide them, the financial resources needed to purchase them and the capitalist relationships within food production that frame their decisions [5,139]. The use of external inputs considerably changes food producing practices as well as agroecosystems in which they are applied. *Inter alia*, some external inputs give rise to unintended consequences (e.g., weed resistance, pollinator decline) that are then stabilized with new external inputs (e.g., stronger herbicide cocktails). This, in turn, reinforces the dependency of food producers on the use of external inputs. The result is that food production is based on a continuous reinvestments in engineered stabilizers rather than tacit knowledge and ecosystems resilience (condition of natural resources). Therefore, if for some reasons (e.g., fossil fuels scarcity, geopolitical tensions, economic crisis) external inputs were not available for food producers: first, it will take a long time for an alternative food production paradigm to become effective (because of, for instance, the need to rebuild the stocks of tacit knowledge and natural resources condition) and second, the outcomes could potentially be far more undesirable than that of a system which never used those stabilizers. Moreover, relying on a limited range of 'stabilizing' external inputs makes the food system particularly vulnerable to disturbances that operate beyond their scope of fixes such as unexpected and non-linear climate change and feedbacks. As high external input systems are capital-intensive, one could think of vulnerability arising from dependence on financial subsidies and the governments that provide them in a similar way.

Vulnerability IV: Latent instability of agri-food markets

When food production is higher than food consumption one might expect that the balancing processes driving the functioning of agri-food markets, i.e., *supply* (B4, Figure 6), *demand* (B5, Figure 6) and *trade* (B6, Figure 6) would equilibrate them. However, many European agri-food markets are imperfect due to governmental support (subsidies) and regulations (e.g., production rules), information and power asymmetries, costs of entry and exit, and inflexibility of natural resources [140–147]. These imperfections either strengthen or weaken the market balancing loops (B4, B5, B6, Figure 6) or create new ones (e.g., *market expansion* balancing loop B7, Figure 7) that sometimes overwhelm the existing ones, leading to inefficiencies or even failure of markets [45]. For instance, one might expect that in the face of rising food production and falling profits, the food producers would reduce or even cease production. While it is true, that the number of food producers tends to decline over time, the EU within CAP offers subsidies to food producers that weaken the balancing feedback loops on agri-food markets and foster the gain around positive feedback loops that drive food production up. As a result food producers stay in unproductive and saturated markets. In other words, subsidies stabilize the viability of food producers and keep food production high, but also make them increasingly reliant on this form of support and the governments that provide them (see Section 3.2) [77,83,148,149]. Furthermore, the balancing loops frequently involve long delays (e.g., length of growing season, duration of transportation and distribution, time to perceive price changes by producers and consumers, etc.), weak responses (e.g., low short-run and high long-run elasticity of demand) or boundedly rational decision making (e.g., *market expansion* balancing loop B7, Figure 7), that predispose the agri-food markets to persistent disequilibrium and instability [45,150–152]. Disturbances such as sudden removal of subsidies, weather- or pest-related crop failures, increased price volatility, food scandals causing sudden drop in food consumption can stimulate and amplify the latent oscillatory behavior of agri-food markets giving rise to undesirable volatility of price of food and/or profits realized by food producers [118,120,153].

Vulnerability V: Striving for efficiency, while losing resilience

The conventional European food system manages its growth and expansion based on ideas of maximizing efficiency realized through *inter alia* scale economies, specialization and technological innovations (i.e., balancing loop of *cost maximization* (B8, Figure 7) that perpetuates the strong reinforcing feedback loop of *efficiency maximization* (R5, Figure 7)). Food producers across Europe experience effects of these processes in many different ways. Scale economies force many small- and medium-scale food producers out of the agri-food business entirely—evidence through declining number of farms and increasing farm size. This trend along with the strong reinforcing spiral of *labor reduction* (R1c, Figure 3) translates into increasingly fewer people in society with knowledge and skills needed to produce food (i.e., further decline in the stock of tacit knowledge, Figure 5). Besides, scale economies drive consolidation and reduce the diversity of scale at which food producers operate. Specialization is apparent, for instance, in the trend towards a single dominant activity on farms and widespread monocultures. Currently, in the EU almost half of the holdings are specialized in cropping and 27% in livestock [154]. Accordingly, as the system specializes, the diversity of organizational forms as well as of crops and animals decreases in the food system. Technical innovations (e.g., application of more and more specific fertilizers, herbicides and pesticides and genetic advances) to a great extent are in hands of few multinational corporations [5]. This narrows down sources of technical innovations as well as the choices available for food producers. For instance, commercial seeds and breeds focus on a few traits in a few crops, which forces food producers to base their production activities on them. The three processes seem to favor each other, so that, for instance, the technical innovations (e.g., promotion of agrochemical use, biotechnology, single crop machinery, etc.) are most (costs) beneficial through scale economies and specialization [5,9,129]. Common feature of all of these processes is that they increase efficiency, but at the same time decrease diversity of different elements in the food system, while diversity is crucial for absorption of shocks and stresses, adaptation, and alternative solutions [5,9,47,155]. Having low diversity in the food system allows disturbances to be augmented in both socio-economic (e.g., food pricing controlled by few) and ecological (e.g., contamination on a single farm can easily effect the entire country) dimensions. Thus, it seems that through strong *efficiency maximization* loop (R5, Figure 7) food producers trade-off short-term productivity against long-term resilience.

In essence, vulnerabilities in the conventional European food system arise if a disturbance strengthens the reinforcing feedback loops and further weakens or delays the balancing loops. For instance, climate change related shocks such as drought, flood or storm, will likely strengthen the *intensification* reinforcing feedback loop (R1b, Figure 3) because of yield losses. Yield losses increase the pressure on food producers to produce more, disregarding the balancing loops of *natural resources degradation and regeneration* (B1, B2 Figure 4), thus further lowering the stock of natural resources condition. When the stock depletes, yield declines, and translates into undesirable outcome of reduced food production and hence food insecurity. Another example is a stress of growing population that demands more resource-intensive animal products (*demand* side of the agri-food market, B5, Figure 6), which most likely strengthens the *efficiency maximization* loop in addition to *intensification* (R1b, Figure 3) and *mechanization* (R1a, Figure 3) loops. As the population grows and demands more animal products (i.e., per capita consumption due to lifestyle rises, Figure 6), food consumption rises. Food producers feel pressure to produce more of both animals and crops, as part of the crop production is redirected to feed for animals. Yet the amount of agriculture land is limited. Food producers are pressured to intensify as well as maximize efficiency. As a result, the disturbance strengthening the reinforcing feedback loops, propagates throughout the system, exacerbating all the vulnerabilities outlined above.

6. Policy Analysis: The Potential Role of Organic Farming

As a whole, the conventional European food system, relying on external inputs and policy stabilizers, reveals resilient features—it provides plentiful and inexpensive food. However, the value judgement of what is resilient or vulnerable to what and over what period of time depends on the beneficiaries. Assessing it through sustainability lenses, we argue that the capitalized, high external input food system is vulnerable to disturbances that operate beyond the system's own boundaries of ontology (i.e., a set of concepts and their relations that are specified in some way), epistemology, or control, such as unanticipated or non-linear climate variability and feedbacks or unpredicted ecological consequences of continuing use of external inputs. Thus, an alternative approach to food system, which does not trade-off long-term resilience for productivity and stability, is called for [5,7–9]. King [16] lists several potential approaches for a resilient food system, including organic and biodynamic farming, permaculture, farmers' markets, community-supported agriculture and community gardens. In Europe organic farming is the fastest growing of all alternatives to the conventional food system, which is regulated at EU level and receives considerable public financial support. However, can organic farming make the European food system more resilient?

In contrast with the conventional European food system, organic farming is a low external input system, in which organic matter cycles and diversification of crops and animals are key concepts. Many meta-analyses and reviews provide evidence for enhanced environmental impact of the organic farming practices in comparison with the conventional system [21,31,36,156,157]. Thus, organic food producers have the potential to address the vulnerability related to worsening conditions of natural resources. Specifically, they are able to recognize the two balancing loops—*natural resource degradation* and *regeneration* (B1 and B2, Figure 4)—and implement practices that strengthen both of them and the important stock of natural resources accumulates, making the system more resilient to disturbances such as climate change. For instance, due to water-holding capacity of soil organic carbon stocks, which is built through common agroecological practices such as diverse and companion cropping, planting green manure and cover crops, and integrating forages and perennials, organically managed farms have been shown to produce higher yields than their conventional counterparts under conditions of severe droughts or excessive rain [158]. In addition, through strategic diversification of crops, organically managed plantations can be also more resilient to pest outbreaks, as commonly a single pest usually damages a particular variety [9].

With regard to trading tacit with standardized knowledge, several studies have shown that organic food producers pay attention to natural cycles in their practices (i.e., balancing loop *loss of tacit knowledge* gains strength (B3, Figure 5)) and hence accumulate much more tacit knowledge than producers in conventional systems [117,159]. In that sense, agroecological practices have potential to lead not only to more natural resources, but build up the human resources as well [117]. Organic food producers may then be better prepared to cope with disturbances over long term.

Organic farming per definition is a low external input system with *inter alia* diversification and nutrient cycling at its heart. It has, thus, potential to preserve higher stocks of natural resources and tacit knowledge as well as to better recognize and operate the balancing loops (B1, B2, Figure 4; B3, Figure 5). Accordingly, organic food producers may be able to escape from being locked into the dangerous dependence on external inputs (R2, Figure 4; R4, Figure 5).

However, implementation of organic food production principles in practice is diverse and ranges from mere 'input substitution' to fundamental 'system redesign' [160]. This implies that there are organic food producers, of which practices diverge only slightly from conventional practices [28]. As organic food producers are not rewarded for continuous improvement, but have to comply just with minimum standards, they are incentivized to simply substitute prohibited with allowed inputs sourced from outside of the system. As a result they will be again locked into the vicious circles creating dependence on external inputs (R2, Figure 4; R4, Figure 5) with all their consequences for resilience of the prevailing food system.

In addition to better environmental outcomes, many studies have found that organic food producers perform better also in socio-economic terms as compared to their conventional counterparts [21,30]. Simply looking at comparisons of organic versus conventional short-term profitability, organic seems to be a promising option to preserve viability of farms. Besides, the organic food system is characterized also by diversity of markets (e.g., specialized organic food stores, farmers' markets and direct farm marketing, food baskets), through which organic food is provided to consumers [20]. These two features—better financial performance and diversity of markets—suggest that potentially the internal market structure of the organic food system is different from the conventional one and that the system can address the vulnerabilities related to socio-economic organization of food production inherent in the latter.

However, there are many signs indicating that organic food system based on certification of food production methods alone, falls into the same reinforcing mechanisms as conventional system and gives up its resilient features for efficiency (R5, Figure 7) and itself is vulnerable.

First, establishing certification put barriers for smaller food producers to enter the sector, because of incurred costs of certification and because it facilitates larger retailers to sell organic products [9]. Hence, the organic food system becomes more and more consolidated and losses its diversity, which has the same consequences for resilience as presented for the conventional European food system.

Second, higher profits realized by organic producers in comparison with conventional farmers are attributed mainly to price premiums paid by consumers and subsidies received from governments [18,20,21]. Some authors point out that the organic producers are becoming more and more dependent on the direct payments [149]. Such dependence makes organic producers increasingly vulnerable to changes in political environment. It is also uncertain, how much of the price premium is received because of willingness to pay or because of the anecdotal unprecedented growth in demand that outpace the growth in supply on the organic market [18], and hence what would happen with profits when the farm-gate prices for organic products fall [20,161].

Third, organic food producers compete with each other based solely on price, which does not internalize all externalities. It means that many of the socially and ecologically progressive attributes of organic produce are neglected in the price of organic food. Such a price-based competition disincentives the organic food producers from continuous improvement of their practices and involves them into the productivity paradigm and the reinforcing spiral of *efficiency maximization* (R5, Figure 7), violating many of the organic principles and reducing its potential to be a viable option for making the European food system more resilient [20,28]. This is evident in the organic 'conventionalization' and 'supermarketization' debate [20,28].

7. Conclusions

In this paper, we have proposed a new way to help policymakers understand the European food system's vulnerabilities and assess whether alternative developments such as organic farming can enhance its resilience. For this, we adopted a system dynamics approach to capture the dynamic complexity of the food system. We identified a number of key systemic vulnerabilities, including the degradation of the natural resource base of food production, the erosion of its tacit knowledge base, its dependence on external inputs and governmental support, the latent instability of the agri-food markets and the strive for efficiency that leads to a loss of diversity in the food system.

We have argued that organic farming has the potential to address these vulnerabilities, but at the same time risks falling into the same systemic pitfalls through a process of conventionalization. More specifically, organic farming as a food system has to be carefully designed and implemented to overcome the contradictions between the dominant socio-economic organization of food production and the ability to implement holistic understanding of organic principles on a broader scale. To make organic farming a viable strategy for reducing the vulnerabilities and enhancing the resilience of the European food system, certification as one of the main intervention proposed by the EU, for instance, will not be sufficient. Certification draws better boundaries around environmental resources and thus limits the negative environmental impacts that agricultural production has. However, it does not interfere with the production growth drivers and thus does not change the nature of any of the feedback mechanisms described in this paper nor does it affect their relative strength, i.e., the extent to which they dominate system's behavior.

Reducing vulnerabilities and increasing resilience of food systems goes beyond intervention engineering. The structural thinking tools developed in this paper provide a basis for an integrated evaluation of interventions, that is, of how interventions acknowledge that accumulation and draining processes cause delays and constraints in food systems' responses to disturbances, that feedback processes cause a reinforcement or dampening of such a response, and that nonlinearity causes an interaction between the response produced by various model components and across model components. The system-oriented approach helps also to characterize the range of synergies and trade-offs between food systems' outcomes that arise from such interventions.

Building on our structural diagram, further research could focus on other outcomes valued by different perspectives. Besides, the structural diagram serves also as a transition between mental models existing in literature and fully operational simulation models with which one could test the system's response to various types and magnitudes of disturbances and interventions. The system dynamics approach captures well the cross-level interactions (e.g., production and consumption) within food systems as long as the individual level is expressed in aggregated terms. Yet cross-scale (i.e., spatially disaggregated) interactions between the biophysical and decision-making, would require a hybrid approach, merging system dynamics with, for example, agent-based modelling [41].

Above and beyond, the understanding of systemic interactions and dynamic complexity of a food system is, however, not enough to identify specific actions and potential policies for increasing the resilience of any particular food system [162,163]. The concrete design and implementation of interventions requires also careful consideration of political agency (e.g., alternative food movements and actors) [14] and negotiation of power relations [164]. This opens up avenues for future research that establish a dialogue between social-ecological systems analyses with, for instance, political ecology ([162]).

Acknowledgments: This paper originates from an EU FP7 funded project TRANSMANGO *"Assessment of the impact of global drivers of change on Europe's food security"*; Grant agreement No. 613532; Theme KBBE.2013.2.5-01. Birgit Kopainsky is supported by the Norwegian Research Council through the project *"Simulation based tools for linking knowledge with action to improve and maintain food security in Africa"* (contract number 217931/F10). We are very grateful to Andreas Gerber for his very helpful feedback on earlier versions of this paper. We would like to thank the reviewers and the editors of this special issue for very useful comments and feedback.

Author Contributions: Natalia Brzezina, Birgit Kopainsky and Erik Mathijs conceived and designed the research; Natalia Brzezina and Birgit Kopainsky performed the research; Natalia Brzezina, Birgit Kopainsky and Erik Mathijs analyzed the data; Natalia Brzezina, Birgit Kopainsky and Erik Mathijs wrote the paper.

Conflicts of Interest: The authors declare no conflict of interest.

Abbreviations

The following abbreviations are used in this manuscript:

AWU   annual work unit
CAP   Common Agriculture Policy EU: European Union
FNS   Food and Nutrition Security
FNVA   farm net value added
NGO   non-governmental organization
R&D   Research and Development
SES   social-ecological systems

References

1. Marten, G.G.; Atalan-Helicke, N. Introduction to the Symposium on American Food Resilience. *J. Environ. Stud. Sci.* **2015**, *5*, 308–320.
2. Hazell, P.; Wood, S. Drivers of change in global agriculture. *Philos. Trans. R. Soc. B Biol. Sci.* **2008**, *363*, 495–515.
3. Swinnen, J.F.M.; Banerjee, A.N.; De Gorter, H. Economic development, institutional change, and the political economy of agricultural protection: An econometric study of Belgium since the 19th century. *Agric. Econ.* **2001**, *26*, 25–43.
4. Kirchmann, H.; Thorvaldsson, G. Challenging targets for future agriculture. *Eur. J. Agron.* **2000**, *12*, 145–161.
5. Hendrickson, M.K. Resilience in a concentrated and consolidated food system. *J. Environ. Stud. Sci.* **2015**, *5*, 418–431.
6. Tansey, G. Food and thriving people: Paradigm shifts for fair and sustainable food systems. *Food Energy Secur.* **2013**, *2*, 1–11.
7. International Assessment of Agricultural Knowledge, Science and Technology for Development. Agriculture at a Crossroads—Global Report. 2009. Available online: http://www.unep.org/dewa/agassessment/reports/IAASTD/EN/Agriculture%20at%20a%20Crossroads_Global%20Report%20(English).pdf (accessed on 1 February 2016).
8. The 3rd SCAR Foresight Exercise. Sustainable Food Consumption and Production in a Resource-Constrained World. 2009. Available online: https://ec.europa.eu/research/agriculture/scar/pdf/scar_feg3_final_report_01_02_2011.pdf (accessed on 1 February 2016).
9. Rotz, S.; Fraser, E.D.G. Resilience and the industrial food system: Analyzing the impacts of agricultural industrialization on food system vulnerability. *J. Environ. Stud. Sci.* **2015**, *5*, 459–473.
10. Godfray, H.C.J.; Crute, I.R.; Haddad, L.; Lawrence, D.; Muir, J.F.; Nisbett, N.; Pretty, J.; Robinson, S.; Toulmin, C.; Whiteley, R. The future of the global food system. *Philos. Trans. R. Soc. Lond. B Biol. Sci.* **2010**, *365*, 2769–2777.
11. Sundkvist, Å.; Milestad, R.; Jansson, A. On the importance of tightening feedback loops for sustainable development of food systems. *Food Policy* **2005**, *30*, 224–239.
12. Ingram, J. A food systems approach to researching food security and its interactions with global environmental change. *Food Secur.* **2011**, *3*, 417–431.

13. EUROSTAT 2014. Population and Population Change Statistics. Available online: http://ec.europa.eu/eurostat/statistics-explained/index.php/Population_and_population_change_statistics (accessed on 20 August 2016).

14. Akram-Lodhi, A.H. *Hungry for Change: Farmers, Food Justice and the Agrarian Question*; Fernwood Publishing: Halifax, NS, Canada, 2013.

15. Stave, K.; Kopainsky, B. A system dynamics approach for examining mechanisms and pathways of food supply vulnerability. *J. Environ. Stud. Sci.* **2015**, *5*, 321–336.

16. King, C.A. Community resilience and contemporary agri-ecological systems: Reconnecting people and food, and people with people. *Syst. Res. Behav. Sci.* **2008**, *25*, 111–124.

17. Stolze, M.; Lampkin, N. Policy for organic farming: Rationale and concepts. *Food Policy* **2009**, *34*, 237–244.

18. International Federation of Organic Agriculture Movements. Organic in Europe: Prospects and Developments. 2016. Available online: http://www.ifoam-eu.org/sites/default/files/ifoameu_organic_in_europe_2016.pdf (accessed on 10 March 2016).

19. Niggli, U. Sustainability of organic food production: Challenges and innovations. *Proc. Nutr. Soc.* **2015**, *74*, 83–88.

20. Darnhofer, I. Contributing to a transition to sustainability of agri-food systems: Potentials and pitfalls for organic farming. In *Organic Farming, Prototype for Sustainable Agricultures*; Bellon, S., Penvern, S., Eds.; Springer: Dordrecht, The Netherlands; Heidelberg, Germany; New York, NY, USA; London, UK, 2014; pp. 439–452.

21. Reganold, J.P.; Wachter, J.M. Organic agriculture in the twenty-first century. *Nat. Plants* **2016**, *2*, 15221.

22. Food Security Information Network. Resilience Measurement Principles. 2014. Available online: http://www.fao.org/fileadmin/user_upload/drought/docs/FSIN%20Resilience%20Measurement%20201401.pdf (accessed on 20 March 2016).

23. Milestad, R.; Darnhofer, I. Building farm resilience: The prospects and challenges of organic farming. *J. Sustain. Agric.* **2003**, *22*, 81–97.

24. Food and Agriculture Organization. Building Resilience for an Unpredictable Future: How Organic Agriculture Can Help Farmers Adapt to Climate Change. 2006. Available online: http://www.fao.org/3/a-ah617e.pdf (accessed on 30 March 2016).

25. Darnhofer, I. Strategies of family farms to strengthen their resilience. *Environ. Policy Gov.* **2010**, *20*, 212–222.

26. Scialabba, N.E.-H.; Müller-Lindenlauf, M. Organic agriculture and climate change. *Renew. Agric. Food Syst.* **2010**, *25*, 158–169.

27. Little Unix Programmers Group. The Role of Agroecology in Sustainable Intensification. 2015. Available online: http://www.snh.gov.uk/docs/A1652615.pdf (accessed on 1 February 2016).

28. Guthman, J. *Agrarian Dreams: The Paradox of Organic Farming in California*; University of California Press: Berkeley, CA, USA; Los Angeles, CA, USA; London, UK, 2004.

29. De Ponti, T.; Rijk, B.; Van Ittersum, M.K. The crop yield gap between organic and conventional agriculture. *Agric. Syst.* **2012**, *108*, 1–9.

30. Crowder, D.W.; Reganold, J.P. Financial competitiveness of organic agriculture on a global scale. *Proc. Natl. Acad. Sci. USA* **2015**, *112*, 7611–7616.

31. Gomiero, T.; Pimentel, D.; Paoletti, M.G. Environmental impact of different agricultural management practices: Conventional vs. organic agriculture. *CRC Crit. Rev. Plant Sci.* **2011**, *30*, 95–124.

32. Palupi, E.; Jayanegara, A.; Ploeger, A.; Kahl, J. Comparison of nutritional quality between conventional and organic dairy products: A meta-analysis. *J. Sci. Food Agric.* **2012**, *92*, 2774–2781.

33. European Commission. Farm Economics Briefs 2013. Organic versus Conventional Farming, Which Performs Better Financially? Available online: http://ec.europa.eu/agriculture/rica/pdf/FEB4_Organic_farming_final_web.pdf (accessed on 10 March 2016).

34. Birkhofer, K.; Bezemer, T.M.; Bloem, J.; Bonkowski, M.; Christensen, S.; Dubois, D.; Ekelund, F.; Fließbach, A.; Gunst, L.; Hedlund, K.; et al. Long-term organic farming fosters below and aboveground biota: Implications for soil quality, biological control and productivity. *Soil Biol. Biochem.* **2008**, *40*, 2297–2308.

35. Bengtsson, J.; Ahnström, J.; Weibull, A.-C. The effects of organic agriculture on biodiversity and abundance: A meta-analysis. *J. Appl. Ecol.* **2005**, *42*, 261–269.

36. Tuomisto, H.L.; Hodge, I.D.; Riordan, P.; Macdonald, D.W. Does organic farming reduce environmental impacts?—A meta-analysis of European research. *J. Environ. Manag.* **2012**, *112*, 309–320.

37. Ericksen, P.J. Conceptualizing food systems for global environmental change research. *Glob. Environ. Chang.* **2008**, *18*, 234–245.

38. Cash, D.W.; Adger, W.; Berkes, F.; Garden, P.; Lebel, L.; Olsson, P.; Pritchard, L.; Young, O. Scale and cross-scale dynamics: Governance and information in a multilevel world. *Ecol. Soc.* **2006**, *11*, 8.

39. Ericksen, P.J. What is the vulnerability of a food system to global environmental change? *Ecol. Soc.* **2008**, *13*, 14.

40. Hammond, R.; Dubé, L. A systems science perspective and transdisciplinary models for food and nutrition security. *Proc. Natl. Acad. Sci. USA* **2012**, *109*, 12356–12363.

41. Kopainsky, B.; Huber, R.; Pedercini, M. Food provision and environmental goals in the Swiss agri-food system: System dynamics and the social-ecological systems framework. *Syst. Res. Behav. Sci.* **2015**, *432*, 414–432.

42. Schlüter, M.; Mcallister, R.R.J.; Arlinghaus, R.; Bunnefeld, N.; Eisenack, K.; Hölker, F.; Milner-Gulland, E.J.; Müller, B.; Nicholson, E.; Quaas, M.; et al. New horizons for managing the enviornment: A review of coupled social-ecological systems modeling. *Nat. Resour. Model.* **2012**, *25*, 219–272.

43. Schlüter, M.; Hinkel, J.; Bots, P.W.G.; Arlinghaus, R. Application of the SES framework for model-based analysis of the dynamics of social-ecological systems. *Ecol. Soc.* **2014**, *19*, 36.

44. Alroe, H.F.; Noe, E. Second-order science of interdisciplinary research a polyocular framework for wicked problems. *Constr. Found.* **2014**, *10*, 65–76.

45. Sterman, J. *Business Dynamics: Systems Thinking and Modeling for a Complex World*; McGraw-Hill Higher Education: Toronto, ON, Canada, 2000.

46. Tendall, D.M.; Joerin, J.; Kopainsky, B.; Edwards, P.; Shreck, A.; Le, Q.B.; Kruetli, P.; Grant, M.; Six, J. Food system resilience: Defining the concept. *Glob. Food Secur.* **2015**, *6*, 17–23.

47. Ingram, J.; Ericksen, P.; Liverman, D. *Food Security and Global Environmental Change*; Earthscan: London, UK; Washington, DC, USA, 2010.

48. Quinlan, A.E.; Berbés-Blázquez, M.; Haider, L.J.; Peterson, G.D. Measuring and assessing resilience: Broadening understanding through multiple disciplinary perspectives. *J. Appl. Ecol.* **2015**, *53*, 677–687.

49. Ford, A. *Modeling the Environment*, 2nd ed.; Island Press: Washington, DC, USA, 2009.

50. System Dynamics Society. Intrdocution to System Dynamics. Available online: http://www.systemdynamics.org/what-is-s/ (accessed on 10 March 2016).

51. Forrester, J. *Industrial Dynamics*; MIT Press: Cambridge, MA, USA, 1961.

52. Lane, D.C. Should system dynamics be desribed as a "hard" or a "deterministic" systems approach? *Syst. Res. Behav. Sci.* **2000**, *17*, 3–22.

53. Ford, J.D. Vulnerability of Inuit food systems to food insecurity as a consequence of climate change: A case study from Igloolik, Nunavut. *Reg. Environ. Chang.* **2009**, *9*, 83–100.

54. Hoffman, M.H.G. Cognitive conditions of diagrammatic reasoning. *Semiotics* **2011**, *186*, 189–212.

55. Barlas, Y.; Carpenter, S. Philosophical roots of model validation: Two paradigms. *Syst. Dyn. Rev.* **1990**, *6*, 148–166.

56. Lane, D.C. With a little help from our friends: How system dynamics and soft OR can learn from each other. *Syst. Dyn. Rev.* **1994**, *10*, 101–134.

57. Vennix, J.A.M. *Group Model Building: Facilitating Team Learning Using System Dynamics*; Wilety: Chichester, UK, 1996.

58. Lane, D.C. Social theory and system dynamics practice. *Eur. J. Oper. Res.* **1999**, *113*, 501–527.

59. Richardson, G.P. *Feedback thought in Social Science and Systems Theory*; University of Pennsylvania Press: Philadelphia, PA, USA, 1991.

60. Burell, G.; Morgan, G. *Sosiological Paradigms and Organisational Analysis: Elements of the Sociology of Corporate Life*; Gower: Aldershot, UK, 1979.

61. Lane, D.C.; Oliva, R. The greater whole: Towards a synthesis of system dynamics and soft systems methodology. *Eur. J. Oper. Res.* **1998**, *107*, 214–235.

62. Lane, D.C. Rerum cognoscere causas: Part I—How do the ideas of system dynamics relate to traditional social theories and the voluntarism/determinism debate? *Syst. Dyn. Rev.* **2001**, *17*, 97–118.

63. Lane, D.C. Rerum cognoscere causas: Part II—Opportunities generated by the agency/structure debate and suggestions for clarifying the social theoretic position of system dynamics. *Syst. Dyn. Rev.* **2001**, *17*, 293–309.

64. A Sustainability Institute Report 2003. Commodity System Challenges Moving Sustainability into the Mainstream of Natural Resource Economies. Available online: http://s3.amazonaws.com/zanran_storage/www.ediblestrategies.com/ContentPages/707629742.pdf (accessed on 12 December 2015).

65. European Commission. The Common Agriculture Policy—A Story to Be Continued. Available online: http://ec.europa.eu/agriculture/50-years-of-cap/files/history/history_book_lr_en.pdf (accessed on 20 April 2016).

66. European Commission. 50 Years of Food Safety in the EU. 2007. Available online: http://ec.europa.eu/food/food/docs/50years_foodsafety_en.pdf (accessed on 20 April 2016).

67. Food and Agriculture Organization. World Outlook and State of Food and Agriculture—1950. Available online: http://www.fao.org/docrep/016/ap638e/ap638e.pdf (accessed on 1 February 1950).

68. Economist Intelligence Unit (EIU). Food Security in Focus: Europe. 2014. Available online: http://foodsecurityindex.eiu.com/Home/DownloadResource?fileName=EIU%20GFSI%202014_Europe%20regional%20report.pdf (accessed on 20 March 2016).

69. EU Agricultural Outlook. Prospects for EU Agricultural Markets and Income 2015–2025. Available online: http://ec.europa.eu/agriculture/markets-and-prices/medium-term-outlook/2015/fullrep_en.pdf (accessed on 1 February 2016).

70. De Vries, B.J.M. *Sustainability Science*; Cambridge University Press: New York, NY, USA, 2013.

71. Brisson, N.; Gate, P.; Gouache, D.; Charmet, G.; Oury, F.X.; Huard, F. Why are wheat yields stagnating in Europe? A comprehensive data analysis for France. *Field Crop. Res.* **2010**, *119*, 201–212.

72. Grassini, P.; Eskridge, K.M.; Cassman, K.G. Distinguishing between yield advances and yield plateaus in historical crop production trends. *Nat. Commun.* **2013**, *4*, 2918.

73. European Commission. EU Agricultural Markets Briefs. Price Developments and Links to Food Security—Price Level and Volatility. 2015. Available online: http://ec.europa.eu/agriculture/markets-and-prices/market-briefs/pdf/05_en.pdf (accessed on 10 March 2016).

74. European Commission. Communication from the Commission to the European Parliament, the Council, the European Economic and Social Committee and the Committee of the Regions—Food Prices in Europe. 2008. Available online: http://eur-lex.europa.eu/legal-content/EN/TXT/PDF/?uri=CELEX:52008DC0821&from=EN (accessed on 1 February 2016).

75. Loopstra, R.; Reeves, A.; Stuckler, D. Rising food insecurity in Europe. *Lancet* **2015**, *385*, 2041.

76. World Health Organization. Obesity Europe. Available online: http://www.euro.who.int/en/health-topics/noncommunicable-diseases/obesity (accessed on 20 March 2016).

77. European Commission. EC Farm Economics Briefs 2011. Income Developments in EU Farms. Available online: http://ec.europa.eu/agriculture/rica/pdf/Brief201101.pdf (accessed on 10 March 2016).

78. European Commission. EC Farm Economics Briefs 2011. EU Production Costs Overview. Available online: http://ec.europa.eu/agriculture/rica/pdf/Brief201102.pdf (accessed on 10 March 2016).

79. European Commission. EU Agricultural Economics Briefs 2013. Structure and Dynamics of EU Farms: Changes, Trends and Policy Relevance. Available online: http://ec.europa.eu/agriculture/rural-area-economics/briefs/pdf/09_en.pdf (accessed on 10 March 2016).

80. European Commission. EU Agricultural Economics Briefs 2014. EU Agricultural Income 2014—First Estimates. Available online: http://ec.europa.eu/agriculture/rural-area-economics/briefs/pdf/003_en.pdf (accessed on 10 March 2016).

81. European Commission. EU Agricultural and Farm Economics Briefs 2015. EU Farm Economics Summary 2012. Available online: http://ec.europa.eu/agriculture/rural-area-economics/briefs/pdf/003_en.pdf (accessed on 10 March 2016).

82. Vrolijk, H.C.J.; De Bont, C.J.A.M.; Blokland, P.W.; Soboh, R.A.M.E. Farm Viability in the European Union Assessment of the Impact of Changes in Farm Payments 2010. Available online: http://edepot.wur.nl/138917 (accessed on 10 March 2016).

83. Matthews, A. FADN Data Highlights Dependence of EU Farms on Subsidy Payments. Available online: http://capreform.eu/fadn-data-highlights-dependence-of-eu-farms-on-subsidy-payments/ (accessed on 10 March 2016).

84. Enjolras, G.; Capitanio, F.; Aubert, M.; Adinolfi, F. Direct payments, crop insurance and the volatility of farm income. Some evidence in France and in Italy. *New Medit* **2014**, *13*, 31–40.

85. Rabobank. Rethinking the F&A Supply Chain. Impact of Agricultural Price Volatility on Sourcing Strategies. 2011. Available online: http://www.boerderij.nl/pagefiles/35979/002_boerderij-download-agd573390d01.pdf (accessed on 10 March 2016).

86. Matthews, A. Recent Trends in EU Farm Incomes 2016. Available online: http://capreform.eu/recent-trends-in-eu-farm-incomes/ (accessed on 10 March 2016).

87. European Environment Agency. Agriculture. Available online: http://www.eea.europa.eu/themes/agriculture/intro (accessed on 20 March 2016).

88. Stoate, C.; Báldi, A.; Beja, P.; Boatman, N.D.; Herzon, I.; van Doorn, A.; De Snoo, G.R.; Rakosy, L.; Ramwell, C. Ecological impacts of early 21st century agricultural change in Europe—A review. *J. Environ. Manag.* **2009**, *91*, 22–46.

89. Rockstrom, J.; Steffen, W.; Noone, K.; Persson, Å.; Chapin, F.S.; Lambin, E.; Lenton, T.M.; Scheffer, M.; Folke, C.; Schellnhuber, H.; et al. Planetary boundaries: Exploring the safe operating space for humanity. *Ecol. Soc.* **2009**, *14*, 32.

90. European Commission. Proposal for a Regulation of the European Parliament and the Council Establishing Rules for Direct Payments to Farmers under Support Schemes within the Framework of the Common Agricultural Policy. 2011. Available online: http://ec.europa.eu/agriculture/cap-post-2013/legal-proposals/com625/625_en.pdf (accessed on 20 March 2016).

91. Pe'er, G.; Dicks, L.V.; Visconti, P.; Arlettaz, R.; Báldi, A.; Benton, T.G.; Collins, S.; Dieterich, M.; Gregory, R.D.; Hartig, F.; et al. TEEB Agriculture policy. EU agricultural reform fails on biodiversity. *Science* **2014**, *344*, 1090–1092.

92. Batáry, P.; Dicks, L.V.; Kleijn, D.; Sutherland, W.J. The role of agri-environment schemes in conservation and environmental management. *Conserv. Biol.* **2015**, *29*, 1006–1016.

93. Erjavec, K.; Erjavec, E. "Greening the CAP"—Just a fashionable justification? A discourse analysis of the 2014–2020 CAP reform documents. *Food Policy* **2015**, *51*, 53–62.

94. Swinnen, J. *The Political Economy of the 2014–2020 Common Agricultural Policy—An Imperfect Strom*; Centre for European Policy Studies (CEPS): Brussels, Belgium, 2015.

95. European Environmental Agency. Food Security and Environmental Impacts. 2012. Available online: http://www.eea.europa.eu/themes/agriculture/greening-agricultural-policy/food-security-and-environmental-impacts (accessed on 20 March 2016).

96. Levers, C.; Butsic, V.; Verburg, P. H.; Müller, D. Drivers of changes in agricultural intensity in Europe. *Land Use Policy* **2016**, *58*, 380–393.

97. La Trobe, H.L.; Acott, T.G.; La Trobd, H.L.; Acotp, T.G. Localising the global food system. *Int. J. Sustain. Dev. World Ecol.* **2000**, *7*, 309–320.

98. Cordell, D.; Neset, T.-S.S. Phosphorus vulnerability: A qualitative framework for assessing the vulnerability of national and regional food systems to the multi-dimensional stressors of phosphorus scarcity. *Glob. Environ. Chang.* **2014**, *24*, 108–122.

99. Neset, T.-S.S.; Cordell, D. Global phosphorus scarcity: Identifying synergies for a sustainable future. *J. Sci. Food Agric.* **2012**, *92*, 2–6.

100. Cordell, D.; Drangert, J.-O.; White, S. The story of phosphorus: Global food security and food for thought. *Glob. Environ. Chang.* **2009**, *19*, 292–305.

101. Wallgren, C.; Mattias, H. Eating energy—Identifying possibilities for reduced energy use in the future food supply system. *Energy Policy* **2009**, *37*, 5803–5813.

102. Pfeiffer, D.A. *Eating Fossil Fuels*; The Wilderness Publications: London, UK, 2003.

103. Olesen, J.E.; Schelde, K.; Weiske, A.; Weisbjerg, M.R.; Asman, W.A.H.; Djurhuus, J. Modelling greenhouse gas emissions from European conventional and organic dairy farms. *Agric. Ecosyst. Environ.* **2006**, *112*, 207–220.

104. The 1st SCAR Foresight Exercise. Agriculture and Environment 2006. Available online: https://ec.europa.eu/research/agriculture/scar/pdf/scar_foresight_environment_en.pdf (accessed on 1 February 2016).

105. ESF/COST Forward Look on European Food Systems in a Changing World. 2009. Available online: http://www.esf.org/fileadmin/Public_documents/Publications/food.pdf (accessed on 1 February 2016).

106. The Government Office for Science. Foresight. The Future of Food and Farming. 2011. Final Project Report. Available online: https://www.gov.uk/government/uploads/system/uploads/attachment_data/file/288329/11-546-future-of-food-and-farming-report.pdf (accessed on 1 February 2016).

107. Tilman, D.; Cassman, K.G.; Matson, P.A.; Naylor, R.; Polasky, S. Agriculture sustainability and intensive production practices. *Nature* **2002**, *418*, 671–677.

108. Ramírez, C.A.; Worrell, E. Feeding fossil fuels to the soil. An analysis of energy embedded and technological learning in the fertilizer industry. *Resour. Conserv. Recycl.* **2006**, *46*, 75–93.

109. Shcherbak, I.; Millar, N.; Robertson, G.P. Global metaanalysis of the nonlinear response of soil nitrous oxide (N2O) emissions to fertilizer nitrogen. *Proc. Natl. Acad. Sci. USA* **2014**, *111*, 9199–9204.

110. Geiger, F.; Bengtsson, J.; Berendse, F.; Weisser, W.W.; Emmerson, M.; Morales, M.B.; Ceryngier, P.; Liira, J.; Tscharntke, T.; Winqvist, C.; et al. Persistent negative effects of pesticides on biodiversity and biological control potential on European farmland. *Basic Appl. Ecol.* **2010**, *11*, 97–105.

111. Mclaughlin, A.; Mineau, P. The impact of agricultural practices on biodiversity. *Agric. Ecosyst. Environ.* **1995**, *55*, 201–212.

112. Food and Agriculture Organization. Livestock's Long Shadow Environmental Issues and Options. Available online: ftp://ftp.fao.org/docrep/fao/010/a0701e/a0701e.pdf (accessed on 20 March 2016).

113. Tamminga, S. Pollution due to nutrient losses and its control in European animal production. *Livest. Prod. Sci.* **2003**, *84*, 101–111.

114. Goodman, D.; Sorj, B.; Wilkinson, J. *From Farming to Biotechnology: A Theory of Agro-Industrial Development*; Basil Blackwell: Oxford, UK, 1987.

115. Lundvall, B.; Johnson, B. The learning economy. *J. Ind. Stud.* **1994**, *1*, 23–42.

116. United Nations Industrial Development Organization. Determinants of Total Factor Productivity: A Literature Review. 2007. Available online: http://www.unido.org//fileadmin/user_media/Publications/Research_and_statistics/Branch_publications/Research_and_Policy/Files/Working_Papers/2007/WP022007%20-%20Determinants%20of%20total%20factor%20productivity.pdf (accessed on 20 March 2016).

117. Morgan, K.; Murdoch, J. Organic vs. conventional agriculture: Knowledge, power and innovation in the food chain. *Geoforum* **2000**, *31*, 159–173.

118. Mitchell, D. *A Note on Rising Food Prices*; Policy Research Working Paper 4682; World Bank: Washington, DC, USA, 2008. Available online: http://elibrary.worldbank.org/doi/abs/10.1596/1813-9450-4682 (accessed on 20 March 2016).

119. Ciaian, P.; Kancs, A. Interdependencies in the energy–bioenergy–food price systems: A cointegration analysis. *Resour. Energy Econ.* **2011**, *33*, 326–348.

120. Abbott, P.C.; Hurt, C.; Tyner, W.E. *What's Driving Food Prices?*; Farm Fundation Issue Report; Farm Fundation: Oak Brook, IL, USA, 2009. Available online: http://www.farmfoundation.org/news/articlefiles/105-FoodPrices_web.pdf (accessed on 20 March 2016).

121. Lipsky, J. Commodity Prices and Global Inflation. In Proceedings of the Council on Foreign Relations, New York, NY, USA, 8 May 2008.

122. Organisation for Economic Co-operation and Development. OECD-FAO Agricultural Outlook 2009–2018. Available online: http://www.oecd.org/berlin/43042301.pdf (accessed on 20 March 2016).

123. Food and Agriculture Organization. Growing demand on agriculture and rising prices of commodities: An opportunity for smallholders in low-income, agricultural-based countries? In Proceedings of the Thirty-First Session of IFAD's Governing Council, Rome, Italy, 14 February 2008.

124. Headey, D.; Fan, S. Anatomy of a crisis: The causes and consequences of surging food prices. *Agric. Econ.* **2008**, *39*, 375–391.

125. Pendell, D.L.; Kim, Y.; Herbel, K. Differences Between High-, Medium-, and Low-Profit Cow-Calf Producers: An Analysis of 2010–2014 Kansas Farm Management Association Cow-Calf Enterprise. 2015. Available online: https://www.agmanager.info/sites/default/files/Cow-Calf_2015_1.pdf (accessed on 20 March 2016).

126. United States Department of Agriculture. *Profits, Costs, and the Changing Structure of Dairy Farming*; Economic Research Report 47; USDA: Washington, DC, USA, 2007.

127. Sgroi, F.; Maria, A.; Trapani, D.; Testa, R.; Tudisca, S. Strategy to increase the farm competitiveness. *Am. J. Agric. Biol. Sci.* **2014**, *9*, 394–400.

128. Karelakis, C.; Abas, Z.; Galanopoulos, K.; Polymeros, K. Positive effects of the Greek economic crisis on livestock farmer behaviour. *Agron. Sustain. Dev.* **2013**, *33*, 445–456.

129. Rasmussen, S. *Production Economics: The Basic Theory of Production Optimisation*, 2nd ed.; Springer: Heidelberg, Germany; New York, NY, USA; Dordrecht, The Nertherlands; London, UK, 2011.

130. Debertin, D.L. *Agricultura Production Economics*, 2nd ed.; Macmillan Publishing Company: New York, NY, USA, 2012.

131. Varian, H.R. *Intermediate Microeconomics*, 8th ed.; WW Norton & Co.: New York, NY, USA; London, UK, 2010.

132. Bragg, L.A.; Dalton, T.J. Factors affecting the decision to exit dairy farming: A two-stage regression analysis. *J. Dairy Sci.* **2004**, *87*, 3092–3098.

133. Foltz, J.D. Entry, exit, and farm size: Assessing an experiment in dairy price policy. *Am. J. Agric. Econ.* **2004**, *86*, 594–604.

134. Ferguson, R.; Hansson, H. Expand or exit? Strategic decisions in milk production. *Livest. Sci.* **2013**, *155*, 415–423.

135. Breustedt, G.; Glauben, T. Driving forces behind exiting from farming in Western Europe. *J. Agric. Econ.* **2007**, *58*, 115–127.

136. Cochrane, W.W. *Farm Price: Myth and Reality*; University of Minnesota Press: St. Paul, MN, USA, 1958.

137. Adger, W.N. Vulnerability. *Glob. Environ. Chang.* **2006**, *16*, 268–281.

138. O'Brien, K.; Leichenko, R.; Kelkar, U.; Venema, H.; Aandahl, G.; Tompkins, H.; Javed, A.; Bhadwal, S.; Barg, S.; Nygaard, L.; et al. Mapping vulnerability to multiple stressors: Climate change and globalization in India. *Glob. Environ. Chang.* **2004**, *14*, 303–313.

139. Magdoff, F. Food as a commodity. *Mon. Rev.* **2012**, *63*, 15–22.

140. Westhoek, H.J.; Overmars, K.P.; van Zeijts, H. The provision of public goods by agriculture: Critical questions for effective and efficient policy making. *Environ. Sci. Policy* **2013**, *32*, 5–13.

141. Harvey, D.; Hubbard, C. Reconsidering the political economy of farm animal welfare: An anatomy of market failure. *Food Policy* **2013**, *38*, 105–114.

142. Ciaian, P.; Swinnen, J.F.M. Market imperfections and agricultural policy effects on structural change and competitiveness in an Enlarged EU. In Proceedings of the XIth Congress of the European Association of Agricultural Economists, 'The Future of Europe in the Global Agri-Food System', Copenhagen, Denmark, 24–27 August 2005.

143. Ciaian, P.; Swinnen, J.F.M. Credit market imperfections and the distribution of policy rents. *Am. J. Agric. Econ.* **2009**, *91*, 1124–1139.

144. Ciaian, P.; Swinnen, J.F.M. Land market imperfections and agricultural policy impacts in the new EU Member States: A partial equilibrium analysis. *Am. J. Agric. Econ.* **2006**, *88*, 799–815.

145. Cohen, B.; Winn, M.I. Market imperfections, opportunity and sustainable entrepreneurship. *J. Bus. Ventur.* **2007**, *22*, 29–49.

146. McCorriston, S. Why should imperfect competition matter to agricultural economists? *Eur. Rev. Agric. Econ.* **2002**, *29*, 349–371.

147. Soregaroli, C.; Sckokai, P.; Moro, D. Agricultural policy modelling under imperfect competition. *J. Policy Model.* **2011**, *33*, 195–212.

148. Matsumoto, A. Do government subsidies stabilize or destabilize agricultural markets? *Contemp. Econ. Policy* **1998**, 452–466.

149. Offermann, F.; Nieberg, H.; Zander, K. Dependency of organic farms on direct payments in selected EU member states: Today and tomorrow. *Food Policy* **2009**, *34*, 273–279.

150. Garrido, A.; Brummer, B.; M'Barek, R.; Meuwissen, M.P.M.; Morales-Opazo, C. *Agriculture Markets Instability: Revisiting the Recent Food Crises*; Routledge: London, UK, 2016.

151. Chavas, J.-P.; Holt, M.T. Market instability and nonlinear dynamics. *Am. J. Agric. Econ.* **1993**, *75*, 113.

152. Meadows, D.L. *The Dynamics of Commodity Production Cycles: A Dynamic Cobweb Theorem*; Wright-Allen Press, Inc.: Cambridge, MA, USA, 1970.

153. Alexandratos, N. Food price surges: Possible causes, past experience, and longer term relevance. *Popul. Dev. Rev.* **2008**, *34*, 663–697.

154. EUROSTAT 2016. Agri-Environmental Indicator—Specialisation. Available online: http://ec.europa.eu/eurostat/statistics-explained/index.php/Agri-environmental_indicator_-_specialisation (accessed on 20 August 2016).

155. Smit, B.; Wandel, J. Adaptation, adaptive capacity and vulnerability. *Glob. Environ. Chang.* **2006**, *16*, 282–292.

156. Mondelaers, K.; Aertsens, J.; Huylenbroeck, G. Van A meta-analysis of the differences in environmental impacts between organic and conventional farming. *Br. Food J.* **2009**, *111*, 1098–1119.

157. Sandhu, H.S.; Wratten, S.D.; Cullen, R. Organic agriculture and ecosystem services. *Environ. Sci. Policy* **2010**, *13*, 1–7.

158. Gattinger, A.; Muller, A.; Haeni, M.; Skinner, C.; Fliessbach, A.; Buchmann, N.; Mäder, P.; Stolze, M.; Smith, P.; El-Hage Scialabbad, N.; et al. Enhanced top soil carbon stocks under organic farming. *Proc. Natl. Acad. Sci. USA* **2012**, *109*, 18226–18231.

159. Seufert, V.; Ramankutty, N.; Foley, J.A. Comparing the yields of organic and conventional agriculture. *Nature* **2012**, *485*, 229–232.

160. Lamine, C. Transition pathways towards a robust ecologization of agriculture and the need for system redesign. Cases from organic farming and IPM. *J. Rural Stud.* **2011**, *27*, 209–219.

161. Smith, E.; Marsden, T. Exploring the "limits to growth" in UK organics: Beyond the statistical image. *J. Rural Stud.* **2004**, *20*, 345–357.

162. Foran, T.; Butler, J.R.A.; Williams, L.J.; Wanjura, W.J.; Hall, A.; Carter, L.; Carberry, P.S. Taking complexity in food systems seriously: An interdisciplinary analysis. *World Dev.* **2014**, *61*, 85–101.

163. Eakin, H.; Winkels, A.; Sendzimir, J. Nested vulnerability: Exploring cross-scale linkages and vulnerability teleconnections in Mexican and Vietnamese coffee systems. *Environ. Sci. Policy* **2009**, *12*, 398–412.

164. Avelino, F.; Wittmayer, J.M. Shifting power relations in sustainability transitions: A multi-actor perspective. *J. Environ. Policy Plan.* **2015**, *18*, 628–649.

The Significance of Consumer's Awareness about Organic Food Products in the United Arab Emirates

Safdar Muhummad, Eihab Fathelrahman and Rafi Ullah Tasbih Ullah

Abstract: Awareness about negative externalities generated by conventional farming is gaining momentum with consumers around the world, opting for alternatively, namely organically, produced food products. Information about consumers' awareness is an essential element for farmers and marketing agencies to successfully plan production that can capture a greater market share. This study discusses effective factors influencing consumers' awareness about the benefits of organic food in the United Arab Emirates. Sample data and ordinary least square (OLS) regression techniques are applied to delineate factors influencing consumers' awareness about organic food. The results from this regression analysis highlight the importance of specific socioeconomic determinants that change awareness about organic food products in United Arab Emirates (UAE) households. This study finds that awareness about organic food is influenced more effective factors such as gender, nationality, and education as well as income, occupation and age. These research findings apply to other economies and societies that have an increasing per capita spending on organic food, but also where people are highly sensitive to information provided about organic food. Therefore, these results are important to these research beneficiaries including food marketing planners, researchers, and agricultural and food policy makers.

Reprinted from *Sustainability*. Cite as: Muhummad, S.; Fathelrahman, E.; Tasbih Ullah, R.U. The Significance of Consumer's Awareness about Organic Food Products in the United Arab Emirates. *Sustainability* **2016**, *8*, 833.

1. Introduction

Traditional or conventional farming is a double-edged sword; on one side it is known for low cost and higher output, but on the other side it is known for its negative human and environmental impacts [1]. The net effects of these factors are not yet known with certainty, but the negative impacts of inorganic farming and food products are well-researched. On the production side, farmers and their families, farm workers such as pesticide applicators, and other field workers are all at a high risk of exposure. Scialabba showed [2] in a study supported by the Food and Agriculture Organization (FAO) that sustainability is also about equity among and between generations. The main contribution of organic agriculture to social well-being is through avoided harm and healthy community development. Avoided harm ranges from loss of arable soil, water contamination, biodiversity erosion, food scares, and pandemics associated with chemical agriculture, as well as the pesticide poisoning of 3 million persons per year resulting in 220,000 deaths, let alone farmers' indebtedness for inputs and suicides (e.g., 30,000 deaths in Maharastra, India, from 1997 to 2005). Lotter (2003) found that demand for organic products is driven by belief that such products are more healthful, tasty, and environmentally friendly than conventional products [3]. Lotter also called for more comparative research to include animal products beside plant products. On the consumption side, the risk to consumers due to the consumption of vegetables at which high levels of synthetic pesticides have been applied is high because consumers are not aware of the health hazards linked to chemical residues in vegetables [4]. In a study by Salazar (2014) titled "Going Organic in Philippines: Social and Institutional Features", the author noticed—through case studies of organic farms and a review of policy developments—that different social arrangements and a supportive environment can be achieved. This is through offering training opportunities, expanding resource access, and organizational support leading to organic farming that expands exponentially in the country [5]. Golderger (2008) showed the presence of a disconnect between farmers and nongovernment organizations in Kenya, the source of organic agriculture organizations which attributed to the delay of farmers' adoption of organic agriculture practices [6].

Organic farming minimizes social, ecological, and biological costs that are associated with conventional farming [7]. Phillip and Dipeolu (2010) [8] defines organic farming to be the farming production management system which promotes agroecosystem health, sustainable biological cycles, and soil biological activity. In other words, organic farming is based on sustainability of the agroecosystem, Food and Agriculture Organization (FAO) [9]. Gil et al. (2000) [10] showed that among the different organic products market segments in Spain, special attention has to be paid to "likely consumers". Consumers represent a potential for market growth, and specific-marketing strategies should be addressed towards them. The authors conclude that a better knowledge of their sociodemographic characteristics is needed. On the demand side, Beharrell and McFie on British (1991) consumers and Al-Taie (2015) on United Arab Emirates consumers showed that such consumers have positive attitudes towards organic products as they perceive them as healthier than conventional alternatives [11,12].

The contribution of conventional agriculture to national development and food security has always been seen as the crux of public policy across the globe, yet in recent decades, organic agriculture has taken over conventional farming in terms of growth. For example, as shown in the Figure 1, organic agricultural land in the world increased from 11 million hectares in 1999 to 43.1 million hectares in 2013 [13].

Similarly, the number of organic producers also increased from 1.8 million in 2011 to 2 million in 2013 [13].

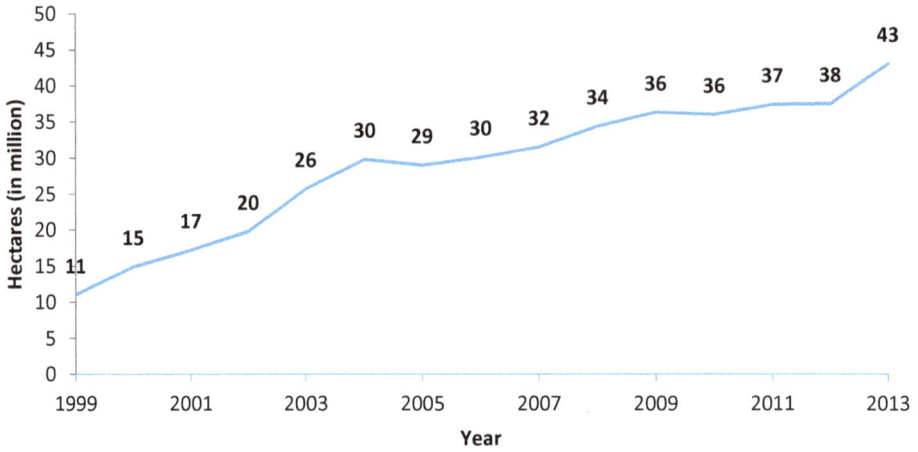

Figure 1. Growth of organic agricultural land in the world (1999–2013). Data Source: Willer et al., (2013) [13].

Zanoli and Naspetti (2002) [14] presented results from an Italian study on consumer perception and knowledge of organic food and related behavior. The authors used the means–end chain model to link attributes of products to the needs of consumers. The authors reported and discussed results on consumer cognitive structures at different levels of experience. In this study, consumers' groups were found to be due to the consumers' frequency of use (experience) of organic products and level of information (expertise). Müller et al. (2015) [15] studied the negative effects on consumers' evaluations of and behavior towards the purchase of certified organic food. The authors used behavioral models and findings from previous studies as well experimental study. The authors, highlighting the importance of consumers' awareness about organic food benefits, argued that when producers' mislabeling of products or consumers' label misinformation is revealed by mass media sources, such mislabeling causes negative effects on consumers' evaluation of and behavior towards the purchase of certified organic food products. These studies agreed that awareness about organic food benefits is an implicit variable. Other explicit variables—such as respondents' education, occupation, and age—were not carefully considered by these previous studies and were not considered to be effective variables to measure the consumers' level of awareness. In this study we argue for the significance of such variables and discuss the relationship between them and the consumers' awareness about organic products.

2. Organic Products Market in United Arab Emirates

In the United Arab Emirates (hereafter UAE), the agricultural sector has never been a major contributor to its gross domestic product (GDP). According to 2012 statistics from the Ministry of Environment and Water (MoEW) [15], agricultural land in the UAE constitutes only 4.6% of its total land. In addition, the value added per worker in the agricultural sector is merely USD 11,795, which is at a large decline since 2000 when this figure was nearly USD 30,000. However, given increasing income and the demand for fresh and healthier food products, agricultural development and food quality are major priorities of the UAE government agricultural policies [16]. These conditions provide an opportunity for organic agriculture to expand in the UAE. Table 1 shows data from 2012 about organic agricultural production in the UAE.

Table 1. Organic agriculture production (2012): selected characteristics in the United Arab Emirates.

Characteritics	Value
Share of organic agricultural land	4.6%
Organic agricultural land total	3905 hectares
Share of organic agriculture in total	0.7%
Number of certified organic farms	34
Number of certified organic crops	More than 62 varieties

Source: Food and Agriculture Organization (FAO) (2013) [17].

Currently, the UAE market for organic food is valued at USD 100–150 million per year [16]. There are 34 certified organic farms producing more than 62 varieties of organically certified crops. While the UAE is one of the leading producers of organic palm dates (more than 1000 tons per year), it also produces organic vegetables, such as beans, beetroot, broccoli, cabbage, chilies, cucumber, eggplant, fennel, carrots, green peas, lettuce, marrows, onions, okra, potatoes, strawberries, sweet corn, and tomatoes [13].

Despite a considerable increase in the demand for organic products around the world [18], organic products are still considered specialty commodities in the food market and occupy a small market share, and there is an overall low market awareness about organic products. A successful campaign for introducing a new product into the market depends on two factors: consumer's awareness about the product and their willingness to pay [19,20]. Determinants influencing consumers' willingness to pay for organic food in the UAE has been addressed elsewhere [19]. The objective of this article is to evaluate factors that effectively influence the consumer's awareness about organic food in the UAE and recommend potential implications and marketing efforts needed to increase awareness about organic food in an effort to increase organic products' market share. The rest of the article is organized as follows. Section 2 describes the organic product market in United Arab Emirates (UAE). Section 3 summarizes lessons learned from previous studies on consumers' awareness about organic food products. Section 4 describes the methods used for data collection and the statistical analysis, as well as explains the selected dependent variable and explanatory variables. Section 5 presents and discusses empirical findings and Section 6 concludes with policy implications based on the application of the study's findings to food policy makers, researchers, and the organic food marketing planners.

3. Consumer's Awareness about Organic Food

Consumer's awareness refers to product characteristics exploration and recognition by consumers. This is the organic product's characteristics/attributes such as nutritional contents, whether the product is certified organic or not, locally produced or imported, country of origin (if it is imported), labeling information including date of expiration, and level of freshness. Furthermore, specific brands are likely to be considered by the consumers to have high quality organic products, so such consideration may affect his or her decision-making about purchasing a specific brand of food products [12,20].

Chouichom et al. (2013) [21] investigated the impact of socioeconomic and demographic variables on organic food purchase behavior in Bangkok. Data was collected from 848 customers using convenient sampling. Since nonrandom sampling was used, the representativeness of the sample was insured by collecting data from customers who shop organic products from supermarkets, as well as from shops in the outskirts of Bangkok. Respondents were divided into three categories: those who have never heard of organic food, those who have heard but never purchased organic food, and those who have both heard and experienced consuming organic food. Then, contingency tables and chi-square tests were used to establish socioeconomic and demographic differences between the three types of consumers. According to the results reported by the authors, there are three main motives to demand organic food in Bangkok: the perceived health benefits, the search for tastier food, and attraction offered by novel food products. Moreover, statistical analysis signified that organic food consumption in Bangkok tends to increase with age, being male, education, and income.

Similarly, Bravo et al. (2013) [20] investigated various motives that encourage organic food consumption in Germany. Using data from the German National Nutrition Survey II, attitude towards organic food purchase is measured through the perceived importance of organic food. Results from the structural equation modeling (SEM) indicated that organic food purchases are influenced by altruistic aspects (i.e., environmental concerns and animal welfare). Additionally, nutritional information and ease of accessibility also positively influences organic food purchases, while price is negatively associated with organic food purchases. Amongst the socioeconomic and demographic variables, the authors reported that respondents from small households, women and older respondents, and those with higher societal status are more likely to purchase organic alternatives. Similar findings are also reported by Meixner et al. (2014) for Russia with the addition that consumers also prefer local foods, and that ease of access to organic food has no effect on consumers' purchasing behavior [22].

Briz and Ward (2009) [23] explored covariates of consumers' awareness in the case of Spain. Using a telephonic interview, the authors collected data from 1003 respondents, all of whom were above 18 years of age. Using multinomial logistic model, the dependent variable, i.e., awareness about organic food, was operationalized as having three categories. Explanatory variables of the model included age, education, income, gender and a general understanding of food nutrition (all are being measured as categorical variables). The authors also controlled for region and size of the city where the respondent belonged. Their results show that consumers' awareness about organic food is influenced only by the level of education, income and the degree by which a consumer is nutrition-conscious. That is, all the stated variables are positively associated with consumers' awareness about organic food. Theoretically, awareness about a product can be measured by using two techniques: brand recall and brand recognition. In brand recall tests, researchers measure the ability of consumers to recall brand names in a particular product category. In a brand recognition test, researchers provide consumers with a list of brands and ask them if they can remember seeing any of the brands before. For empirical studies, both techniques can be used to provide simple descriptions (i.e., average, range of response, mode, and standard deviation) or to provide more detail by using factor analysis [24]. Kesse-Guyot et al. (2013) [25] profiled organic food consumers in a large sample of French adults using cluster analysis and found that a sizable group of regular organic food consumers had shown a healthy profiles compared to nonorganic food consumers. However, the authors suggested further studies to analyze organic food intake in relation to health markers. A study that considered the demographic portrayal of organic vegetable consumption within the United States (U.S.) by Dettmann et al. (2007) [26] found that the organic market sector in the U.S. is one of the fastest growing food sectors, with growth rates in organic food sales averaging 18% per year between 1998 and 2005. The largest segment within the organic market was found to be fresh produce, comprising 36% of retail sales in 2005. The study applied a logistic model framework and a Heckman two-stage model to depict the relationship of organic vegetable expenditures as a ratio of total household vegetable expenditures. The authors found that race, education level, and household income consistently influenced the consumer's decision to purchase organic vegetables.

In brief, previous studies discussed and analyzed the importance of consumers' awareness about organic food products at various settings and conditions. These conditions varied from local organic products to the overall organic food products including imported products. These previous studies agreed that the socioeconomic/demographic factors play significant roles on determining the consumers' decision to buy organic product. In this study we confirm previous studies summarized above. Furthermore, this study addresses and analyzes the influence of non-income and non-occupation status variables such as gender, education, and nationality as we argue that they could be more relevant to changes on the dependent variable (i.e., awareness about organic food products). Variables such as household income and occupation status are important to make the consumer demand an effective demand, so their inclusion in the econometric model specification is critical. However, this study's objective is to further analyze the significance of each of such on the non-income related explanatory (independent) variables towards the overall awareness of consumers about organic food products.

4. Data and Methods

The study is based on a sample survey of 300 respondents from the United Arab Emirates (UAE). The descriptive statistics of the respondents in the sample indicates that the typical respondent (median) of the head household in the sample is male, in the age category between 30 and 39 years old, married with children, completed his college education, originally non-national, works for the private sector, has income between 10,000 and 15,000 UAE Dirhams (exchange rate is 1 USD = 3.65 UAE Dirhams), and lives with a family of four members. The data was collected through a well-structured questionnaire, containing questions on awareness about organic food and sociodemographic information of the respondents. The questionnaire for this study was designed to solicit the respondents about their level of awareness about organic food products, organic food market characteristics, and their ranking for the importance of organic food features such as their perception of whether they consider organic food product to be chemical-free, fresh and healthy, better-tasting or not, was produced used best agricultural production practices, among many other attributes. The questionnaire in this study was divided into two parts. Part one asked questions in relation to the respondents' perception and attitudes towards organic products in UAE, including questions about the frequency of buying organic food, and positive and negative attributes of buying organic food. For example, the questionnaire asked if the respondents considered organic food to be expensive, whether they still purchase the products due to the benefits they draw from consumption of organic food. This part also included questions in relation to respondent's level of trust of organic food certification in UAE and other places. The second part of the questionnaire included questions about the respondents' socioeconomic or demographic variables. These questions enable the research to draw conclusions about the correlations between such explanatory variables on one side and the research-dependent variable, namely awareness about organic food.

Awareness can be measured by asking simple descriptive questions from the respondents, using factor analysis (FA)/principle component analysis (PCA).

To depict factors influencing consumer's awareness about organic food in the UAE, we used the following:

$$AW_i = \beta_0 + \beta_1 AG_i + \beta_2 GR_i + \beta_3 NY_i + \beta_4 ED_i + \beta_5 MI_i + \beta_6 ES_i + \beta_7 HS_i + \epsilon_i \quad (1)$$

The dependent variable is awareness about organic food (AW), which is an index constructed from various awareness indicators asked from respondents. Based on similar a principal component analysis by Qendro (2015) [27] and Petrrescu et al. (2015) [28], we have asked respondents questions like, "Do you consume organic food because you think it is chemical free?" Responses were recorded for a total of 17 such questions in the form of 0 (which implies no) and 1 (which implies yes). Each of the 17 questions highlighted a key characteristic of organic food, such as freshness, taste, and nutrients. To construct our dependent variable, we have adopted the simple counting procedure which adds responses to all questions (Lea, and Worsley, 2005) [29] for adopting a similar methodology to develop the index of awareness variables. On the explanatory variables side, a set of variables were included to represent the respondents' socioeconomic characteristics. These variables are namely Age, Gender, Nationality, Education, Monthly income, Employment status, and Household size.

As a robustness check of our awareness measure, we have also constructed another measure of awareness about organic food. One of the questions in the survey asked respondents, "How much have you heard or read about organic food?" Responses to the questions were recorded on a 4 point Likert scale (i.e., 1 = nothing at all, 2 = a little, 3 = some, and 4 = a lot). We have considered options 1 and 2 as respondents who are unaware or less aware about organic food and the remaining two options as respondents who possess awareness about organic food. Since the dependent variable thus constructed is a dichotomous one, the estimation technique adopted for the purpose is binary logistic model. The results obtained are given in Appendix 6, which show that none of the explanatory variables assume signs different from the main model, nor are these much different in magnitude. As a further robustness check, the awareness variable is also specified as standard normal scores (Appendix 6) and the results resemble the original ones.

AG is the age of the respondent, which is a multinomial variable having five categories: 1 signifies age younger than 20 years of age, 2 between 20 and 29 years of age, 3 between 30 and 39 years of age, 4 between 40 and 49 years of age, and

5 greater or equal to 50 years of age. A dichotomous variable representing gender is named GR, where 1 represents male and 0 female. The dichotomous variable, NY representing nationality, where 1 represents a UAE resident and 0 otherwise. The respondent's education level, ED, has six categories, where 1 represents the lowest level of education and 6 the highest. MI is the monthly income of the respondents and has 7 categories, where 1 is the lowest income level and 7 is the highest income level. Meanwhile, ES is a dichotomous variable representing the employment status of the respondent, where 1 represents those who are actively employed (e.g., government employees, self-employed, and private sector employees) and 0 represents those who are not active participants in the job market (e.g., retired, housewives, student, unemployed, and others). Finally, HS represents the number of people living in the household, including the respondent.

Since the dependent variable in Equation 1 is a continuous variable, the appropriate estimation method is ordinary least squares (OLS), provided that the basic Gauss–Markov properties are satisfied. The model is thus tested for multi-collinearity, heteroskedasticity, and misspecification. All tests confirmed that the model does not suffer from any of the above stated sources of biases. The following section presents and discusses the results.

5. Results and Discussion

To study the relationship between the consumers' awareness about organic food and their socioeconomic characteristics. The ordinary least square (OLS) regression was selected to illustrate and predict such a relationship. The estimated outputs of Equation 1 are presented in Table 2. The equation was estimated using the OLS method and the results were checked for multicollinearity (using variance inflating factor criteria), heteroskedasticity (using White's heteroskedasticity test), and misspecification errors (using Ramsey's RESET test) (given in Appendix 6). Since the results satisfy these tests, we can safely conclude that our results are free from any econometric bias. The best fit curve of this model is depicted in Figure 2.

The results show that of all the explanatory variables analyzed in our study, only gender, nationality, and education influence awareness about organic food positively. Because none of the other explanatory variables are statistically significant, we ignore them in the subsequent discussion. However, these variables are important for the overall econometric model design and fitness representing the dataset. An explanation for why gender is a significant explanatory variable is that male respondents have more awareness about organic food than their female counterparts. This finding is in line with previous research cited in the previous section of this article. These studies noted that the reason behind more organic food awareness among male members is because overall, male members are relatively more educated in developing countries where these studies are conducted. The correlation matrix illustrated in Appendix 6 shows the correlation between the explanatory variables (independent variables) themselves. No evidence found of highly significance correlation between these variables which supports selection of these variables to be possible influential variables on the level of the respondents' awareness of organic products in the study sample.

Table 2. Awareness about organic food.

Variable	Coefficient	Std. Error	T-Statistics	p-Value
Age	−0.189	0.274	−0.691	0.490
Gender	1.055 **	0.538	1.959	0.051
Nationality	1.631 **	0.731	2.232	0.027
Education	0.658 **	0.277	2.376	0.018
Monthly income	−0.134	0.145	−0.919	0.359
Employment status	−0.025	0.647	−0.039	0.969
Household size	−0.013	0.063	−0.213	0.832
Constant	3.449 **	1.793	1.924	0.055

** signifies statistical significance at 5 percent level.

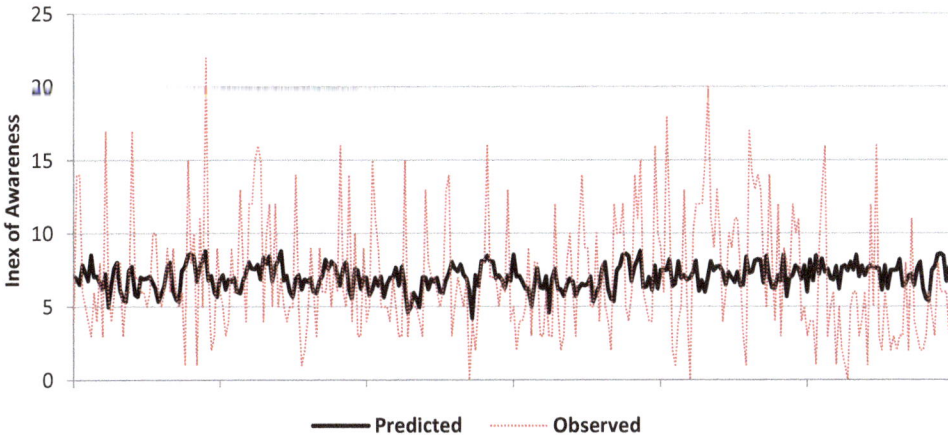

Figure 2. Predicted awareness index against observed levels of awareness about organic agriculture in United Arab Emirate.

The results also found that being a national of the UAE increases awareness about organic food. An explanation for this is that the majority of non-UAE nationals are immigrants who are low-paid workers lacking education and circles that can provide them awareness about organic food.

The positive and significant impact of education on awareness about organic food implies that more education increases awareness about organic food. There

187

is a strong connection between education and awareness of organic food benefits, including the environmental benefits of production (the supply side) and consumer health benefits (the demand side). These results confirmed previous studies' findings and highlight the significance of more relevant (to the level of awareness about organic food) explanatory (independent) variables in the study such as the consumer's gender, level of education, and nationality. Furthermore, study findings show that such respondents' effective explanatory (independent) variables play a significant role on increasing awareness about organic products. Policy makers, researchers, and marketing analysts may consider such significant variables when designing their organic products' market share.

6. Conclusions and Recommendations

A key challenge faced by the modern world is food security. Conventional farming may be considered a success in the quest for food security, but it comes at the cost of the overall well-being of the environment and ecosystem. Organic farming is a relatively new system of farming; it avoids costs associated with conventional farming and it is considered beneficial for the planet as a whole. Because organic farming is relatively new, the perceived benefits associated with it are not well-known. It will take time and effort for its benefits to become common knowledge.

Increasing awareness about the benefits of organic food compared to conventionally-produced food is relatively expensive. Therefore, awareness about the advantages of organic agriculture has not been well-considered in previous studies as an approach for sustainable production and consumption. In order to encourage organic farming and the production of organic food, the market for organic food will have to be established, so that organic farmers can get an outlet for their products. However, the establishment of a market for organic food depends on knowing what factors influence consumer awareness about organic food. Once such factors are ascertained, farmers will be better equipped to market their organic products and capture a larger share of the vegetable market. This study is an attempt to identify factors that determines consumers' awareness about organic food. This study utilizes sample data and contemporary econometric techniques to investigate factors that influence consumers' awareness about organic food. Furthermore, this study implements an ordinary least square (OLS) analysis of consumers' income, occupation status, and age as determinants of the level of consumers' awareness about organic food products. Furthermore, the study expanded the analysis to include more relevant explanatory variables such as gender, education, and nationality in order to better explain the influence of such non-income variables on consumers' awareness about organic food products. Income and occupation status are effective demand variables. Both variables contribute to the overall fitness of the regression model. However, specific variables such as gender, respondents' level of education, and nationality were found to be more significant.

This study finds that awareness about organic food is influenced by more relevant factors such as gender, nationality, and education as well as income, occupation status, and age. Therefore, efforts to successfully expand awareness about organic foods should consider nonconventional socioeconomic/demographic characteristics of consumers when segmenting markets. Understanding of the significance of the influential factors that determine the consumers' decision of buying organic food helps the policy makers, researchers, and marketing planners focus their efforts and expand them in the areas of providing the consumers with more information in relationship to the products' characteristics and designing awareness activities about organic food production, consumption, and health benefits. Such effort will support the organic farmers' in gaining larger market shares for their products.

Future studies may consider the need for comprehensive framework to support organic products marketing including policy formulation, research, education, and outreach to support the organic food production. Furthermore, future studies may also consider changes of consumers' behavior in response to expanding use of information through advanced information technology such as social media.

Author Contributions: All the authors contributed equally to this work.

Conflicts of Interest: The authors declare no conflict of interest.

Appendix A

Table A1. Awareness about Organic Food (Logistic Regression).

Variable	Coefficient	Std. Error	Wald	Significance
Age	−0.052	0.311	0.028	0.867
Gender	0.334	0.286	1.368	0.242
Nationality	0.093	0.366	0.065	0.799
Education	0.059	0.458	0.016	0.898
Monthly income	−0.382	0.324	1.395	0.237
Employment status	−0.069	0.323	0.045	0.832
Household size	−0.003	0.033	0.011	0.918
Constant	0.838	0.722	1.349	0.245

Appendix B

Table B1. Awareness about Organic Food (Awareness Specified as Standard Normal Scores).

Variable	Coefficient	Std. Error	t-Statistics	p-Value
Age	−0.046	0.066	−0.691	0.490
Gender	0.255	0.130	1.959	0.051
Nationality	0.395	0.177	2.232	0.027
Education	0.159	0.067	2.376	0.018
Monthly income	−0.032	0.035	−0.919	0.359
Employment status	−0.006	0.157	−0.039	0.969
Household size	−0.003	0.015	−0.213	0.832
Constant	−0.861	0.434	−1.984	0.048

Appendix C. Diagnostic Tests

Table C1. Collinearity test.

Variable	Tolerance	Variance Inflation Factor
Age	0.842	1.188
Gender	0.941	1.062
Nationality	0.610	1.639
Education	0.879	1.138
Monthly income	0.687	1.455
Employment status	0.657	1.522
Household size	0.938	1.067

Table C2. Heteroskedasticity test.

White Heteroskedasticity Test			
F-Statistics	1.414	Probability	0.167
R-Squared	15.338	Probability	0.167

Appendix D

Table D1. Correlation Matrix.

		Age of the Respondent (Original)	Gender of the Respondent	Nationality of the Respondent	Education of the Respondent (Original)	Income of the Respondent (Original)	Employment Status of the Respondents	Household Size
Age of the respondent (Original)	Pearson Correlation	1	0.032	−0.209 **	−0.247 **	0.096	0.387 **	−0.054
	Sig. (2-tailed)		0.585	0.000	0.000	0.120	0.000	0.356
Gender of the respondent	Pearson Correlation	0.032	1	−0.068	−0.031	0.056	0.250 **	−0.077
	Sig. (2-tailed)	0.585		0.237	0.588	0.369	0.000	0.187
Nationality of the Respondent	Pearson Correlation	−0.209 **	−0.068	1	−0.107	0.442 **	−0.329 **	0.137 *
	Sig. (2-tailed)	0.000	0.237		0.063	0.000	0.000	0.018
Education of the respondent (original)	Pearson Correlation	−0.247 **	−0.031	−0.107	1	0.040	−0.254 **	−0.119 *
	Sig. (2-tailed)	0.000	0.588	0.063		0.523	0.000	0.041
Income of the respondent (Original)	Pearson Correlation	0.096	0.056	0.442 **	−0.040	1	0.138 *	0.097
	Sig. (2-tailed)	0.120	0.369	0.000	0.523		0.026	0.119
Employment status of the respondents	Pearson Correlation	0.387 **	0.250 **	−0.329 **	−0.254 **	0.138 *	1	−0.074
	Sig. (2-tailed)	0.000	0.000	0.000	0.000	0.026		0.216
Household Size	Pearson Correlation	−0.054	−0.077	0.137 *	−0.119 *	0.097	−0.074	1
	Sig. (2-tailed)	0.356	0.187	0.018	0.041	0.119	0.216	

** Correlation is significant at the 0.01 level (2-tailed). * Correlation is significant at the 0.05 level (2-tailed).

References

1. Lumpkin, H. Organic Vegetable Production: A Theme for International Agricultural Research. In Proceedings of the Seminar on Production and Export of Organic Fruit and Vegetables in Asia, Bangkok, Thailand, 3–5 November 2003.
2. Scialabba, N.E.-H. Organic Agriculture's Contribution to Sustainability. In Proceedings of the USDA Organic Farming Systems Research Conference, Washington, DC, USA, 29 April 2013.
3. Lotter, D.W. Organic agriculture. *J. Sustain. Agric.* **2003**, *21*, 59–128.
4. Coulibaly, O.; Cherry, A.; Nouhoheflin, T.; Aitchedji, C.; Al-Hassan, R. Vegetable producer perceptions and willingness to pay for biopesticides. *J. Veg. Sci.* **2007**, *12*, 27–42.
5. Salazar, R.C. Going organic in the Philippines: Social and institutional features. *Agroecol. Sustain. Food Syst.* **2014**, *38*, 199–229.
6. Goldberger, J.R. Diffusion and adoption of non-certified organic agriculture: a case study from semi-arid Makueni District, Kenya. *J. Sustain. Agric.* **2008**, *32*, 531–564.
7. Koocheki, A. Organic farming in Iran. In Proceedings of the 6th IFOAM-Asia Scientific Conference, Benign Environment and Safe Food, Yangpyung, Korea, 7–11 September 2004.
8. Phillip, B.; Dipeolu, A. Willingness to pay for organic vegetables in Abeokuta, South West Nigeria. *Afr. J. Food Nutr. Sci.* **2010**, *10*, 11.
9. Food and Agriculture Organization (FAO) of the United Nations. Organic Agriculture: What Are the Environmental Benefits of Organic Agriculture? Available online: http://www.fao.org/organicag/oa-faq/oa-faq6/en/ (accessed on 1 October 2015).
10. Gil, J.M.; Gracia, A.; Sanchez, M. Market segmentation and willingness to pay for organic products in Spain. *Int. Food Agribus. Manag. Rev.* **2000**, *3*, 207–226.
11. Beharrell, B.; MacFie, J. Consumer attitudes to organic foods. *Br. Food J.* **1991**, *93*, 25–30.
12. Al-Taie, W.A.; Rahal, M.K.; AL-Sudani, A.S.; AL-Farsi, K.A. Exploring the Consumption of Organic Foods in the United Arab Emirates. *SAGE Open* **2015**.
13. Willer, H.; Lernoud, J.; Home, R. The World of Organic Agriculture. Statistics and Emerging Trends 2013: Summary. Available online: http://www.organic-world.net/yearbook/yearbook-2013.html (accessed on 22 August 2016).
14. Zanoli, R.; Naspetti, S. Consumer motivations in the purchase of organic food: A means-end approach. *Br. Food J.* **2002**, *104*, 643–653.
15. Müller, C.E.; Gaus, H. Consumer response to negative media information about certified organic food products. *J. Consum. Policy* **2015**, *38*, 387–409.
16. Ministry of Environment and Water (MoEW). *UAE State of Green Economy*; Ministry of Environment and Water (MoEW): Dubai, UAE, 2014.
17. Food and Agriculture Organization (FAO). FAOSTAT. Available online: http://faostat.fao.org/site/377/default.aspx#ancor (accessed on 10 October 2015).
18. Aryal, K.P.; Chaudhary, P.; Pandit, S.; Sharma, G. Consumers' willingness to pay for organic products: A case from Kathmandu valley. *J. Agric. Environ.* **2009**, *10*, 15–26.
19. Muhammad, S.; Fathelrahman, E.; Ullah, R.U. Factors Affecting Consumers' Willingness to Pay for Certified Organic Food Products in United Arab Emirates. *J. Food Distrib. Res.* **2015**, *46*, 37–45.

20. Bravo, C.P.; Cordts, A.; Schulze, B.; Spiller, A. Assessing determinants of organic food consumption using data from the German National Nutrition Survey II. *Food Qual. Preference* **2013**, *28*, 60–70.

21. Chouichom, S.; Liao, L.M.; Yamao, M. General View Point, Perception and Acceptance of Organic Food Products among Urban Consumers in the Thai Marketplace. In *Sustainable Food Security in the Era of Local and Global Environmental Change*; Springer: Berlin, Germany, 2013; pp. 187–201.

22. Meixner, O.; Haas, R.; Perevoshchikova, Y.; Canavari, M. Consumer Attitudes, Knowledge, and Behavior in the Russian Market for Organic Food. *Int. J. Food Syst. Dyn.* **2014**, *5*, 110–120.

23. Briz, T.; Ward, R.W. Consumer awareness of organic products in Spain: An application of multinominal logit models. *Food Policy* **2009**, *34*, 295–304.

24. Keller, K.L. Conceptualizing, measuring, and managing customer-based brand equity. *J. Mark.* **1993**, *57*, 1–22.

25. Kesse-Guyot, E.; Peneau, S.; Mejean, C.; de Edelenyi, F.S.; Galan, P.; Hercberg, S.; Lairon, D. Profiles of organic food consumers in a large sample of French adults: results from the nutrinet-santé cohort study. *PLoS ONE* **2013**, *8*, e76998.

26. Dettmann, R.L.; Dimitri, C. Organic consumers: A demographic portrayal of organic vegetable consumption within the United States. In Proceedings of the EAAE International Marketing and International Trade of Quality Food Products Meeting, Bologna, Italy, 8–10 March 2007.

27. Qendro, A.-E. Albanian and UK Consumers' Perceptions of Farmers' Markets and Supermarkets as Outlets for Organic Food: An Exploratory Study. *Sustainability* **2015**, *7*, 6626–6651.

28. Petrescu, D.C.; Petrescu-Mag, R.M. Organic Food Perception: Fad, or Healthy and Environmentally Friendly? A Case on Romanian Consumers. *Sustainability* **2015**, *7*, 12017–12031.

29. Lea, E.; Worsley, T. Australians' organic food beliefs, demographics and values. *Br. Food J.* **2005**, *107*, 855–869.

The Resilience of a Sustainability Entrepreneur in the Swedish Food System

Markus Larsson, Rebecka Milestad, Thomas Hahn and Jacob von Oelreich

Abstract: Organizational resilience emphasizes the adaptive capacity for renewal after crisis. This paper explores the sustainability and resilience of a not-for-profit firm that claims to contribute to sustainable development of the food system. We used semi-structured interviews and Holling's adaptive cycle as a heuristic device to assess what constitutes social and sustainable entrepreneurship in this case, and we discuss the determinants of organizational resilience. The business, Biodynamiska Produkter (BP), has experienced periods of growth, conservation and rapid decline in demand, followed by periods of re-organization. Our results suggest that BP, with its social mission and focus on organic food, meets the criteria of both a social and sustainability entrepreneurial organization. BP also exhibits criteria for organizational resilience: two major crises in the 1970s and late 1990s were met by re-organization (transformation) and novel market innovations (adaptations). BP has promoted the organic food sector in Sweden, but not profited from this. In this case study, resilience has enhanced sustainability in general, but trade-offs were also identified. The emphasis on trust, local identity, social objectives and slow decisions may have impeded both economic performance and new adaptations. Since the successful innovation Ekolådan in 2003, crises have been met by consolidation rather than new innovations.

Reprinted from *Sustainability*. Cite as: Larsson, M.; Milestad, R.; Hahn, T.; von Oelreich, J. The Resilience of a Sustainability Entrepreneur in the Swedish Food System. *Sustainability* **2016**, *8*, 550.

1. Introduction

In a world undergoing constant change, the perspective of resilience offers a framework for facilitating sustainable development. Resilience is "the capacity of a system to absorb disturbance and reorganize while undergoing change so as to still retain essentially the same function, structure, identity, and feed-backs" [1], (p. 4). For an ecosystem, this can involve dealing with overfishing or pollution, while for a society, it involves an ability to deal with all kinds of disturbances. For a company, it can involve handling increased competition or fluctuating demand. Since the publication of Holling's [2] article "Resilience and Stability of Ecological Systems," ecosystem resilience has been rather well researched. The resilience of social systems, e.g., [3], as well as that of organizations and firms, e.g., [4,5], is gaining interest, as is the resilience of social-ecological systems, such as food systems [6,7].

The resilience approach focuses on the dynamic interplay between periods of gradual and sudden change and considers ways to adapt to change [8]. Resilience is thus a conceptual framework for understanding how persistence and transformation coexist in living systems, including human societies and food systems. Central to this paper is resilience thinking as a means of understanding the dynamics in a food business organization.

The role of entrepreneurs is often emphasized in discussions about sustainable development. One example from policy is the United Nations (UN) initiative "Supporting Entrepreneurs for Sustainable Development" [9], aiming to develop practical tools to help social and environmental entrepreneurs to scale up and inspire new entrepreneurial ventures to deliver social and environmental benefits. An example from social science is a special issue, Entrepreneurship and Sustainable Development, of the *Journal of Business Venturing*, where attention is paid to social and environmental entrepreneurs, e.g., [10].

According to Avery and Bergsteiner [11], operating on sustainable principles can enhance business performance and resilience. This is certainly an entrepreneurial challenge: "the choice to adopt a more sustainable strategy, one that research and practice show leads to higher resilience and performance over the long term, remains in the hands of each executive team" [11], (p. 13).

If strategies for environmental sustainability may enhance business resilience to external turbulences, the reverse causal order is not true: strategies for increasing business resilience may not benefit environmental sustainability unless this is the focus. For example, since the last century until 2012, the fossil fuel sector has been one of the most resilient globally and adapted to oil crises, but it has never contributed to environmental sustainability.

Business resilience means the ability to renew itself after crisis. Such adaptive capacity and renewal is the "natural consequence of an organization's innate resilience" [4], (p. 54). In other words, resilience is not just stability (not undergoing change), but successfully adapting to external influences [8]. A business that strives for sustainable development should contribute to social, environmental and economic development in its production while at the same time remaining flexible and resilient [12]. Sustainability requires resilience: if a sustainability entrepreneurship cannot adapt to change, it may be unable to contribute to desired changes in the long run. A company's efforts to be resilient, however, are not necessarily supported by government action [5].

In this paper, we explore the resilience of a business with a dual social and environmental orientation, a business that claims to contribute to the sustainable development of the food system. Two research questions are explored: (1) What constitutes social and sustainability entrepreneurship in this case? (2) What determines the organization's resilience? We approach these questions through a case study of a business that has experienced periods of growth, conservation, release (caused by a rapid decline in demand) and re-organization. This makes it an interesting case from the perspective of organizational resilience.

2. Entrepreneurship and Resilience

2.1. Entrepreneurs and Entrepreneurship

Entrepreneurs and entrepreneurship have been examined by scholars in several disciplines and have therefore been described in different ways. Entrepreneurship should promote a sense of ownership, long-term commitment, learning and strategic thinking and should facilitate decision making under uncertainty [13]. Most research focuses on economic aspects of entrepreneurship [14], but the role of entrepreneurs in achieving sustainable development, an aspect of entrepreneurship that is of interest in this paper, has received increasing interest, e.g., [9,10,15]. Some authors focus on entrepreneurship's social aspects. Social entrepreneurs, or those engaged in community business entrepreneurship, can contribute to economic, as well as social development [16,17]. Thake and Zadek [18] describe these businesses and organizations as community-based social entrepreneurs. Social entrepreneurs often adopt a not-for-profit form, but this is not always the case. A social entrepreneurship organization, what Bacq and Janssen [19] call a "social entrepreneurial venture" (SEV), must meet three criteria: (1) its social mission must be explicit and central; (2) its business idea, that is, the productive activity of goods or services generating an income, must go hand in hand with its social mission; and (3) the legal framework does not define SEVs, e.g., the social mission and purpose should not be defined by or limited to being a not-for-profit organization.

Besides social entrepreneurship, various concepts have been used to describe environmentally- oriented forms of entrepreneurship [12]. Social entrepreneurs "tailor their activities to be directly tied with the ultimate goal of creating social value" [17], (p. 22), and environmental entrepreneurs tailor theirs to environmental concerns [20,21]. Ecopreneurship, according to Schaper [20,21], uses more sustainable business practices; the motivations and orientations of green entrepreneurs have been investigated by Walley et al. [22], who found that many green start-ups are driven by economic motives as much as they are by wider sustainability goals.

Steyaert and Katz [23], (p. 193) foresee a "new multiverse of entrepreneurship" including "social, cultural, ecological, civic (…)" entrepreneurship. A suitable label for the business studied here might be, following Parrish and Tilley [24], sustainability entrepreneur. Schaltegger and Wagner [25] argue that while the literature on social entrepreneurship has focused on societal goals, seeing economic profit as a means only, the literature on environmental entrepreneurship has focused on integrating environmental goals with the business case as a strategy to make profits. Sustainable entrepreneurship, they argue, integrates all dimensions of sustainability and treats profits as both a means and an ends. Parrish and Tilley [24] indicate that sustainability entrepreneurship supports sustainable development in a broad sense and generates a reinforcing cycle of benefits to the entrepreneur, to other people and to the surrounding environment.

Central to entrepreneurship is social capital. Trust, reciprocity and common norms are central aspects of social capital [26]. Trust is crucial for establishing an entrepreneurial society, one "in which individuals in all kinds of organizations and in all aspects of life behave in an entrepreneurial manner" [13], (p. 34). Social capital and trust can be viewed as constituting a comparative advantage that an entrepreneur or a group of entrepreneurs enjoy over competitors. Trust reduces transaction costs and makes co-operation easier [27,28]. Trust improves relations with staff and clients and can also be an outcome of a long-term perspective on environmental and social responsibility [11].

2.2. Resilience and the Adaptive Cycle

Folke [8], (p. 259) argues that resilience "incorporates the idea of adaptation, learning and self-organization in addition to the general ability to persist disturbance". Thus, the adaptive capacity of an organization is part of its resilience. To build resilience in social-ecological systems (SESs), diversity in the decision-making structure is critical and can be applied locally [28,29], as well as regionally and globally [30]. To enhance resilience, social capital and trust have been shown to be important [31,32].

Resilience is a descriptive term, retaining essentially the same function is "good" only to the extent that this function is desirable [33]. Since increasing the organically-grown acreage in Sweden is a politically decided goal (states as 20% organically-grown acreage), the resilience of the organic food system has normative connotations [34]. Resilience and adaptation are used in what follows to describe the socio-economic situation in our case study. Detailed empirical studies are needed because it is only after an extreme crisis that the degree of an organization's resilience becomes fully visible [35]. Lengnick-Hall *et al.* [36] argue that organizational resilience is enhanced by learning and innovation among individual employees, which in turn can be promoted by human resource management. Learning and innovation are also emphasized in the reconstruction/reorganization phase (see Figure 1) by Linnenluecke *et al.* [35].

During periods of political or economic turbulence or environmental change, resilient communities and organizations are well prepared and quick to recover from the challenges they face [35]. Diversified production can absorb shocks and in practice acts as insurance for the local economy. Sources of organizational resilience include wisdom, perceptual stance and contextual integrity [37], which also provides

197

authenticity [38]. Resilience is eroded when rigidities are built up in organizations, resulting in low capacity for adaptation and renewal when crises emerge.

Holling [39] illustrated these dynamics as an "adaptive cycle", a description that should be thought of as a metaphor, a heuristic or conceptual model that can facilitate the understanding of complex systems, including organizations (see Figure 1). Initially, it was used to study the dynamics of ecosystems where ecologists had observed that periods of exploitation, for example, the rapid growth that follows a forest fire, were followed by periods of slower growth and accumulation of energy and structure. The adaptive cycle added the release and reorganization phases.

Figure 1 illustrates how longer periods of stability and the slow accumulation of resources alternate with shorter periods of turbulence [40]. Periods of turbulence are characterized by rapid reorganization and are sometimes used to illustrate what Schumpeter [41] called creative destruction, when turbulence breaks down structures, in turn creating opportunities for innovation. During the reorganization phase in Figure 1, there is a possibility that the organization might transform and take an unexpected turn and develop in a new direction. This phase is unpredictable, and there is a risk that capital could be drained. The adaptive cycle can be applied to different systems and different scales. Resilience at a higher scale, e.g., a large corporation, may require transformations of smaller units of this corporation, just like resilience of the biosphere requires transformation of the food and energy systems [42]. The adaptive cycle is used below to illustrate periods of growth and crisis in an organically-certified business in Sweden.

3. Case-Study Approach: Materials and Methods

3.1. The Case Study

Biodynamiska Produkter (Biodynamic Products; BP) is a not-for-profit foundation "providing consumers with organic and biodynamic food of high quality, in a way that enables farmers to continue developing" [43]. This dual aim is meant to provide favorable long-term conditions for organic/biodynamic farmers. BP consists of four parts: a fruit and vegetable box scheme (Ekolådan), a wholesaler, a trading company and two production units (an organic fruit farm and an organic market garden in Sweden). BP is a fully-certified organic business situated in Järna, 40 kilometers southwest of Stockholm. This area is the most important anthroposophic cluster in Sweden, of which BP has been an integral part since its start in 1966. Anthroposophy is based on the philosophical teachings of Rudolf Steiner (b. 1861, Austria) and entails biodynamic agriculture, education, healthcare and more. Järna today is the anthroposophical center of all the Nordic countries. Anthroposophically-inspired businesses and organizations in Järna include Waldorf Schools and the Steiner College, the Vidar Clinic, Ekobanken (a not-for-profit bank), several farms and market gardens, food-processing concerns (e.g., Saltå Kvarn) and food distributors, including BP. The trading division of BP co-operates with similar organic mid-scale initiatives in Europe that together (*i.e.*, not in the conventional retail network) source items, such as coffee, bananas and other fruits from overseas. Since these products come to ports in the Netherlands, BP has one employee there, as well. The wholesaler, which supplies restaurants, kitchens, organic shops and its own box scheme, buys all produce and processed products from Sweden, Europe and overseas.

The most recent and most important (in terms of publicity) part of BP is its box scheme, Ekolådan, studied in detail in this paper. The boxes delivered by Ekolådan contain only organic or biodynamic produce, but the box itself does not carry any visible label - neither Sweden's organic certification, KRAV [44], nor the European Union's (EU) organic label, nor the Demeter label. Demeter is the international organization certifying biodynamic products, which is a special type of organic food [45]. The Swedish branch of Demeter is located in Järna. After buying the fruit and vegetables from farmers, Ekolådan/BP alone has control over the rest of the value chain until the boxes are delivered to consumers' doors. Every box also contains a newsletter, including the names and locations of all of the farmers that contributed the produce for the box. In this way, Ekolådan wants to establish a relationship between consumers and producers. Since Ekolådan's establishment in 2002/2003, the chain of actors between the producers/farmers via the foundation (BP) and Ekolådan to final consumers has been fairly stable. Some producers have been added over time, and the number of customers has varied; those customers include individuals, households and offices. BP and Ekolådan share the same economic accounts and workforce (Figure 2), but BP is the overarching business and decision-making unit. Thus, Ekolådan is a part of BP and is not run as a business of its own.

BP and Ekolådan are interesting for several reasons. First, they claim to support an environmentally-friendly food chain. Second, BP has a long history of experiencing ups and downs. Ekolådan has also experienced turbulence, but during a shorter period. Third, Ekolådan is different from all other actors on the box scheme market in Sweden in that it is run by a not-for-profit foundation.

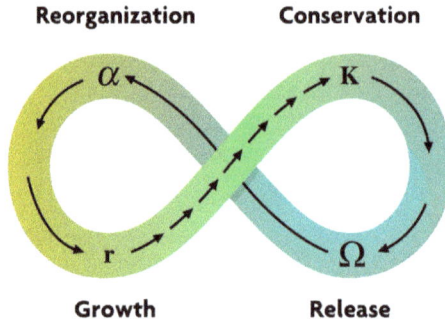

Figure 1. The dynamics of ecological, economic or social systems can be described in terms of an adaptive cycle with four phases: growth, conservation, release and reorganization. Long periods of stability and slow accumulation of structure (short arrows, from exploitation or growth to conservation, r to K) alternate with shorter, more turbulent periods of release of resources that create opportunities for innovation and reorganization (long arrows, when a crisis moves the system, K to Ω to α, and finally, to some new r phase). If the new r phase is fundamentally different from the previous r phase, a transformation has occurred. Modified from Holling [46].

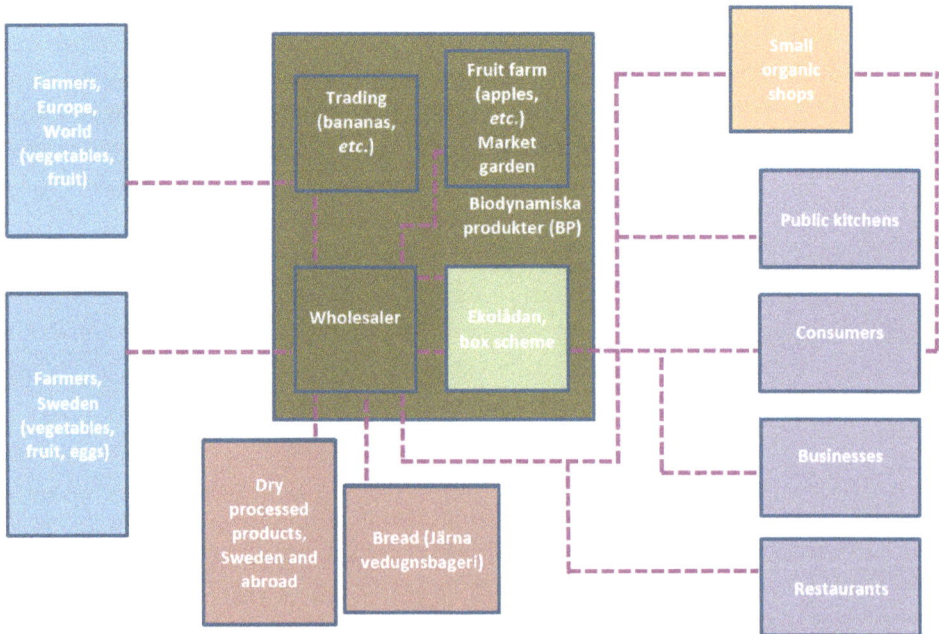

Figure 2. BP's stakeholder network. Modified from Milestad and von Oelreich [46].

3.2. Methods

Having identified the central respondent for this case study, we used a snowball sampling technique [47], whereby the initial respondent was asked to identify other relevant respondents, including people he thought would present a perspective different from his own. The key respondent (R1) has multiple roles within BP and Ekolådan. He is the chairperson of BP and responsible for project development, as well as for BP's banana imports. We conducted three interviews with him. In addition, we chose to interview five more respondents: the current newsletter writer of Ekolådan, who was also a former quality manager and was responsible for the box contents (R2); Ekolådan's senior purchaser (R3); one biodynamic farmer with long-term ties to Ekolådan (R4); the person responsible for customer service (R5); and BP's founder and former CEO (R6) (see Table 1).

Table 1. The respondents and number of interviews.

Respondent	Position at Ekolådan/BP	Number of interviews *
R1	Chair of BP's board	3
R2	Newsletter writer, former quality manager	1
R3	Senior purchaser	1
R4	Farmer delivering to BP and Ekolådan	2
R5	Responsible for customer service	1
R6	Founder and former CEO of BP	1

* Not counting follow-up interviews conducted by telephone or e-mail.

Semi-structured interviews were conducted [48]. Of nine interviews, six took place in person and three by phone. All but two interviews were carried out in 2014; the last two took place in May and June 2015. An interview guide was used, but room was allowed for additional questions and explanations. When needed, follow-up questions were asked via e-mail or telephone. The interviews focused on resilience and adaptation, trust and economic sustainability. Some questions were of a general character (e.g., when the company/initiative was started and by whom; how suitable partners were identified initially). Other questions were more specific, including how the food chain is organized (e.g., the importance of trust and how trust is built in the business); on the balance between quality differentiation, volume and economic performance (e.g., the qualities the company uses to differentiate their products from other products); on business logics (e.g., what actors are seen as strategic partners); communication (e.g., on the views of biodynamic and organic, growth, quality, *etc.*); handling change (e.g., what the main critical phases had been; what have been the main barriers to growth). The interviews were recorded and all interviews, except two, were fully transcribed. The last two were partly transcribed. The transcripts and notes were analyzed thematically. All respondents, except for Respondent 6, currently have a professional relationship with the business.

4. Social and Sustainability Entrepreneurship in BP and Ekolådan

4.1. BP as a Social Entrepreneurship Organization

Following Bacq and Janssen [19], a social entrepreneurial venture (SEV) must meet three criteria (Table 2). Criteria 1 and 2 are reflected in BP's charter (Table 2, Box 1). BP has consistently built its own independent and complete organic food chain, from farmer to consumer. BP's approach has privileged independent control over growth. One respondent (R1) described this as *"klein aber mein"* (German expression meaning "small but mine"). Over the years, BP has experienced different legal frameworks without shifting its focus. After a few years as a private company, BP became a foundation in 1974 (R6). This was not a purpose of its own, but merely a way of raising capital to develop the business without risking a violation of BP's core purpose, as stated in the charter, that is to support biodynamic activities (see Box 1). BP supported organic/biodynamic food production when it was a private company and continues to do so today as a foundation (R6); hence, the third criterion is also fulfilled (Table 2). R1 described it as "exciting to keep it as a foundation—[there's] no private owner, [and] we have no real requirements regarding profits, except for our own, so we must make a profit to be able to develop ... What's not re-invested is given away ... You don't have a profit-hungry owner chasing you. You can dare to engage in projects that business-wise are high risk, to say the least, if you find them culturally valuable". Another respondent expressed similar thoughts: "a fundamental value" of Ekolådan that distinguishes it from other actors is that "we're not here to make money" (R2).

Box 1. The BP charter.

BP's charter states that its aim is to "trade in and contribute to processing of produce that, to the extent possible, has been produced according to the biodynamic methods developed by Rudolf Steiner. The Foundation also aims in its operations to fulfil an alternative within business ... Any profits from the foundation's operations, after allocations for consolidation and development of the operations have been made, will be handed over as a gift to anthroposophical activities, prioritizing research, education, and information in biodynamic operations. Operations shall be run in continuous co-operation with the consumers and producers of the company's products, and its operations shall mirror the consumers' demand for biodynamic foods and the producers' ability to grow and sell these". [49] (Authors' translation.)

Table 2. Criteria for classification as a social entrepreneurial venture. SEV, social entrepreneurial venture.

Criteria Defining a Social Entrepreneurship Venture *	Biodynamic Products
1. Its social mission must be explicit and central.	"Any profits from the foundation's operations ... will be handed over as a gift ... " **
2. Its market orientation must be consistent with its social mission.	"Operations shall be run in continuous co-operation with the consumers and producers of the company's products, and its operations shall mirror the consumers' demand for biodynamic foods and the producers' ability to grow and sell these". **
3. The legal framework does not define SEVs; they can be found in the private for-profit sector and in the public sector.	Both structures, private company and foundation, have been shown to suit BP's purposes.

* Bacq and Janssen [19]; ** According to BP's charter [49].

The social aspects of entrepreneurship were also emphasized by the producer delivering to BP. On the importance of social responsibility, respondent R4, a biodynamic farmer, said: "at our farm, production is environmentally sound, but we also create job opportunities and we make a large social contribution in hosting some 50 pupils at different periods over the year". This practice could be expanded so that each farm hosts one or two persons with special needs and provides housing for elderly individuals, R4 reasoned. "What I want to see is that a well-functioning biodynamic farm not only is a farm producing products but also is a place where social responsibility is exercised" (R4). Thus, the social imperative was present both at the farm level and at BP itself.

The business studied was characterized by a high degree of trust and social capital. The importance of trust was mentioned by almost all of the respondents: "To stand for what we say we do is of utmost importance" (R5); "I have never been at a company where so much is built on trust as here" (R2). R1 remarked, "I think that we live out of trust ... people think we are honest ... it is our only capital. And being such a small company on such a tough market, I think we would have been gone long ago without it". He continued, "Sometimes we're a bit slow and not always very professional, but everyone knows—we are always there".

One example of this trust, or perhaps lack of professionalism, is the attitude towards written contracts. R3 usually told producers that "we can have [formal] contracts if you want," but "I've never used contracts as long as I have been here". The view that contracts are overrated was shared by respondent R4, who thought it was better to "try to be as accurate as possible" than to promise and not be able to live up to something. The only time contracts were used was when payments were made in advance (R1).

4.2. Sustainability Entrepreneurship

Although contested [50], organic food is generally seen as environmental friendly, and support for organic farming has found a place in environmental policy both in Sweden and in the EU. BP has a history as an early adopter of new trends, of introducing new organic products and of catalyzing the organic market as such in Sweden. According to R3, "BP has been around for 40 years, and we have increased and decreased in size over the years. We have adapted and learnt continuously ... For example, we were first in introducing [organic] coffee; now everyone has coffee, and our brand is very small. We were first with Ekolådan [as an organic box scheme in Stockholm]. We built Änglamark [the organic brand of Coop, a large retailer], before they started to run it themselves. We packed and distributed plenty of products to different retailers, which we don't any more. We imported pears and apples and other fruit from Latin America, until all the others started doing it; now we have quit. We adapt all the time". Another respondent described BP as leading its competitors: "This company has always been ahead ... we were first in organic meat ... first in serving retailers ... first with organic bananas ... but then the large [actors] enter the scene" (R1).

With its focus on both environmental and social sustainability, BP fits well with how Schaltegger and Wagner [25], (p.224) describe sustainability entrepreneurship: "attempts not only to contribute to sustainable development of the organization itself, but also to create an increasingly large contribution of the organization to sustainable development of the market and society as a whole, [requiring] substantial sustainability innovations". Other criteria for sustainability entrepreneurship, given by a literature review, are also fulfilled by BP (Table 3); the last criteria being resilience, which is analyzed below.

Table 3. Summary of the findings concerning the constituents of sustainability entrepreneurship in the case of Biodynamic Products (BP).

Sustainability Entrepreneurship Attribute	Reference	Findings in BP
Ethical business case related to social and environmental sustainability.	Bacq and Janssen [19]; Schaltegger and Wagner [25].	BP is clearly value-driven, based on its charter.
High level of trust within staff and towards clients.	Avery and Bergsteiner [11]; Schaltegger and Wagner [25].	Trust-building is a key characteristics of the whole business model.
Promoting a cause beyond the success of the business.	SEEDS [9]; Parrish and Tilley [24].	BP did not scale up itself, but helped forming a niche, which inspired other entrepreneurs to expand the organic market.
Long-termism.	Gibb [13]; Schaltegger and Wagner [25].	The charter of BP emphasizes long-term strategies over short-term success.
High adaptive capacity, including renewal after crises (resilience) and innovation.	Schaltegger and Wagner [25].	Both BP and Ekolådan have survived crises by innovation.

5. The Resilience of BP and Ekolådan

We now describe important phases in the history of BP and Ekolådan through the resilience lens, using the adaptive cycle. Doing this highlights the organization's dynamics, strengths and weaknesses and allows us to discuss them. We describe BP's history in three phases (1966–early 1980s, late 1980s–2002, 2003–2015), of which the last focuses on Ekolådan.

5.1. Biodynamic Products, 1966–Early 1980s (Cycle I)

Phase r to K (growth), Cycle I: BP was started by R6 in 1966 as a private company that bought products from producers and delivered to a large number of stores that specialized in healthy food (R6). The producers were not certified by any organization, neither KRAV, nor Demeter had been established in Sweden at the time. Instead, BP, or rather, R6 himself, guaranteed that the products were cultivated without any use of "poison" (indeed, retailers advertised the produce as "grown without poison"). Demand was greater than supply, and BP advertised for new producers interested in converting to "non-poisonous production". However, BP could not assist farmers financially to convert, and this impeded the growth of the market. As a way of generating the capital needed for the expansion of BP and to guarantee that the foundational idea of BP of supporting organic agriculture would not be violated over time, the private company was converted into a foundation in 1974. R6 formulated the foundation's charter himself (see Box 1) and, together with a few others, formed an interim board of directors (R6). This board contacted potential donors and asked for a contribution to facilitate the expansion and the survival of the business. The fundraising was successful, and donors included other businesses with a mutual interest in the growing sector: farmers and retailers, as well as individuals engaged in what later became organic and biodynamic farming. In 1974, Stiftelsen Biodynamiska Produkter (Biodynamic Products Foundation; BP) was established with R6 as its first CEO.

Phase K to Ω (release), Cycle I. Profitability remained poor after the transformation to a foundation and in 1975/1976, R6 reluctantly left the position as CEO. The board wished to see more professional management, and an external CEO was recruited. In retrospect, R6 said the decision had been the right one.

Phase Ω to α (re-organization), cycle I: After the transformation to a foundation and the change of CEO, both investments and profitability increased, and this increased margin was needed to support the external activities mentioned in the charter (see Box 1). Until 1978, R6 was employed by BP as responsible for sales and producer contacts. He was a member of the board until 1980, but since then, he has had no formal contact with the foundation (R6) (Figure 3).

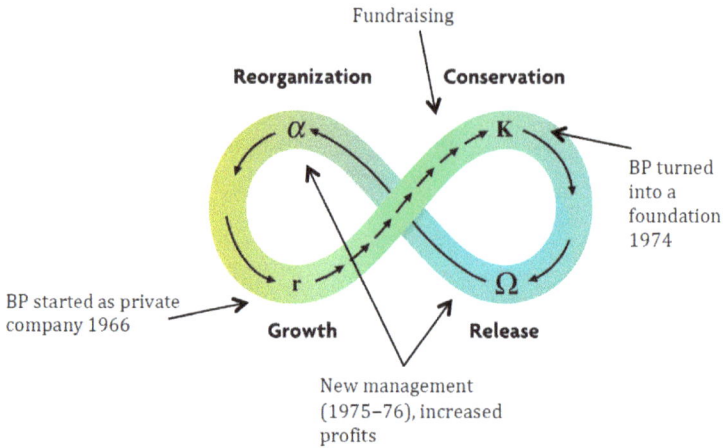

Figure 3. The development of BP (1966–early 1980s) mapped onto Holling's adaptive cycle.

5.2. Biodynamic Products, Late 1980s–2002 (Cycle II)

Phase r to K, Cycle II: In the late 1980s, Coop, one of the largest retailers in Sweden, asked BP for help establishing its own organic brand. The retailer lacked a large network of organic producers, something that BP had. For BP, this was an opportunity to distribute products to a larger group of customers and to enlarge the organic market, albeit under the retailer's brand. It was a difficult decision for BP, because it would mean that Coop, with an increasing range of organic products, would start competing with an important group of BP customers, namely health stores. If BP turned down the retailer's offer, the retailer would develop its own supply chain and perhaps compete with BP and decrease its market shares, according to R1. Retailers were set to enter the organic food sector more substantially. BP opted to engage in the collaboration; Coop became the first major retailer in Sweden to realize the potential of organic food. In the early 1990s, other retailers followed, and BP was contracted by SABA, a large wholesaler, to deliver organic produce throughout Sweden. Thanks to this arrangement, BP could sell large quantities of fruit and vegetables from a range of European producers.

Phase K to Ω, Cycle II: This expansion phase was followed by some backlashes. In the mid-1990s, SABA wanted to introduce its own list of organic products with their own supply chains. Retailers who wanted to buy organic produce straight from BP were told to buy from SABA instead. Otherwise, they would not have access to SABA's main list of products, SABA threatened. As a result, BP lost customers and had to close its packaging unit and lay off a number of employees (R1). At about the same time, in the second half of the 1990s, Coop ceased selling BP's coffee brand without notice. This was yet another indicator of the risks of relying too heavily on retailers. Meanwhile, a reaction from the health stores facing increased competition from retailers was to stop selling food that required refrigerated display.

Phase Ω to α, Cycle II: Early in the 21st century, Saltå Kvarn, a mid-sized organic mill, importer and distributor of flour, muesli, pasta, dried fruit, and so on, approached BP. Owing to the geographical proximity and joint history in the biodynamic movement, Saltå Kvarn initiated negotiations with BP around merging the two organizations into one private company. BP hesitated for several reasons. One related to the aim in BPs charter of financially supporting organic/biodynamic production. Would this continue in the future if new owners joined the company? Another concern for BP was that Saltå Kvarn aimed to increase sales of its own products to retailers. BP, having had several bad experiences collaborating with retailers, was more reluctant about this development. BP eventually rejected the merger (Figure 4).

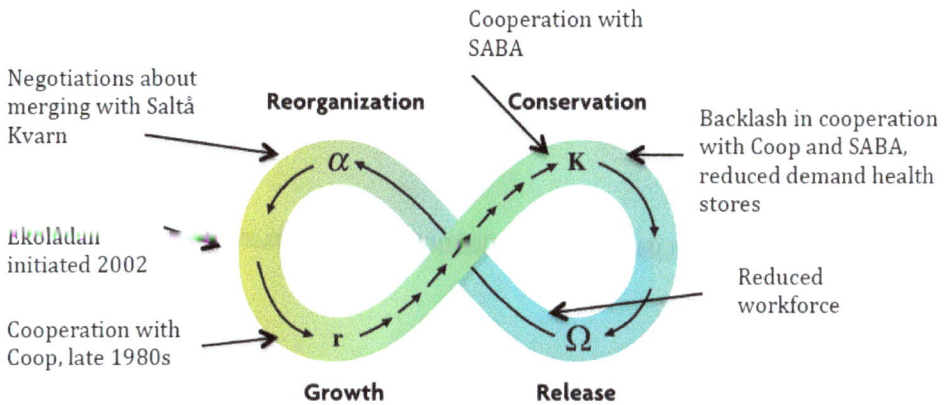

Figure 4. BP's development during the period of late 1980s–2002 through the resilience lens/Holling's adaptive cycle, starting with cooperation with Coop.

In 2002, volumes were lower than they had been in the early 1990s, when BP had its own packaging operation and before many health stores had stopped selling fruit and vegetables that need refrigeration. The supply offered by major retailers was poor, and many organic/biodynamic farmers were on the verge of closing their businesses. The idea emerged to organize a food chain without retailers. This was an opportunity for BP, on the one hand, to offer farmers stable demand at stable prices and, on the other hand, to offer consumers organic/biodynamic products. The innovation that emerged was the box scheme Ekolådan. After the introduction of Ekolådan, BP's development has fluctuated along with Ekolådan's ups and downs, since it has become the dominant part of BP. Hence, the third cycle focuses on Ekolådan.

5.3. Ekolådan, 2003–2015 (Cycle III)

The first Ekolådan boxes were delivered in autumn 2003. R1, chairperson of BP, recruited an external project leader who brought the idea from the U.K. and who contacted and recruited R2 (R1; R2).

Phase r to K, Cycle III: After a modest start with 13 boxes delivered in October 2003, the word spread, and the number of boxes delivered rapidly increased (R2). The growth phase can partly be explained by good timing. Ekolådan was the first of its kind in the Stockholm region (R2). Early on, Ekolådan received frequent media coverage (R2), which catalyzed interest. Hardly any resources at all were spent on advertising. Instead, the news was spread by word of mouth and through presentations about Ekolådan at events (R1). The increased demand called for a larger organization. The number of employees rose from 10 to more than 30 (see Figure 5) in Ekolådan alone. The number of boxes delivered peaked in 2008–2009, at some 4500 per week (R1) (see Figure 5).

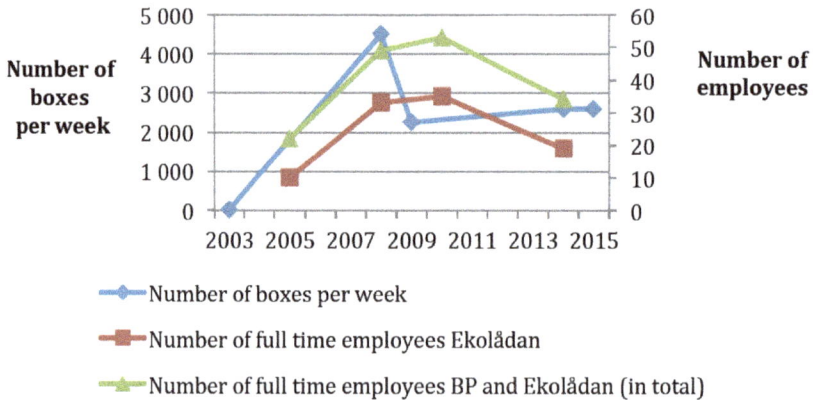

Figure 5. The number of boxes delivered weekly and the number of BP and Ekolådan employees. Sources: R2 (2003 and 2009), R1 (2008 and 2014), R5 (2015).

Phase K to Ω, Cycle III: In 2008–2009, three unrelated events occurred that affected both Ekolådan and BP in a most concrete way. First, on a macro level, the global financial crisis hit the market with full force. Customers unsubscribed from Ekolådan, citing their reduced income (R2). Second, on a local level, competitors now were on par with Ekolådan. BP had perhaps been the first to market an organic box scheme in the Stockholm region, but other actors soon copied the box concept or developed similar products. Retailers introduced web stores with home delivery. Others introduced home delivery of ready-made food bags with recipes and matching ingredients. There were niche deliveries for large or small households, for low carb/high fat diets, for vegetarians, and so on. Several actors claimed

208

to have "as much organic food as possible," thereby competing for Ekolådan's customers (R2). Finally, large retailers also improved their supply of organic fruits and vegetables. When Ekolådan was launched, these retailers had very few organic fruits and vegetables, and what they had was expensive and of poor quality: "The organic [food you found] was hidden in a corner at Konsum [a Coop supermarket]. It was old and wrinkled and labelled KRAV [organic], and it cost twice as much as these other shining vegetables and fruits" (R2). However, a few years later, the major retailers could compete with specialty retailers, offering large quantities of attractively-displayed organic fruit and vegetables at competitive prices. Thus, one of Ekolådan's unique selling points had been lost. The increased competition was a more severe blow for Ekolådan than the global financial crisis was (R1).

The drop in demand was severe and sharp. The number of boxes delivered fell by 50% in 2009, plummeting from 4500 per week to 2250 (see Figure 5). In the period 2009–2012, Ekolådan ran a deficit, and the BP board discussed the possibility of closing the division altogether (R2).

Phase Ω to α, Cycle III: The reduced number of boxes called for a smaller work force. The number of Ekolådan and BP employees combined had increased from 22 in 2005 to 49 in 2008, peaking at 53 in 2010 before decreasing to 34 in 2014 (see Figure 5). During the decline and consolidation period, the chair of the board took over responsibility for staff issues and was in fact the one who laid off personnel (R1). According to R1, "We experienced the growth as rather organic. We probably fooled ourselves and hired too many people too fast. We probably should have been more cautious".

Phase α to r, Cycle III: The decline in demand was eventually halted. The period 2012–2014 was characterized by consolidation. The number of employees now correlated with incoming orders, and an increase in demand was seen for the first time since the crisis. By 2014, some 2600 boxes were being delivered per week (R1), and that number stabilized in 2014–2015 (R5; see Figure 5). The increase from the post-crisis low of 2250 boxes per week to today's 2600 corresponds to 16% growth in absolute terms. However, in relation to the market and to competitors, it was losing market shares. Thus, Ekolådan might still be in a challenging period characterized by reorganization and tough competition. We return to this question in the Discussion Section (Figure 6).

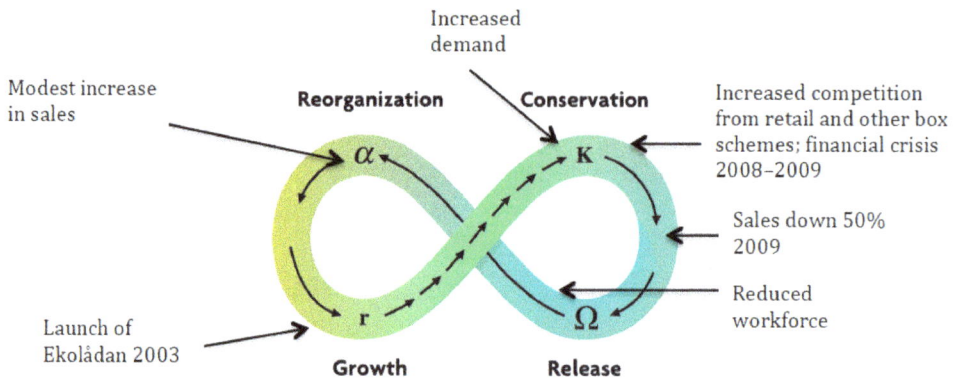

Figure 6. Ekolådan's development from 2003 to 2015 through the lens of Holling's adaptive cycle.

5.4. On a Conservative Business Profile

Over the years, BP has experienced phases of continuous growth followed by periods of crisis. Lately, Ekolådan has increased its sales, but it has not kept pace with the market for organic food. R6 reflected on this and emphasized that BP has been managed in an economically-responsible fashion, noting his own respect for the financial management. He did wish, however, that BP had been more growth oriented and bolder in its choices. R1 was ambiguous when it came to the rather conservative business profile of BP. On the one hand, he acknowledged that a more aggressive attitude could have benefitted BP, but this would have required more capital for marketing and product development. One of BP's weaknesses, according to R1, is that the foundation is specialized: it carries only organic products and has a limited assortment. Customers who desire a larger range of products might turn to another wholesaler. Overall, R1 believed it had been the right decision to remain a foundation, but BP's lack of capital was a disadvantage and meant that BP "can be seen as slow" by others.

In the early days of Ekolådan, the managers found the slow decision making frustrating. Nevertheless, "in retrospect, one can see that perhaps it was good to stay calm" (R2). Slow decision making might have benefitted Ekolådan during the crisis inasmuch as at the moment when Ekolådan was "almost at a closing down scenario," it kept going and focused on improving its operations and becoming more efficient (R1). One respondent (R5) saw a danger in adapting Ekolådan to a changing environment. "I believe it is very important to protect our trademark and not try to follow the latest trend ... this is what we stand for, and this is what we deliver, and we will continue to do so". Thus, the organizational form as a foundation implies built-in resistance. A foundation reacts more slowly than a private company does. "A generational shift is about to take place in the house ... and with that I believe there will be changes ... What is good is the foundation. You can't do just anything; the fundamental idea must be kept" (R3).

Another issue that emerged in the interviews was the problem BP had communicating its values to consumers, despite the fact that each Ekolådan box was accompanied by a newsletter with an editorial. Initially, when Ekolådan was introduced, the founders had chosen not to mention Järna or Demeter and to distance themselves from anthroposophy. When competition increased, and there arose a need to distinguish Ekolådan from its competitors, its identity as a foundation in Järna became something worth communicating. What had previously been seen as "ugly German labels" and "anthroposophically old" was now seen as "genuine" (R2). For example, "Today I believe that Järna still stands for something trustworthy, something that has a quality" (R2). However, the values BP stands for are not communicated by the organic labels. According to R1, the importance of organic labelling is "close to zero for Ekolådan". Since KRAV is used by mainstream producers, "labelling in general has lost its meaning for us niche traders". One respondent asked, "Why should we have the same labelling as Chiquita?" (R1). In other words, being a small cutting-edge company, why should Ekolådan use the same label as a mainstream multinational company only certifying a small part of its products? Thus, the values must be communicated in another way, and BP has not yet found such a channel. For example, no resources at all were being spent on advertising during our study period.

6. Discussion and Conclusions

6.1. Social and Sustainability Entrepreneurship

Our results suggest that this case meets the criteria for both a social and an entrepreneurial venture [19] (Table 2) and the broader framework of sustainability entrepreneurship (Table 3). BP has a strong commitment for creating social value, and the business has a clear environmental profile, to support environmentally-friendly food production and consumption. BP could be an environmental entrepreneur [20,21], but sustainability entrepreneurship fits better, since this term integrates all aspects of sustainability, including innovation [25].

By supporting other businesses in its network, BP supports not only organic development, but also, and not least, economic development on the local and regional levels. It can therefore be argued that BP is well suited to contribute to the sustainable development of the region in general and of the food system in particular. However, BP has not managed to profit from the organic food boom in Sweden, which they had contributed to by several innovations. Even if making profits has never been the main objective, our interviews reveal some disappointment concerning the economic performance. The reasons for this seem to be related to both BPs social goal of using

211

surplus to support organic farming rather than investing in the business and the strong focus on resilience.

6.2. The Resilience of BP

BP exhibits several of the characteristics that may determine organizational resilience: renewal and innovation; flexible organizational culture; high level of trust; strong local identity; and value-based authentic trademark (Table 4). While resilience generally supports, and provides a more evolutionary understanding of, sustainability (e.g., handling crises by innovations and adaptations), this case study also suggests some trade-offs between resilience and sustainability. We identified three possible trade-offs or complex relations: trust, value-driven local identity and slowness related to the organizational form. First, trust and social capital build resilience. For BP, common norms, specifically, a shared understanding of the importance of environmentally-friendly food production and the practice of not relying on written contracts, point to the importance of trust and reciprocity. However, in some cases, the high level of trust has been used to justify a lack of business orientation, e.g., it took several years to reduce the level of staff after the drastic decline in sales of Ekolådan, which threatened the sustainability of the whole box scheme (Figure 5).

Second, concerning complex relations, the value-driven local identity enhances the trademark and consumer loyalty, but also conservatism; this neither promotes resilience (innovation and adaptation) nor sustainability (in terms of business orientation and a more "aggressive attitude", as noted by R1). Third, the organizational form of a foundation entails more internal control than that of a listed company. Focusing on controlling the local value chain has increased the robustness and adaptive capacity of the overall community. This is supported by Johannisson [51]. Nevertheless, BP is subject to tough competition, which sometimes compels timely adaptations. However, decision making in a foundation can be slow, and this could reduce the company's resilience ([4], p. 54) since adaptation opportunities are not realized. On the other hand, one respondent argued that slow decisions probably saved Ekolådan from going out of business.

Olsson et al. [29] argue that social transformation is needed when moving from a less desired trajectory to one where the capacity to manage for sustainability is strengthened. Regarding BP, transforming a private company into a non-profit foundation could be viewed as an attempt to increase the organization's resilience by shifting to a more stable organizational form. In a study of the transformation of the dairy industry in Australia, Sinclair et al. ([7], p. 371) argue that "in some situations adaptation is an inadequate response to changing conditions and a transformation is required". They view adaptation and transformation as different degrees of change along a continuum where transformation represents the highest level. Deliberate transformations are often carried out with the intention of obtaining a specific goal [52]. In this case, one could argue that transforming the private company into a foundation was a deliberate transformation, at least in organizational terms, intended to reach the goal of supporting biodynamic agriculture and increasing the organizational resilience of BP.

Table 4. Summary of findings concerning the constituents of organizational resilience and how these criteria are met by Biodynamic Products (BP).

Organizational Resilience	Reference	Findings in BP
Ability to renew itself and self-organize after crisis	Hamel and Välikangas [4]; Folke [8]	BP has shown capacity for self-organization and renewal after crises in the mid-1970s and late 1990s.
Adaptation is sometimes not enough, and transformation is required	Sinclair *et al.* [7]	BP has, over time, both adapted (in terms of products) and transformed (in organizational form) depending on conditions.
The importance of control and independence	Milestad [6]	BP has privileged independence and control over growth.
Learning and innovation, especially in the α phase	Linnenluecke *et al.* [35]; Lengnick-Hall *et al.* [36]	BP has innovated continuously, introducing organic coffee, fruits and meat in Sweden, as well as the box scheme Ekolådan.
Organizational culture and working environment that stimulates flexibility and innovation	Lengnick-Hall *et al.* [36]	BP fosters trust, informality and flexibility with employees, promoting experimentation.
Value-based contextual integrity and wisdom generates authenticity, trust and loyalty	Kantur and İşeri-Say [37]; Cameron *et al.* [38]; Aldrich and Meyer [31]; Bernier and Meinzen-Dick [32]	BP enjoys a strong local trademark and integrity as idealistic biodynamic enthusiasts. This provides authenticity, trust and loyalty among customers: "They know what we stand for".

In a socio-economic system, accumulated capital (e.g., skills, productivity, networks, mutual trust) is developed and integrated during the progression from r to K in an adaptive cycle (e.g., Figure 1) [40]. Farmers delivering products to Ekolådan considered themselves fairly economically diversified, relatively stable and well prepared to tackle change and turbulence [53]. In terms of Figure 1, they have moved from r to K. If or when change arises, the supplying farmers are thus arguably prepared to handle the release phase in Ω. Of interest for the development of Ekolådan and BP is, of course, the larger development of organic food in Sweden. In the period 2008–2014, the total increase in sales amounted to more than 150%, and demand increased steadily even during the financial crisis [54]. During the last few years, demand has increased at an accelerating rate. In 2013, the increase was 13% [55]; in 2014, it was 38% [56]; and in 2015, the increase in organic sales was 39% [57]. This can be contrasted with the development of Ekolådan: sales have increased by 16% from 2009 to today. This result is somewhat contradictory. On the one hand, it is a sign of an increased interest in BP's and Ekolådan's products (organic food), and BP has clearly played an important role in the market's development in Sweden. On the other hand, BP has not managed to capitalize on this increased demand to which it has contributed. Our results suggest that this is due to a combination of competition from conventional retail, the cautious attitude internally and the difficulty BP has in reaching consumers who now face multiple organic options. The growth in the number of delivered boxes suggests that Ekolådan has consolidated and recovered from the 2009 crisis. However, Ekolådan underperforms *vis-à-vis* the market and is vulnerable to new competition since the concept is easy to copy.

6.3. Conclusions

Sustainability entrepreneurship means, "promoting a cause beyond the success of the business" (Table 3), and for BP, controlling the company and the food chain has been prioritized over increasing volumes. BP has contributed to local and regional sustainable development for almost 50 years. Over the years, the market has grown more rapidly than BP itself, and their sales have not increased at the same rate as the market has developed. The organization catalyzed the expansion of organic fruit and vegetable sales and led the way by introducing a range of other organic products in Sweden. However, there are trade-offs between sustainability entrepreneurship and organizational resilience, which the BP case highlighted. Sustainability requires transformation after crisis, not merely adaptation, which can be illustrated with the adaptive cycle. While a case study cannot be generalized, we find the application of the adaptive cycle of business development and its implications for transformative innovation clarifying. Case studies have a role to play in knowledge building around adaptive cycles in business development and the business ecosystem that is required for this to be successful. This case study should assist in building a stronger dynamic business theoretical model that can address the lack of dynamic models in this area. The existing models are very profit-driven and lack sustainability as an integral aspect to their modelling.

Acknowledgments: We would like to thank the two anonymous reviewers for their valuable suggestions and comments. We are also most thankful to Studio Flygar for help with the illustrations. Part of the research for this article was carried out within the Healthy Growth project, funded by the FP7 European Research Area (ERA)-Net project Core Organic II.

Author Contributions: Markus Larsson conceived and designed the case study with Rebecka Milestad. Markus Larsson, Rebecka Milestad and Jacob von Oelreich performed the case study; Markus Larsson analyzed the data; Markus Larsson and Thomas Hahn wrote the paper, with some contributions from Rebecka Milestad and Jacob von Oelreich.

Conflicts of Interest: The authors declare no conflict of interest.

References

1. Walker, B.H.; Holling, C.S.; Carpenter, S.; Kinzig, A. Resilience, adaptability, and transformability in social-ecological systems. *Ecol. Soc.* **2004**, *9*, article 5. Available online: http://www.ecologyandsociety.org/vol9/iss2/art5/ (accessed on 20 May 2016).
2. Holling, C.S. Resilience and Stability of Ecological Systems. *Annu. Rev. Ecol. Syst.* **1973**, *4*, 1–23.
3. Adger, N.; Hughes, T.; Folke, C.; Rockström, J. Social-Ecological Resilience to Costal Disasters. *Science* **2005**, *309*, 1036–1039.
4. Hamel, G.; Välikangas, L. The Quest for Resilience. *Harv. Bus. Rev.* **2003**, *September*, 52–63.
5. Doeksen, A.; Symes, D. Business Strategies for Resilience: The Case of Zeeland's Oyster Industry. *Sociol. Ruralis* **2015**, *55*, 325–342.
6. Milestad, R. Building Farm Resilience. Prospects and Challenges for Organic Farming. Ph.D. Thesis, The Swedish University of Agricultural Science, Uppsala, Sweden, 4 April 2003.
7. Sinclair, K., Curtis, A.; Mendham, E.; Mitchell, M. Can resilience thinking provide useful insights for those examining efforts to transform contemporary agriculture? *Agric. Hum. Values* **2014**, *31*, 371–384.
8. Folke, C. Resilience: The emergence of a perspective for social-ecological systems analyses. *Glob. Environ. Chang.* **2006**, *16*, 253–267.
9. SEED. *Turning Ideas into Impact: Setting the Stage for the Next 10 Years of Green and Inclusive Growth through Entrepreneurship*; SEED: Berlin, Germany, 2015.
10. Hall, J.K.; Danke, G.A.; Lenox, M.J. Sustainable development and entrepreneurship: Past contributions and future directions. *J. Bus. Ventur.* **2010**, *25*, 439–448.
11. Avery, G.C.; Bergsteiner, H. Sustainable leadership practices for enhancing business resilience and performance. *Strateg. Leadersh.* **2011**, *39*, 5–15.
12. Larsson, M.; Andersson, A.; Enberg, S. Trust and Resilience—A Case Study of Environmental Entrepreneurs in Järna. Available online: http://www.vaxteko.nu/html/sll/slu/ekologiskt_lantbruk/EKL47/EKL47AMETA.HTM (accessed on 8 June 2016).
13. Gibb, A. *Effective Policies for Small Business. A Guide for the Policy Review Process and Strategic Plans for Micro, Small and Medium Enterprise Development*; OECD/United Nations Industrial Development Organization (UNIDO): Paris, France, 2004.
14. Lundström, A.; Stevenson, L. *Entrepreneurship Policy: Theory and Practice*; Springer: Boston, MA, USA, 2005.
15. Parrish, B.D. Sustainability-driven entrepreneurship: Principles of organization design. *J. Bus. Ventur.* **2010**, *25*, 510–523.

16. Johnstone, H.; Lionais, D. Depleted communities and community business entrepreneurship: Revaluing space through place. *Entrep. Region Dev.* **2004**, *16*, 217–233.

17. Abu-Saifan, S. Social Entrepreneurship: Definition and Boundaries. *Tech. Innov. Manag. Rev.* **2012**, *2*, 22–27. Available online: http://timreview.ca/article/523 (accessed on 9 June 2016).

18. Thake, S.; Zadek, S. *Practical People, Noble Cause: How to Support Community-Based Social Entrepreneurs*; New Economics Foundation: London, UK, 1997.

19. Bacq, S.; Janssen, F. The multiple faces of social entrepreneurship: A review of definitional issues based on geographical and thematic criteria. *Entrep. Region. Dev.* **2011**, *23*, 373–403.

20. Schaper, M. The essence of ecopreneurship. *Greener Manag. Int.* **2002**, *2002*, 26–30.

21. Shaper, M. *Making Ecopreneurs: Developing Sustainable Entrepreneurship*; Gower Publishing: Farnham, UK, 2010.

22. Walley, L.; Taylor, D.; Greig, K. Beyond the visionary champion: Testing a typology of green entrepreneurs. In *Making Ecopreneurs: Developing Sustainable Entrepreneurship*; Schaper, M., Ed.; Gower Publishing: Farnham, UK, 2010.

23. Steyaert, C.; Katz, J. Reclaiming the space of entrepreneurship in society: Geographical, discursive and social dimensions. *Entrep. Region Dev.* **2004**, *16*, 179–196.

24. Parrish, B.D.; Tilley, F. Sustainability entrepreneurship: Charting a field in emergence. In *Making Ecopreneurs: Developing Sustainable Entrepreneurship*; Shaper, M., Ed.; Gower Publishing: Farnham, UK, 2010.

25. Schaltegger, S.; Wagner, M. Sustainable entrepreneurship and sustainability innovation: Categories and interactions. *Bus. Strategy Environ.* **2011**, *20*, 222–237.

26. Pretty, J.; Ward, H. Social Capital and the Environment. *World Dev.* **2001**, *29*, 209–227.

27. Fukuyama, F. Social capital, civil society and development. *Third World Q.* **2001**, *22*, 7–20.

28. Pretty, J. Social Capital and the Collective Management of Resources. *Science* **2003**, *302*, 1912–1914.

29. Olsson, P.; Folke, C.; Hahn, T. Social-ecological transformation for ecosystem management: The development of adaptive co-management of a wetland landscape in southern Sweden. *Ecol. Soc.* **2004**, *9*, article 2. Available online: http://www.ecologyandsociety.org/vol9/iss4/art2/ (accessed on 20 May 2016).

30. Dietz, T.; Ostrom, E.; Stern, P.C. The Struggle to Govern the Commons. *Science* **2003**, *302*, 1907–1912.

31. Aldrich, D.P.; Meyer, M.A. Social Capital and Community Resilience. *Am. Behav. Sci.* **2015**, *59*, 254–269.

32. Bernier, Q.; Meinzen-Dick, R. *Networks for Resilience-The Role of Social Capital*; International Food Policy Research Institute: Washington, DC, USA, 2014.

33. Hahn, T.; Nykvist, B. Are adaptations self-organized, autonomous and harmonious? Assessing the social-ecological resilience literature. *Ecol. Soc.* **2016**, submitted.

34. Swedish Environmental Protection Agency. Sweden's Environmental Objectives. Available online: http://www.miljomal.se/sv/Environmental-Objectives-Portal/ (accessed on 10 June 2016).

35. Linnenluecke, M.K.; Griffiths, A.; Winn, M. Extreme weather events and the critical importance of anticipatory adaptation and organizational resilience in responding to impacts. *Bus. Strategy Environ.* **2012**, *21*, 17–32.

36. Lengnick-Hall, C.A.; Beck, T.E.; Lengnick-Hall, M.L. Developing a capacity for organizational resilience through strategic human resource management. *Hum. Resour. Manag. R.* **2011**, *21*, 243–255.

37. Kantur, D.; İşeri-Say, A. Organizational resilience: A conceptual integrative framework. *J. Manag. Organ.* **2012**, *18*, 762–773.

38. Cameron, K.S.; Dutton, J.E.; Quinn, R.E.; Wrzesniewski, A. Developing a discipline of positive organizational scholarship. In *Positive Organizational Scholarship. Foundations of a new Discipline*; Cameron, K.S., Dutton, J.E., Quinn, R.E., Eds.; Berrett-Koehler Publishers: San Francisco, CA, USA, 2003; pp. 361–370.

39. Holling, C.S. Resilience of ecosystems; local surprise and global change. In *Sustainable Development of the Biosphere*; Clark, W.C., Munn, R.E., Eds.; Cambridge University Press: Cambridge, UK, 1986; pp. 292–317.

40. Holling, C.S. Understanding the Complexity of Economic, Ecological, and Social Systems. *Ecosystems* **2001**, *4*, 390–405.

41. Schumpeter, J. *Capitalism, Socialism and Democracy*; Harper and Row: New York, NY, USA, 1950.

42. Folke, C.; Carpenter, S.; Walker, B.; Scheffer, M.; Chapin, T.; Rockström, J. Resilience Thinking: Integrating Resilience, Adaptability and Transformability. *Ecol. Soc.* **2010**, *15*, article 20. Available online: http://www.ecologyandsociety.org/vol15/iss4/art20/ (accessed on 9 June 2016).

43. Ekolådan. Ekolådans rötter (The Roots of Ekolådan). Available online: www.ekoladan.se (accessed on 7 July 2014).

44. KRAV. Available online: www.krav.se (accessed on 10 June 2016).

45. Svenska Demeterförbundet (The Swedish Demeter Association). Available online: www.demeter.nu (accessed on 10 June 2016).

46. Milestad, R.; von Oelreich, J. *Full Case Study Report: Ekolådan-Sweden*; KTH Royal Institute of Technology: Stockholm, Sweden, 2015.

47. Biernacki, P.; Waldorf, D. Snowball sampling: problems and techniques of chain referral sampling. *Sociol. Method Res.* **1981**, *10*, 141–163.

48. Kvale, S.; Brinkmann, S. *Interviews: Learning the Craft of Qualitiative Research Interviewing*; SAGE Publications Ltd.: London, UK, 2009.

49. BP's Charter. Available online: http://web05.lansstyrelsen.se/stift/StiftWeb/FoundationDetails.aspx?id=1000180 (accessed on 10 June 2016).

50. Johansson, B. *Är eko reko? Om ekologisk lantbruk i Sverige (Is Organic Fair? Organic Agriculture in Sweden)*; The Swedish Research Council Formas: Stockholm, Sweden, 2003.

51. Johannisson, B. Entrepreneurship—The making of new realities. In *De Glesa Strukturerna i den Globala Ekonomin (The Thin Structures of the Global Economy)*; The Royal Academy of Agriculture and Forestry: Stockholm, Sweden, 2002.

52. O'Brien, K. Global environmental change II: From adaptation to deliberate transformation. *Prog. Hum. Geogr.* **2012**, *36*, 667–676.
53. Axelsson, B. A Conventions Theory Analysis of Farmers in the Ekolådan Distribution Network—Justifications and Conventions. Master's Thesis, Stockholm University, Stockholm, Sweden, 31 May 2012.
54. Ekoweb. Ekologisk livsmedelsmarknad 2015 (Organic Food Market 2015). Available online: http://www.ekoweb.nu/?p=11363 (accessed on 8 January 2016).
55. KRAV. Market Report 2014. Available online: http://www.krav.se/forsaljning-2013 (accessed on 8 January 2016).
56. KRAV. Market Report 2015. Available online: http://www.krav.se/marknadsrapport-2015/forsaljning (accessed on 8 January 2016).
57. Ekoweb. Ekologisk Livsmedelsmarknad 2016 (Organic Food Market 2016). Available online: http://www.e-pages.dk/maskinbladet/1180/ (accessed on 24 April 2016).

Section 3:
Beyond Organic: Shaping Future Farming and Food Systems

An Interpretive Framework for Assessing and Monitoring the Sustainability of School Gardens

Francesco Sottile, Daniela Fiorito, Nadia Tecco, Vincenzo Girgenti and Cristiana Peano

Abstract: School gardens are, increasingly, an integral part of projects aiming to promote nutritional education and environmental sustainability in many countries throughout the world. In the late 1950s, FAO (Food and Agriculture Organization) and UNICEF (United Nations Children's Fund) had already developed projects to improve the dietary intake and behavior through school and community gardens. However, notwithstanding decades of experience, real proof of how these programs contribute to improving sustainability has not been well-documented, and reported findings have mostly been anecdotal. Therefore, it is important to begin a process of collecting and monitoring data to quantify the results and possibly improve their efficiency. This study's primary goal is to propose an interpretive structure—the "Sustainable Agri-Food Evaluation Methodology-Garden" (SAEMETH-G), that is able to quantifiably guide the sustainability evaluation of various school garden organizational forms. As a case study, the methodology was applied to 15 school gardens located in three regions of Kenya, Africa. This application of SAEMETH-G as an assessment tool based on user-friendly indicators demonstrates that it is possible to carry out sustainability evaluations of school gardens through a participatory and interdisciplinary approach. Thus, the hypothesis that the original SAEMETH operative framework could be tested in gardens has also been confirmed. SAEMETH-G is a promising tool that has the potential to help us understand school gardens' sustainability better and to use that knowledge in their further development all over the world.

Reprinted from *Sustainability*. Cite as: Sottile, F.; Fiorito, D.; Tecco, N.; Girgenti, V.; Peano, C. An Interpretive Framework for Assessing and Monitoring the Sustainability of School Gardens. *Sustainability* **2016**, *8*, 801.

1. Introduction

School gardens are increasingly part of projects related to the promotion of environmental and nutritional education in many countries throughout the world. In school garden projects, fruits and vegetables are grown in areas around or near the school, sometimes providing a small-scale staple food source, as well as other complementary activities. However, this is not a new approach; already in the

1950s, FAO (Food and Agriculture Organization) and UNICEF (United Nations Children's Fund) had begun the "Applied Nutrition Projects" meant to improve nutrition through school and community gardens. Numerous other interventions by government and non-government organizations followed, aiming to spread the development of a "garden culture". In what are commonly considered the developed countries, a "garden-based learning" (GBL) approach has prevailed, where gardens are laboratories for learning science, environmental studies, as well as topics such as art and literature.

In the report, "Revisiting garden-based-learning in basic education", published by FAO and UNESCO in 2004 [1], the authors document how there was already a strong movement in the 1800s tied to scholastic gardens, both in Europe (especially in Austria), as well as in North America. At the beginning of the 20th century, the great American horticulturalist, Liberty Hyde Bailey wrote:

"... to open the child's mind to his natural existence, develop his sense of responsibility and of self dependence, train him to respect the resources of the earth, teach him the obligations of citizenship, interest him sympathetically in the occupations of men, touch his relation to human life in general, and touch his imagination with the spiritual forces of the world" [2]

emphasizing how experiential learning, ecological literacy, and environmental awareness, as well as technical agricultural subjects, could all be integrated within a garden. It is interesting to note how some of the key principles of sustainable development, such as inter- and intra-generational equity and the interrelation between multifaceted aspects, have already been mentioned in relation to a school curriculum almost 80 years before in the 1987 Brundtland report (known as "Our Common Future") [3].

In the southern hemisphere, the tableau is more variable: the origins of school gardens are less documented and quite often not institutionalized in official school curricula. In these cases, their design focused on the principal aim, which was not always achieved [3], of supplying food for school meals and improving the children's nutrition and health. Similarly to what happened previously in developed countries [4], youths who live in urban areas (but not only) have less and less experience with natural ecosystem complexity and are becoming strangers to the source of the food that they consume, with evident nutritional imbalances that cause important health problems, such as obesity [5–7].

By putting together these considerations, we can see how the perception of school gardens is still evolving and represents a response to the increasingly pressing needs for greater food security, environmental protection, more secure livelihoods, and better nutrition [8]. A school garden is both a sustainable action by itself, as well as a generator of other sustainable actions [9].

From the many analyses carried out on various projects [10–13], it has clearly emerged that for a school garden to be successful, some key "active ingredients" are always needed [14]. The school garden must be designed and carried out together with the local community and must correspond to the socio-cultural and environmental place, particularly for crop choices and garden management. Successful school garden projects do not just aim to involve the school's children, but also the school's directors, teachers, and parents, or rather school garden programs can and must have multiplying effects, encouraging the creation of private gardens in the case of school-age children, as described by Drescher [3]. Furthermore, regarding a successful program's objectives, gardens must build ties and synergies between learning, nutrition, health, agriculture, and sustainability [15].

One of the most interesting aspects of school gardens is their ease of realization; they can be developed both in rural and urban contests, with limited financial investments and manual labor needs. Furthermore, the potential use of domestic organic waste for compost provides the opportunity to institute an efficient use of limited resources and to close the nutrient cycle. This benefits the environment and forms a sustainable system [16–18]. Furthermore, another important contribution to sustainability comes from the large variety of crops, including those belonging to the local germoplasm [19], that can be found in school gardens and that create systems that are much more diversified in respect to the widespread agricultural models, even the small scaled ones [20]. Finally, to reduce environmental risks, the crops are almost always cultivated in conditions that reduce the necessity for external inputs to a minimum (for example the creation of compost, use of legume species, and crop rotation) and that maximize quality yields. All of this shows how school gardens can be a new gymnasium for sustainable education [21]. Thus, they should be proposed as more than an educational objective, but as the very method where the message, as well as the structure, practices, and the entire educational system are all congruent [22]. In fact, in the last few years we have rediscovered an interest in education that includes nature activities (excursions in parks, observations of the wild flora and fauna, etc.), the impact of every-day life (education on waste, recycling, separated trash collection, home water-use, energy saving), and even agriculture and animal husbandry. The underlying theme is education on the relationships between humans and ecosystems, which was already delineated by Stapp in 1969 [23]. This widespread and growing attention towards environmental themes comes from a need to feel that one is making a contribution, and is fundamental because it is directed towards new generations, and to solving conflicts between the current model of development dominant in the population and the limits imposed by the finiteness of Earth's ecosystem [21].

In line with what has been sustained until now, it is possible to synthesize the objectives of the current school gardens as (1) reaching a better understanding of biological processes, sustainable agricultural practices, and environmental sensibility; (2) providing better information regarding healthy food choices, favoring the assumption of a varied diet, and guaranteeing irrigation water and sanitary services; and (3) reducing the cost of food and providing a safety net for the poor, giving them the possibility to cultivate their own food. Notwithstanding more than 50 years of experience regarding healthy food with school garden programs, the evidence that these gardens contribute in an integrated way to sustainability, with nutritional, educational, and economic results is not well documented and is largely anecdotal. Although many quantitative and qualitative studies have shown positive outcomes in the areas of food behavior (especially for vegetable intake) [4,24], and academic performance (especially for disruptive students) [25], there is the need to learn from these programs in a more structured way and to collect data to improve their efficiency and quantify the results obtained in terms of sustainability. The lack of an integrated evaluation of school gardens undermines the multifaceted contribution that they produce for society.

This work's objective is to evaluate the environmental, social, and economic sustainability of school gardens by applying an interpretive structure called the "Sustainable Agri-Food Evaluation Methodology-Garden" (SAEMETH-G), derived from an analogous model built for small scale agro-food systems [26]. SAEMETH-G situates itself within the studies that aim to translate the general principles of sustainability into practical and operational tasks for small agricultural systems by directly involving the users [27,28].

2. Materials and Methods

2.1. Geographical Location and Selected School Gardens

The study was carried out in Kenya in the counties of Embu, Muranga, and Nakuru (Figure 1). School gardens are widespread in these three counties, fulfilling educational, as well as community, needs in a regional context where agriculture is one of the principal sources of livelihood for the population. School gardens play a fundamental role in maintaining an awareness of how agriculture works; outside of school gardens, agriculture is almost completely absent from the school curriculum and the majority of young people who complete their primary and secondary education did not receive any training for an agricultural career.

School gardens initiatives are carried out by several different participants: they can come directly from government institutions or through agricultural extension officers, NGOs, foreign donors, or directly as a teachers' initiative. In this area, the local section of the Ministry of Agriculture is very active: numerous school, family, and community gardens have been formed thanks to the support and training provided by the Ministry (4K-Club), which is also working to promote the principles of organic farming.

Many local NGOs, including PICE (Progressive Initiatives for Community Empowerment), and NECOFA (Network for Ecofarming in Africa), in collaboration with foreign NGOs and associations, are operating in the three counties with the primary objective of educating the local community and sustainably using the existing human and natural resources to improve economic and social wellbeing. In Nakuru County alone, there are 90 vegetable gardens promoted by the Slow Food Foundation for Biodiversity through the project "10,000 gardens in Africa" [29].

Figure 1. Map of Kenya and the counties (underlined in red) where the school gardens included as case studies are located.

For the selection of the 15 case studies, we used a qualitative targeted sampling procedure [30], individuating the most representative school gardens of the various counties. In total, 15 gardens were selected with sizes that varied between 90 and 5000 m^2. The 15 gardens show different forms of interaction and participation among students, teachers, and the local community. Table 1 shows an overview of the selected school gardens.

225

Table 1. Overview of the selected school gardens.

County	Initiative	School	Locality	Size m^2	Participant	Start Date
Nakuru 1	Slow food	Primary School	Langa Langa town	3030	80	2012
Nakuru 2	ONG Necofa	Primary School	Village di Tayari	90	40	2012
Nakuru 3	Slow food	Primary School	Village di Kangawa	500	48	2011
Nakuru 4	ONG Ygep	Secondary school	Village di Temoyetta	3500	156	2010
Nakuru 5	ONG Necofa	Primary School	City of Elburgon	1000	52	2005
Muranga 1	Agricultural extension officer	Primary School	Village of Karega	1500	30	2010
Muranga 2	Agricultural extension officer	Primary School	Village of Nyako	2450	30	2010
Muranga 3	Agricultural extension officer	Primary School	Village of Ngungugu	375	30	2009
Muranga 4	Agricultural extension officer	Primary School	Village of Kiganjo	2450	30	2010
Muranga 5	Agricultural extension officer	Primary School	Village of Thika Greens	3000	40	2010
Embu 1	Local agriculture ministry	Primary School	City of Embu	1500	26	2010
Embu 2	School teachers	Primary School	City of Embu	5000	35	2004
Embu 3	Local agriculture ministry	Primary School	Village of Manyatta	4000	32	2003
Embu 4	School teachers	Secondary school	City of Runyenjes	4000	22	2012
Embu 5	Local agriculture ministry	Secondary school	City of Embu	2000	16	2011

2.2. SAEMETH-G Method: Dimensions, Components, and Indicators of Sustainability

The SAEMETH-G method has been developed as an attempt to make the concept of sustainability operative in school gardens, taking into consideration the triple bottom line of social, environmental, and economic sustainability. The sustainability assessment framework's construction was based on an interdisciplinary dialogue among a team of five Kenyan and 10 Italian experts, including the authors of the present work.

The 15 experts were both the heads of theoretical projects (professors, teachers, and researchers) but, most of all, of practical school garden projects (agronomists, managers of cooperative development projects, NGOs) addressing social, environmental, and economical themes. The team was composed to include different school garden stakeholders.

The construction of the framework moved across three levels of increasing complexity: first the selection of the sustainability dimensions; then, the individuation of the components; and, finally the choice of proper indicators as described in Table 2. Three focus groups were organized to support the exchange among research participants across the three levels of the framework elaboration. The socio-cultural, agro-environmental, and economic dimensions of sustainability, already selected for the SAEMETH framework for small agri-food system [26], were considered to also be well-suited for school gardens.

Regarding the weight of the dimensions, the outcome of the exchange among the research stakeholders, reached during the first focus group, was to attribute an equal importance (equal weight = maximum 100 for each measurement) to each of the three dimensions in the total measure of sustainability.

Table 2. Dimension numbers (Level 1), components (Level 2), and indicators (Level 3) of school garden sustainability.

Level 1: Dimension	Socio-Cultural	Agro-Environmental	Economic
Level 2: Component	Internal relationships External relationships	Biodiversity Culture/terroir Farming practices Productive process Energy	External input Selling
Level 3: Indicator (number of indicators)	19	22	9

The definition of the components and the attribution of weights to the components (Level 2) of the various dimensions with the equal weights system led to the following outcome:

- for the social-cultural dimensions: two components were selected (internal and external relationships) with a weight equal to 50;
- for the agro-environmental dimensions: five components were selected (biodiversity, culture/terroir, farming practices, productive process, energy) with a weight equal to 20; and
- for the economic dimensions: two components were selected (external input, products sold) with a weight of 50.

This structure reflects the trade-offs made between the considered objectives and the priorities emphasized by the research team starting directly from the proposals of the different stakeholders [27]. By following the approach used for the formulation of SAEMETH [26], and already successfully applied by Van Calker et al. [31] and by Meul et al. [32], the research team tried to mediate the subjectivity of the school garden sustainability components in order to have a framework that allows data collection to be standardized and results to be comparable.

The selection, test, and refinement of the indicators were the most challenging part in terms of time and debate. Various indicators were tested for each component as well as various maximum and minimum values for these indicators. This pilot phase involved three school gardens (one for each county). Quantitative and qualitative data were considered for the indicator selection. For each of the chosen indicators, we have defined a minimum threshold (0 = for the worst situations) and a maximum (10 = the best situations); the reference values are, in some cases, derived through the best techniques available, in other cases through the results of an ad hoc questionnaire and through the proposals of experts. Finally, a set of indicators was agreed upon for the assessment of the 15 selected school gardens for the socio-cultural (Table 3), agro-environmental (Table 4), and economic (Table 5) dimension.

Table 3. Indicators and definition for Level 3 relative to the socio-cultural dimension.

Level 1 Dimension	Level 2 Component	Level 3 Indicator	Indicator Definition	Data Type *	Indicator Weight **
Socio-Cultural	Internal relationships	Decision-making structure	Transparency and clarity between the producers	b	5
		Organization of the group	Presence/absence of an organization of producers	b	5
		Involvement of younger generations	% of young people per the total product	a	5
		Role of younger generations	% of young people pursuing the strategy of garden management	b	5
		Involvement of women	% of women per total of product	a	5
		Role of women	% of women pursuing the strategy of garden management	b	5
		Use of the products	Rediscovery of historical recipes	a	5
		Contribution to the diversification of the diet	The garden allows you to diversify the diet	b	5
		Knowledge is transferred to the population in the garden	Sharing decisions and choices	a	5
		Participation of the producers	How often the group meets	a	5
	External relationships	Vertical transmission of knowledge	Recognition of the role of older generations	a	5.55
		Relationships with public and private institutions	Improvement of the relationships with public institutions and private entities and the possibility of influencing public policy	b	5.55
		Relationships with the local network	There has been an improvement in the local population	a	5.55
		Communication	Knowledge is transmitted to the population in the garden	b	5.55
		Communications systems	Social networks are used to promote the garden	b	5.55
		Events	Participation in events related to the Food Network	a	5.55
		Transmission of knowledge	The group transfers knowledge to children	b	5.55
		Relationship with suppliers	There is a direct relationship	b	5.55
		History and territory	The garden has strengthened the area's history	b	5.55

* a = quantitative data; b = questionnaire; ** the weights sum up to 100 for each Level 1 dimension.

Table 4. Indicators and definition for Level 3 relative to the agro-environmetal dimension.

Level 1 Dimension	Level 2 Component	Level 3 Indicator	Indicator Definition	Data Type *	Indicator Weight **
Agro-environmental	Biodiversity	Number of species	% diversification of products	a	6.66
		Number of local varieties/breeds	% of local varieties/breeds grown	a	6.66
		Varieties/Race	Number of varieties/breeds	a	6.66
	Culture/terroir	Systems	Traditional practices affecting orchard management	b	5
		Deforestation	Slash-and-burn	b	5
		Type of fences	Type of material used for the fences	b	5
		Traditional tools	Use of traditional tools for cultivation	b	5
	Farming practices	Seeds	% in-house production of propagation material	a	1.81
		Forest and woody plants	% in-house production of propagation material	a	1.81
		Rotations	% crop rotations	a	1.81
		Intercropping	% intercropping with other plant species	a	1.81
		Green manure	Using green manuring	a	1.81
		Composter	Compost is created	b	1.81
		Organic fertilization	% use of natural fertilizers	a	1.81
		Fertilization	% use of synthetic chemical fertilizers	a	1.81
		Defense products	% use of synthetic chemical pesticides	a	1.81
		Natural defense products	% use of natural pesticides	a	1.81
		Irrigation	Water conservation and an efficient use of resources	b	1.81
	Productive process	Transformation	Rediscovery or experimentation with transformed products	a	10
		Conservation	Improvement of conservation quality	a	10
	Energy	Water source	Type of water used for irrigation	b	10
		Renewable energy	Use of renewable energy sources	a	10

* a = quantitative data; b = questionnaire; ** the weights sum up to 100 for each Level 1 dimension.

Table 5. Indicators and definition for Level 3 relative to the economic dimension.

Level 1 Dimension	Level 2 Component	Level 3 Indicator	Indicator Definition	Data Type *	Indicator Weight **
Economic	External input	Buying seeds-seedlings-saplings	% products bought	a	7.14
		Buying forest plants	% products bought for forest plants	a	7.14
		Buying compost	% compost bought	a	7.14
		Buying chemical fertilizers	% products bought for the chemical fertilizer	a	7.14
		Buying chemical herbicides/pesticides	% products bought for the chemical defense	a	7.14
		Buying natural pesticides/herbicides	% products bought for the natural defense	a	7.14
		Land	Type of contract that regulates the possession of the garden	b	7.14
	Selling	Selling products	% of products sold on total	a	25
		Type of sales	Commercial network used	b	25

* a = quantitative data; b = questionnaire; ** the weights sum up to 100 for each Level 1 dimension.

2.3. Collection and Statistical Elaboration of the Data

The data were collected for each garden during two visits, lasting about three hours each, (interviews were conducted with at least 30% of the people involved) including a meeting with the project manager in loco conducted by an expert trained in our method. The training program of the expert, carried out in Italy, included a theoretical part with lessons on how to obtain the information on a specific indicator as well as training in the field developed in gardens located in the cities of Turin and Palermo (Italy). At the end of the training period, the expert was well versed in asking for and verifying responses in a standardized way. English was used as the reference language. The interviewer was always accompanied by a translator, who translated the question into Swahili or, where necessary, into the local dialect.

Once the data had been collected, they were first elaborated and viewed graphically, similarly to SAEMETH [26], by putting dimensions, components, and indicators together so that they could be analyzed both singularly and as a whole, considering different scales of analysis (the 15 school gardens, a single school garden, a single dimension, a single component). For the information regarding Level 1 (dimension), the data have been visibly grouped together in a bar graph. For Level 2, a radar chart shows all of the components of total sustainability together, independently of their size. This operation is made possible by the equal-weights approach regarding the size pertaining to each one. This tool supports school gardens coordinator to conceive of their achievements in a holistic way. The indicator values of the analyzed systems are positioned along the axes of a radial diagram scaled from 0 to 100, from the worst (0) to the best (100); therefore, the external ring of the diagram represents the optimal values measured for each component. Furthermore, for Level 2, a principal components analysis (PCA) was performed in order to show the behavior of the components in the school garden's sustainability assessment. Kaiser-Meyer-Olkin and Barlett's sphericity tests were used to test and analyze the appropriateness of the PCA. For an easier interpretation of the PCA results, varimax rotation was applied. For Level 3, a cluster analysis was used in order to show the trend of the 50 indicators in relation to the 15 school gardens. Ward's method of hierarchical clustering with squared Euclidean distance was applied to explore the sample grouping. All of these statistical analyses have been performed with SPSS software 13.0 (SPSS Inc., Chicago, IL, USA).

3. Results

3.1. Application of SAEMETH-G: Level 1—Dimensions

The bar graph (Figure 2) shows the total sustainability of value each school garden. Only the Iruguini Garden (Muranga 1) exceeds the threshold of 200, showing positive values (the minimum sustainability threshold was equal to 50 for each single dimension—defined by the research team) for all of the three dimensions. All of the other school gardens registered a total sustainability value comprised between 150 and 190. In the case of the gardens of Embu (1, 2, and 4) and Nakuru (2 and 4), the agro-environmental dimension is less than 40, indicating problems relative to the cultivation techniques adopted.

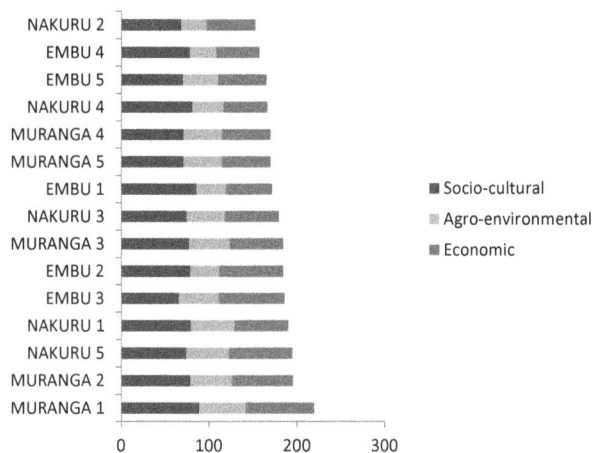

Figure 2. The score relative to the sustainability dimensions (Level 1) for each case study.

Furthermore, it shows how socio-cultural and economic sustainability are positive elements for all of the analyzed school gardens. In particular, the values reached by the socio-cultural scale underline how important the school garden is as a gymnasium for interpersonal relationships inside and outside students' school journey.

3.2. Application of SAEMETH-G: Level 2—Components

The radar graphs (Figure 3) show the distribution of the various components (expressed as percentages) in each garden, aggregated according to geographical location. It is one of the possible result representations. This way of gathering the data has been selected in order to look for the presence of a trend within the territorial context defined by the county (homogeneity of climate conditions, ethnicity know-how in garden management).

In all of the three territories, and in all of the gardens, the internal relationships and culture and regional components reach values above 70%. It is particularly interesting to note that, in addition to the indicators regarding the involvement of women and youths, those that regard the internal relationship component, and most of all, the indicator for diet diversification, reached elevated thresholds. In contrast, external relationships were seen to be lacking, showing a certain difficulty by the schools to communicate their own activities to the outside world through any means of communication.

Regarding the components of the agro-environmental dimensions, no particularly virtuous situations are to be seen, with the exception of the biodiversity component, which reaches values near 80% (most of all in Nakuru and Muranga 1).

All of the schools taken into consideration consumed all of what they grew in their gardens (in the school canteen and/or events), so that the sales component was 0. Even if the schools tended to own the land and not use synthetic products, the acquisitions component (in particular seeds and forest plants) was moderately elevated. In particular, the Embu 3 and Muranga 3 garden were the least self-sufficient.

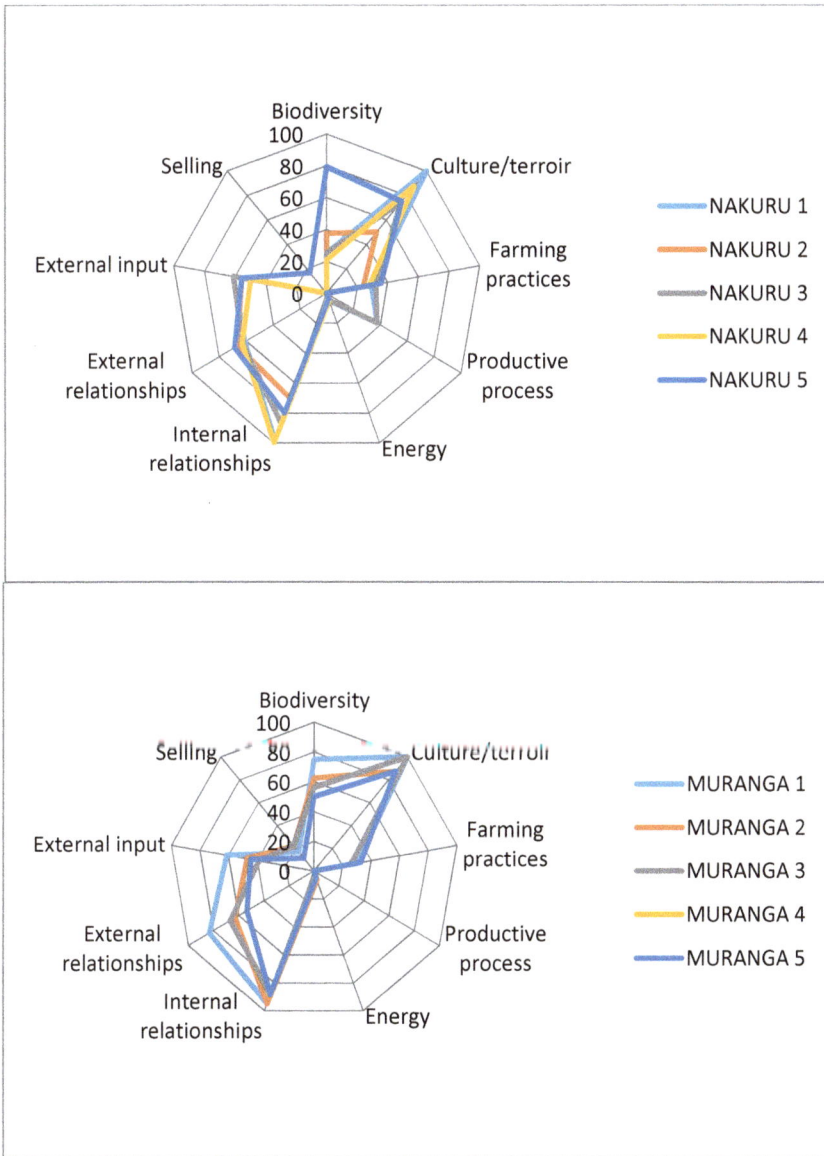

Figure 3. *Cont.*

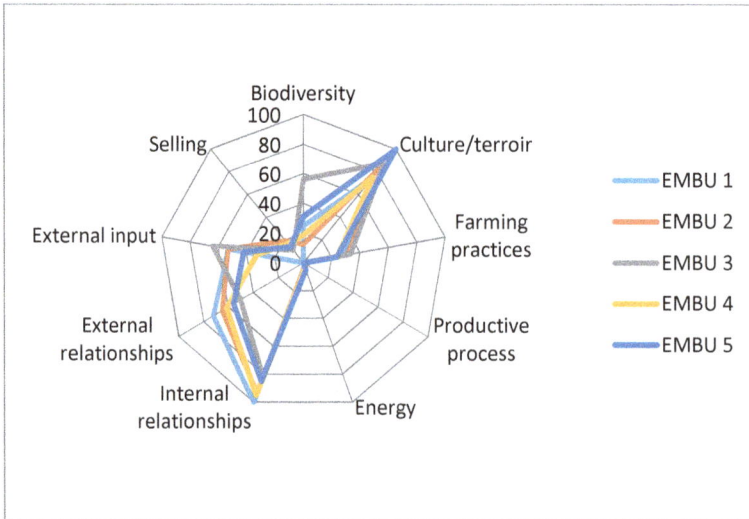

Figure 3. Each radar chart reports data coming from a county where five different school gardens have been analyzed; the differently colored lines evidence the sustainability value (ranging from 0 to 100) for the nine components.

3.3. Principal Components Analysis (Level 2) and Cluster Analysis (Level 3)

With the aim of evaluating which of the Level 2 components were the most influential in determining the value of sustainability, a PCA was carried out with the data relative to all of the gardens in all of the geographic locations.

The primary purpose of the PCA is to reduce the nine components (representatives of analyzed phenomenon as derived from their articulation into indicators) in some latent variables by performing a linear transformation of the variables. Therefore, the variable with higher variance (highlighted in bold) is drawn on the first axis, the second on the second axis, and so on. In order to reduce the complexity, the main (for variance) among the new latent variables (factors) is usually analyzed.

As can be observed in Table 6, Factor 1 is explained by the indicators aggregated in an input and acquisition process. Factor 2 is connected to the agro-environmental indicators, grouped into Biodiversity and Agricultural Practices, while Factor 3 emphasizes aspects that are more socio-cultural and regard generational exchange such as culture, region, and internal relationships. The fact that these three principal components are represented by factors included in the three considered dimensions (socio-cultural, agro-environmental, and economic) clearly shows that the indicators chosen by the stakeholders for these representations are reliable and demonstrates the relevance of all three dimensions (Figure 4).

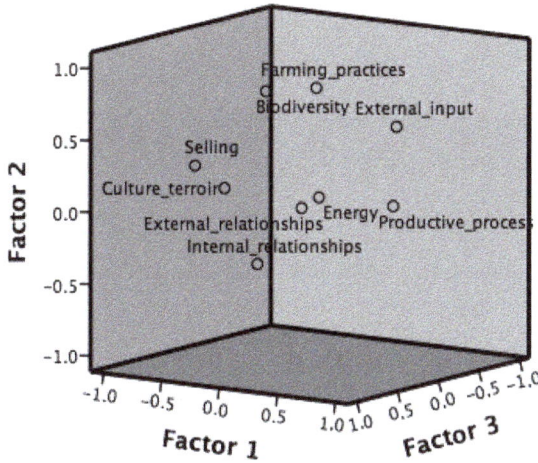

Figure 4. Graphic representation of components performed by using the data of all the gardens analyzed through the PCA (principal component analysis). The environmental dimension is well represented by the components on Factor 2, while Factors 1 and 3 are explained by components that belong indistinctly to socio-cultural, environmental, and economic dimensions.

Table 6. The rotated component matrix where the factor loadings were obtained by performing a PCA (principal component analysis).

Component	Factor				
	1	2	3	4	5
Internal_relationships	0.156	−0.270	**0.863**	0.163	0.133
External_relationships	−0.021	−0.013	0.072	**0.986**	−0.012
Productive_process	**0.842**	0.103	0.171	−0.079	−0.004
Biodiversity	−0.495	**0.717**	−0.169	0.241	0.223
Culture_terroir	−0.134	0.229	**0.857**	−0.065	−0.088
Farming_practices	0.133	**0.845**	0.113	−0.155	−0.031
Energy	0.106	0.069	0.041	−0.011	0.979
External_input	**0.644**	0.590	−0.152	0.320	0.030
Selling	−0.785	0.238	0.287	−0.015	−0.379

Extraction method: principal component analysis. Rotation method: varimax with Kaiser normalization. The factor loadings with the highest positive impact on factor expression are typed in bold. Factor 1 is well explained by the agro-environmental and economic dimensions, Factor 2 from agro-environmental dimension, and Factor 3 from both the socio-cultural and the agro-environmental dimensions.

Finally, to show the homogenous presence able to characterize the sustainability of the analyzed gardens, a cluster analysis (Figure 5) was carried out taking into consideration all of the Level 3 elements (indicators). The cluster analysis is a multivariate statistical analysis technique able to logically group the countings in order to minimize the differences inside the groups and to maximize the differences among groups.

The analyzed gardens were aggregated according to the geographic areas, with the exception of the Nakuru 4 and Embu 5 gardens. This kind of analysis made it possible to show how the similar climate conditions and the cultural component of the local populations had a particular influence on every-day actions. In fact, the Muranga county gardens, all aggregated into a single cluster, were able to strongly influence the practices and processes characteristic of the Kikuyu culture, the ethnicity dominant in the area [33].

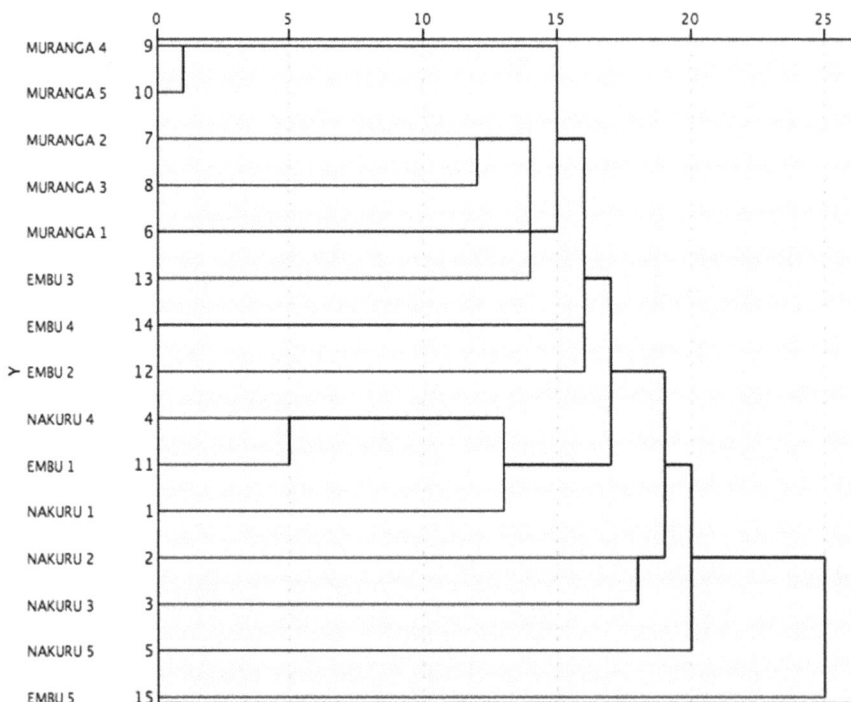

Figure 5. Dendrogram showing homogenous groups of the school gardens case studies. The Ward's method of hierarchical clustering produces a cluster which evidences homogenous groupings per county with respect to values accounting for all of the selected indicators. Classes of similar school gardens are grouped all together.

4. Discussion

SAEMETH-G is an interpretative framework for assessing and monitoring the sustainability of school gardens, inspired by the SAEMETH method [26] for evaluating small-scale crops. It is based on the parity of socio-cultural, agro-environmental, and economic dimensions of sustainability (in terms of weight due to an acknowledgement of the same level of relevance) with the aim of facilitating a synthetic vision of the school gardens, as a multifaceted learning tool [34].

Its application on the 15 case studies in Kenya has demonstrated the functionality of a method proposed for a wide-spectrum and integrated evaluation of school gardens as a mix of quantitative and qualitative data. For some components and indicators of the agro-environmental and economic dimensions in particular, a quantitative approach has been possible (for example, the evaluation of the lifecycle and of the calculation of net margins). However it is clear that there are several constraints to exactly measuring some features of school gardens: the cost of the harvest and analysis of the data might be high and, furthermore, quite often these data are unavailable. The choice of using a large number of qualitative indicators, on one hand, penalizes the possibility of a precise analysis of a single indicator but, on the other hand, allows for a wider perspective of the capacity of an environmentally-, economically-, and most of all socially-sustainable system inside a school curriculum [9]. The method has shown a substantial flexibility and, thus, can also be applied to different models of school gardens (managed by teachers, by local agricultural officers, by local NGOs). In fact, the analyzed gardens include all of the aspects of experiential education tied to local knowledge even though they represent different school garden programs. There are good reasons to believe that the garden micro-system as a sustainable action (albeit with variable margins of improvement according to the specific situation). It is a way of working on the approach children have toward sustainability [9]. Thus, it is hoped that these kinds of learning laboratories will continue to spread.

According to the ecological principal of interdependence [35], in which variations in an ecosystem's components produce variations of other components, virtuous processes at a scholastic level can also generate changes at the family and community level [36].

The participation of all of the subjects involved in the realization and management of school gardens together with experts, including those outside the academic world, in various phases of the methodology has, furthermore, carried out a fundamental role in the development of SAEMETH-G, the importance of which has already been shown in other works, such as that of Olsson and Folke [37].

Aside from analyzing single SAEMETH-G gardens, the methodology was seen to be useful for comparing different systems at a school level, and could also be applied in the future to analyze school garden in other geographical contexts as well as other kinds of gardens such as community gardens. It could also become a supporting tool for monitoring the sustainability performance of school gardens over time by identifying possible spaces for improvement.

However, it is important to strengthen this research with other case studies with the objective of better understanding the importance of the synergies between the components and indicators in order to further refine the point-system criteria. In addition, the analysis of these synergies could reduce the number of indicators and make the method more widely applicable for explaining complex systems in simple ways. At the moment the framework is also still weighed down by the remarkable training that is necessary for those carrying out the data collection.

5. Conclusions

Among the different instruments that can favor an interdisciplinary and every-day approach to sustainability, caring for a school garden has revealed itself to be particularly effective across different nations and cultures. Furthermore, many different school subjects can be involved in the educational activities connected to it and it can play a fundamental role in bridging the gap perceived by new generations between the production and consumption of food.

Clearly, the school garden offers a "learning space" that is potentially more innovative and experiential than traditional school contexts. It should also be recognized that the school garden experience is not always as easy to implement as it seems because it requires not only adequate space and tools, but also teachers with appropriate skills (theoretical and practical management).

SAEMETH-G has shown itself to be an analysis method that is sufficiently flexible to be applied to models that are managed in different ways, even if based on approaches with a similar foundation. Additionally, even though the selected case studies received a good sustainability score on average, the method has contributed to showing the necessity of intervening in the training and productive processes with the aim of improving some fundamental aspects of sustainability, most of all in the agro-environmental field. Gardens may suffer when practical and theoretical skills are lacking, particularly when training is unavailable. In such cases, school-garden experiences can be improved by providing the teachers who manage the gardens (and activities related to them) with more scientific and informative support and by designing the school garden to be a complete agro-ecological system, complete and as independent as possible from external inputs and the associated negative externalities. In this way, the resulting garden can become a key part of a systemic education that supports an understanding of local and global issues.

Acknowledgments: We would like to thank all of the producers, professionals, NGO workers, colleagues and collaborators that actively participated in the research project. The authors are also grateful to Cassandra Funsten (mother-tongue English speaker and graduate student in Palermo's Agriculture Department) for her fundamental assistance editing the English language in the final version of the paper.

Author Contributions: Cristiana Peano and Francesco Sottile designed the research, interpreted results and wrote the paper. Daniela Fiorito, Vincenzo Girgenti and Nadia Tecco collected data, collaborated in the literature review, performed research and checked results. All authors read and approved the final manuscript, analyzed the data and participated jointly to the discussion. All authors have read and approved the final manuscript.

Conflicts of Interest: The Authors declare no conflict of interest.

References

1. Desmond, D.; Grieshop, J.; Subramanium, A. *Revisiting Garden-based Learning in Basic Education*; Food and Agriculture Organisation of the United Nations: Rome, Italy, 2002.

2. Bailey, L.H. *The Nature Study Idea*; McMillan: New York, NY, USA, 1909.

3. Dresher, A.W. Improving child nutrition and agricultural education through the promotion of School Garden Programs. In *First Draft Concept Note on School Gardens*; FAO: Rome, Italy, 2002.

4. Blair, D. The child in the garden: An evaluative review of the benefits of school gardening. *J. Environ. Educ.* **2009**, *40*, 15–38.

5. Cheng, M.H. Asia-Pacific faces diabetes challenge. *Lancet* **2010**, *375*, 2207–2210.

6. Frerichs, L.; Brittin, J.; Sorensen, D.; Trowbridge, M.J.; Yaroch, A.L.; Siahpush, M. Influence of school architecture and design on healthy eating: A review of the evidence. *Am. J. Public Health* **2015**, *105*, e46–e57.

7. Harmon, A.; Lapp, J.L.; Blair, D.; Hauck-Lawson, A. Teaching Food System Sustainability in Dietetic Programs: Need, Conceptualization, and Practical Approaches. *J. Hunger Environ. Nutr.* **2011**, *6*, 114–124.

8. FAO. *World Food Summit Five Years Later, Related Events. School and Hospital Gardens*; FAO: Rome, Italy, 2002. Available online: http://www.fao.org/WorldFoodSummit/ sideevents/papers/Y6824e.htm (accessed on 3 April 2016).

9. Varela-Losada, M.; Vega-Marcote, P.; Pérez-Rodríguez, U.; Álvarez-Lires, M. Going to action? A literature review on educational proposals in formal Environmental Education. *Environ. Educ. Res.* **2016**, *22*, 390–421.

10. Oluoch, M.O.; Pichop, G.N.; Silué, D.; Abukutsa-Onyango, M.O.; Diouf, M.; Shackleton, C.M. Production and harvesting systems for African indigenous vegetables. Chapter 5. In *African Indigenous Vegetables in Urban Agriculture*; Shackleton, C.M., Pasquini, M.W., Drescher, A.W., Eds.; Earthscan: London, UK, 2009.

11. De Neergard, A.; Drescher, A.W.; Kouame, C. *Urban and Peri-Urban Agriculture in African African Indigenous Vegetables in Urban Agriculture*; Earthscan: London, UK, 2009; pp. 36–64.

12. Chen, H.M.; Huang, Y.L. The theory and application of horticultural therapy. *J. Chin. Soc. Hortic. Sci.* **2005**, *51*, 135–144.

13. Tenkouano, A. The nutritional and economic potential of vegetables. In *State of the World 2011: Innovations that Nourish the Planet*; Starke, L., Ed.; Norton: New York, NY, USA, 2011; pp. 27–38.

14. Skinner, E.A.; Chi, U. The Learning-Gardens Educational Assessment Group, Intrinsic Motivation and Engagement as "Active Ingredients" in Garden-Based Education:

Examining Models and Measures Derived from Self-Determination Theory. *J. Environ. Educ.* **2012**, *43*, 16–36.

15. Keatinge, D.H.; Chadha, M.L.; Hughes, J.A.; Easdown, W.J.; Holmer, R.J.; Tenkouano, A.; Yang, R.Y.; Mavlyanova, R.; Neave, S.; Afari-Sefa, V.; et al. Vegetable gardens and their impact on the attainment of the Millennium Development Goals. *Biol. Agric. Hortic.* **2012**, *28*, 71–85.

16. BMVBS/BBR (Bundesministerium für Verkehr, Bau und Stadtentwicklung/Bundesamt für Bauwesen und Raumordnung). *Städtebauliche, Ökologische und Soziale Bedeutung des Kleingartenwesens. Forschungen Heft 133 (Importance of Allotment Gardening for Urban Planning, Ecological and Social Aspects)*; BMVBS/BBR: Berlin, Germany, 2008. (In German)

17. Torquebiau, E. Are tropical agroforestry home gardens sustainable? *Agric. Ecosys. Environ.* **1992**, *41*, 189–207.

18. Seidu, R.; Heistad, A.; Amoah, P.; Drechsel, P.; Jenssen, P.D.; Stenström, T.A. Quantification of the health risk associated with wastewater reuse in Accra, Ghana: A contribution toward local guidelines. *J. Water Health* **2008**, *6*, 461–471.

19. Galluzi, G.; Eyzaguirre, P.; Negri, V. Home gardens: Neglected hotspots of agrobiodiversity and cultural diversity. *Biodivers. Conserv.* **2010**, *19*, 3635–3654.

20. Gari, J.A. *Agrobiodiversity Strategies to Combat Food Insecurity and HIV/AIDS Impact in Rural Africa*; FAO: Rome, Italy, 2003.

21. Sterling, S. Educazione Sostenibile. 2006. Available online: http://www.animamundi.it/HOME.html (accessed on 11 August 2016). (In Italian)

22. Lorenzini, G. Educazione Ambientale: I Principi e le Pratiche Biologi Italiani Giugno. 2012. Available online: http://www.ceeauniba.net/files/EdAmb_principi_e_pratiche.pdf (accessed on 11 August 2016).

23. Stapp, W.B. The concept of environmental education. *Environ. Educ.* **1969**, *1*, 30–31.

24. Hermann, J.; Parker, S.; Brown, B.; Siewe, Y.; Denney, B.; Walker, S. Afterschool gardening improves children's reported vegetable intake and physical activity. *J. Nutr. Educ. Behav.* **2006**, *38*, 201–202.

25. Ruiz-Gallardo, J.-R.; Verde, A.; Valdés, A. Garden-based learning: An experience with "at risk" secondary education students. *J. Environ. Educ.* **2013**, *44*, 252–270.

26. Peano, C.; Tecco, N.; Dansero, E.; Girgenti, V.; Sottile, F. Evaluating the sustainability in complex agri-food systems: The SAEMETH framework. *Sustainability* **2015**, *7*, 6721–6741.

27. Astier, M.; Speelman, E.N.; Lopez-Ridaura, S.; Masera, O.R.; Gonzalez-Esquivel, C.E. Sustainability indicators, alternative strategies and trade-offs in peasant agroecosystems: Analysing 15 case studies from Latin America. *Int. J. Agric. Sustain.* **2011**, *9*, 409–422.

28. De Mey, K.; D'Haene, K.; Marchand, F.; Meul, M.; Lauwers, L. Learning through stakeholder involvement in the implementation of MOTIFS, an integrated assessment model for sustainable farming in Flanders. *Int. J. Agric. Sustain.* **2011**, *9*, 350–363.

29. Slow Food Foundation for Biodiversity. Available online: http://www.fondazioneslowfood.com/en/ (accessed on 2 January 2016).

30. Patton, M.Q. Qualitative Research. In *Encyclopedia of Statistics in Behavioral Science*; John Wiley & Son: New York, NY, USA, 2005.

31. Van Calker, K.; Berentsen, P.; de Boer, I.; Giesen, G.; Huirne, R. An LP-model to analyze economic and ecological sustainability on Dutch dairy farms: Model presentation and application for experimental farm de Marke. *Agric. Syst.* **2004**, *82*, 139–160.

32. Meul, M.; van Passel, S.; Nevens, F.; Dessein, J.; Rogge, E.; Mulier, A. MOTIFS: A monitoring tool for integrated farm sustainability. *Agron. Sustain. Dev.* **2008**, *28*, 321–323.

33. Dewees, P.A. *Household Economy, Trees and Woodland Resources in Communal Areas of Zimbabwe Mimeo*; Oxford Fresty Institute, University of Oxford: Oxford, UK, 1992; p. 85.

34. Brunotts, C.M. School Gardening—A Multifaceted Learning Tool. An Evaluation of the Pittsburgh Civic Garden Centers Neighbors and Schools Gardening Together. Unpublished Master's Thesis, Duquesne University, Pittsburgh, PA, USA, 1998.

35. Kelly, J.G.; Ryan, A.M.; Altman, B.E.; Stelzner, S.P. Understanding and changing social systems: An ecological view. In *Handbook of Community Psychology*; Rappaport, J., Seidman, E., Eds.; Kluwer Academic/Plenum: New York, NY, USA, 2000; pp. 133–159.

36. Ozer, E.J.; Weinstein, R.S.; Maslach, C.; Siegel, D. Adolescent AIDS prevention in context: The impact of peer educator and classroom characteristics on the effectiveness of a school-based, peer-led program. *Am. J. Community Psychol.* **1997**, *25*, 289–323.

37. Olsson, P.; Folke, C.; Hahn, T. Social-ecological transformation for ecosystem management: The development of adaptive co-management of a wetland landscape in southern Sweden. *Ecol. Soc.* **2004**, *9*, 2.

Integrated Multi-Trophic Recirculating Aquaculture System for Nile Tilapia (*Oreochlomis niloticus*)

Puchong Sri-uam, Seri Donnuea, Sorawit Powtongsook and Prasert Pavasant

Abstract: Three densities of the sex-reversed male Nile tilapia, *Oreochromis niloticus* (20, 25, 50 fish/m^3) were cultivated in an integrated multi-trophic recirculating aquaculture system (IMRAS) that involves the ecological relationship between several living organisms, i.e., phytoplankton, zooplankton, and aquatic plants. The results indicated that, by providing proper interdependency between various species of living organisms, the concentrations of ammonia, nitrite, nitrate, and phosphate in the system were maintained below dangerous levels for Nile tilapia throughout the cultivation period. The highest wet weight productivity of Nile tilapia of 11 ± 1 kg was achieved at a fish density of 50 fish/m^3. The aquatic plants in the treatment tank could effectively uptake the unwanted nitrogen (N) and phosphorus (P) compounds with the highest removal efficiencies of 9.52% and 11.4%, respectively. The uptake rates of nitrogen and phosphorus by aquatic plants could be ranked from high to low as: *Egeria densa* > *Ceratophyllum demersum* > *Vallisneria spiralis* and *Vallisneria americana* > *Hygrophila difformis*. The remaining N was further degraded through nitrification process, whereas the remaining P could well precipitate in the soil sediment in the treatment tank.

Reprinted from *Sustainability*. Cite as: Sri-uam, P.; Donnuea, S.; Powtongsook, S.; Pavasant, P. Integrated Multi-Trophic Recirculating Aquaculture System for Nile Tilapia (*Oreochlomis niloticus*). *Sustainability* **2016**, *8*, 592.

1. Introduction

A rapid increase in the world population has accelerated the demand for food, and this leads to challenges in providing an adequate supply of nutrients via intensive agriculture. Typical agricultural systems, particularly aquaculture systems, are mono-cultured where the target aquaculture species such as fish or shrimp is cultivated in a batch mode. The system has to be large enough that the left over feed and wastes from the culture are being naturally treated with several types of microorganisms inhabiting the system. However, it is quite common to have a high-density culture where the system has to be fed with a large quantity of feed and in certain cases with extra aeration. In this case, problems always arise when the waste cannot be adequately treated, resulting in an unsuitable living conditions for the culture. This can lead to stress which negatively affects the growth and the productivity of the system. One attempt to deal with this problem is to have a recirculating system where the unwanted waste is taken out of the culture tank and treated very effectively elsewhere. Recirculating aquaculture system (RAS) is an integrated closed cultivation system where the circulation between the cultivation and treatment tanks helps maintain the quality of the water. This clean water allows a better control of disease [1] and promotes a better growth of aquatic animals which enhances the productivity of the system. In addition, the treatment tank can also act like a holding basin when the cultivation tank needs to be emptied for maintenance. This ability to collect clean water eliminates the need of water from the external irrigation system and prevents unnecessary contamination which might come from external sources. In RAS, wastewater from an aquatic culture containing major nutrients such as nitrogen and phosphorus compounds is not only treated by typical nitrification and denitrification processes in the biological filters [2–6] or integrated biofloc systems [7–11], but also by the uptake of vegetable/ornamental or aquatic plants [12–16]. By providing a proper balance between these various species, this RAS shares a common important concept with an integrated multi-trophic system, which is the synergistic relationship between the living organisms that helps promote the sustainability and the economics of the whole system. In multi-trophic systems, such waste will be used as feed for other organisms, e.g., aquatic plants, simulating the symbiotic relationship in natural ecosystems. The design of this multi-trophic aquaculture is quite important as this will affect the economics of the aquaculture. Well selected food-chain-like organisms enable the farmers to generate more income from by-products that can be harvested from the system. There is a therefore a clear need to develop the multi-trophic recirculating aquaculture system (MRAS) prototype as an integrated closed loop system for Nile tilapia-plankton-aquatic plant cultivation in Thailand to ensure future success of the system and to help guarantee the security of the food supply for the quickly increasing global population.

In this work, this multi-trophic recirculating aquaculture system was based on Nile tilapia (*Oreochromis niloticus*) as the major species. Due to their rapid growth rate and high resistance to disease, Nile tilapia is one of the most farmed aquatic animals [17–21]. Moreover, they require relatively low oxygen for survival [22–24] and a natural surface oxygen transfer is generally adequate for their effective growth. This excludes the need for surface aeration which is a major electricity cost for the system. However, typical culture practice for Nile tilapia still does not incorporate the concept of RAS which renders the culture system susceptible to several environmental disturbances such as water quality, natural drought, etc. Moreover, Nile tilapia excrete waste in the form of nitrogen and phosphorus compounds, which if not treated properly can exert negative effects on the environment [15,25–28]. Nitrogen and phosphorus excreted from Nile tilapia are used as the feed for microalgae which in this case is *Chlorella* sp. This algal species could grow reasonably well in a tropical climate and therefore could be cultured with minimal maintenance. The biomass of *Chlorella* sp. is fed to *Moina macrocopa* tanks which can be more easily harvested when compared to *Chlorella* sp., whose small size leads to harvesting difficulties. The remaining nitrogen and phosphorus waste are used to grow ornamental aquatic plants (*Egeria densa, Ceratophyllum demersum, Hygrophila difformis, Vallisneria spiralis* and *Vallisneria americana*) which can also be harvested, where the clean water is recirculated back to the fish tank. With this configuration, the system will more effectively utilize the feed and the farmers will benefit from having a variety of products apart from the major aquacultural species (the economics of the culture system were further improved using the combination of phytoplankton or microalgae cultivations and zooplankton). The symbiotic mechanism of the various organisms constitutes the novel concept of the integrated multi-trophic recirculating aquaculture system (IMRAS) which is the main focus of this research.

2. Materials and Methods

2.1. System Setup

In this work, a duplicate cultivation of sex-reversed male Nile tilapias (*Oreochromis niloticus*) was carried out in the control and treatment systems. In the control cultivation, fish were cultured in the oval shape opaque fiber (diameter 0.8 m, depth 0.4 m, working volume of 200 L) where the water was not circulated and not treated (representing typical cultivation practice). On the other hand, the treatment system consisted of a series of tanks connected together as shown in Figure 1. This system, called an integrated multi-trophic recirculating aquaculture system (IMRAS), included a fish tank (Section 2.2), phytoplankton tank (Section 2.3), zooplankton tank (Section 2.4) and aquatic plants tank (Section 2.5). The water in the aquatic plants tank was pumped to the fish tank using a submersible pump. An overflow conduit was installed from the fish tank to the phytoplankton tank and the aquatic plants tank. A valve was provided to allow a partial overflow of the water from the fish tank to the phytoplankton tank. This valve remained open until the phytoplankton tank was filled up, at which point the valve was shut and the tank was then operated in a batch mode for the cultivation of *Chlorella* sp. as described in Section 2.3. Once the stationary growth phase was reached, the phytoplankton culture was transferred to the zooplankton tank as feed for *Moina macrocopa* as described in Section 2.4, and the overflow valve was turned on again. A part of the water from the fish tank continuously overflowed to the aquatic plants tank before being pumped back to the fish tank to finish the cycle. The water pumping rate was set at 700 mL/min which is equivalent to a recirculation with a hydraulic retention time of one day.

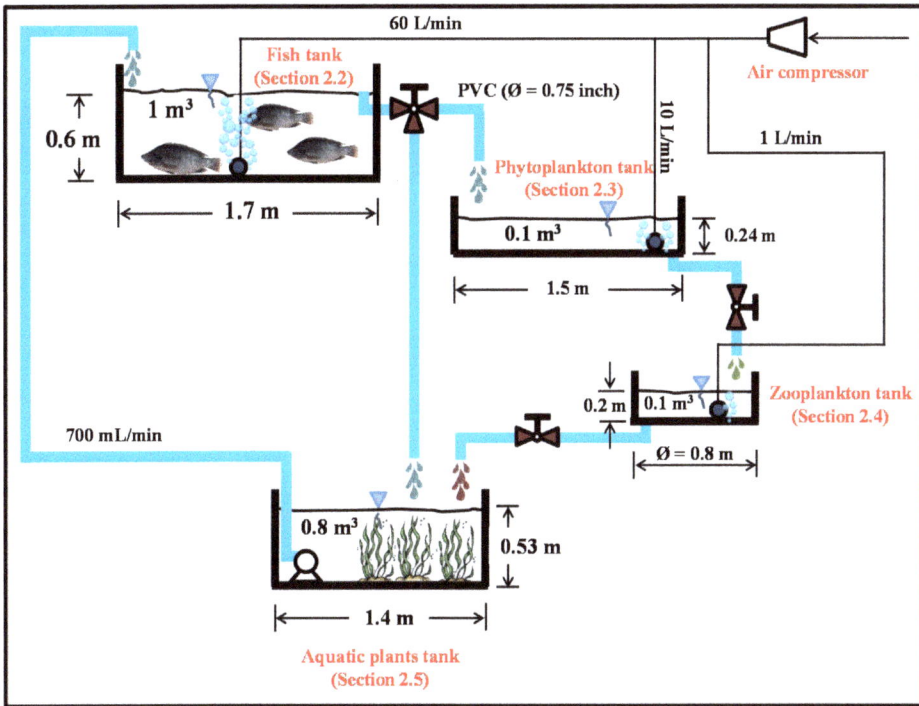

Figure 1. Integrated multi-trophic recirculating aquaculture system (IMRAS) setup.

During the experiment, the growth rates of Nile tilapia, *Chlorella* sp., *Moina macrocopa*, and aquatic plants, along with the water quality such as concentration of ammonia, nitrite, nitrate, phosphate, alkalinity, temperature and dissolved oxygen (DO), were measured following the standard methods for water and wastewater analysis [29].

2.2. Fish Tank (Nile Tilapia)

The fish tank was made from fiber glass with a working volume of 1000 L (dimension: length 1.7 m, width 1 m, depth 0.6 m). An air compressor (LP100, Resun) was used to provide dissolved oxygen at a level greater than 5 mg/L and also to promote liquid circulation. This level of dissolved oxygen was reported to be enough for growth of this fish [30].

The experiment started with three different fish stockings, i.e., 20, 25 and 50 fish/m^3, with an initial average weight of 2 g/fish. This was meant to examine the effect of fish density on the final productivity, where 50 fish/m^3 represents a typical high density culture whilst 20 and 25 fish/m^3 a low density culture. Feeding was provided twice a day (morning and evening) each at 5% of the total fish weight.

The commercial feed composition was 28% crude protein (as provided by Charoen Pokphand Foods PCL, Thailand: nitrogen and phosphorus content were 5.24% and 1.14% of dry weight matter). This amount of protein was recommended as suitable for Nile tilapia by Ribeiro et al. [31]. Note that during the experiment, the weight of all fish was recorded at every 28 days. The experiment was carried out for 112 days before harvest.

At the end of the experiment, the weight gain (g), daily weight gain (g/d), feed conversion ratio (FCR) and survival rate (%) are calculated as follows:

$$\text{Weight gain } (g) = \text{Final wet weight } (g) - \text{Initial wet weight } (g) \qquad (1)$$

$$\text{Daily weight gain } (g/d) = \frac{\text{Weight gain } (g)}{\text{Cultivation time } (d)} \qquad (2)$$

$$\text{FCR} = \frac{\text{Total amount of fish feed fed } (g)}{\text{Total wet weight gain } (g)} \qquad (3)$$

$$\text{Survival rate } (\%) = \frac{\text{Total number of fish at final } (fish)}{\text{Total number of fish at initial } (fish)} \times 100 \qquad (4)$$

In addition, Nile tilapia sample was analyzed for its moisture content, dry weight matter and chemical compositions in order to calculate nitrogen and phosphorus mass balances.

2.3. Phytoplankton Tank (Chlorella sp.)

Chlorella sp. was cultivated with the modified M4N medium [32,33] where KNO_3 and K_2HPO_4 were omitted as N and P sources were obtained from the fish excrete. The culture tank was made from transparent glass with a working volume of 100 L (length 1.5 m, width 0.28 m, depth 0.24 m). The initial biomass density was 0.01 g/L (10^6 cell/mL) with a continuous aeration rate of 0.1 vvm (10 L/min). A sample was collected once a day to measure dry weight. *Chlorella* sp. was harvested as it entered the stationary growth phase (generally after 4 days of cultivation) and was used as a feed for *Moina macrocopa*.

2.4. Zooplankton Tank (Moina macrocopa)

The zooplankton tank was made from fiber glass with a working volume of 100 L (diameter 0.8 m, depth 0.2 m). *Chlorella* sp. as harvested in Section 2.3 was used in the cultivation of *Moina macrocopa* with an initial concentration of 0.1 g/L. An aeration rate of 0.01 vvm (1 L/min) was supplied at the center of the tank in order to increase the level of dissolved oxygen in water and also to prevent cell precipitation. *Moina macrocopa* generally spent 4 days to reach its stationary phase in which it was harvested with 150 µm plankton net. The culture water after cell removal was sent to be treated in the aquatic plants tank (Section 2.5).

2.5. Aquatic Plants Tank

The aquatic plants tank was the major component of the treatment because most nitrogen and phosphorus compounds were removed here and used for the growth of the aquatic plants. Moreover, aquatic plants could capture the suspending sediment from the fish tank, which helped maintain not only the level of nitrogen and phosphorus, but also the clarity of the water in the system. The aquatic plants tank was made from fiber glass with a total working volume of 800 L (length 1.4 m; width 1.2 m; depth 0.53 m). The tank was operated under outdoor condition to utilize sunlight as an energy source for the growth of the aquatic plants.

In this tank, soil was filled at the bottom at the height of 5 cm, where the water depth was 48 cm. Several aquatic plants, i.e., *Hygrophila difformis*, *Vallisneria spiralis*, *and Vallisneria americana*, each with initial fresh weight of 100 g were planted and distributed evenly in the soil, whereas *Egeria densa* and *Ceratophyllum demersum* (100 g each) were floated on the water surface. Aquatic plants were harvested every 14 days such that the remaining weight of each plant was equal to the initial fresh weight (100 g each). The harvested aquatic plants were analyzed for their dry weights, moisture contents, nitrogen and phosphorus balances.

3. Results and Discussion

3.1. Growth of Nile Tilapia

The integrated multi-trophic recirculating aquaculture system (IMRAS) was operated without replacing the water (fresh clean water was regularly added to replenish water lost by evaporation) for 336 days or three fish crops and the growth of the fish was demonstrated in Figure 2. The results indicate that the fish in the low density system (20 fish/m^3) grew at a faster rate when compared with those from the higher fish densities (i.e., 25 and 50 fish/m^3). However, the system with the fish density of 50 fish/m^3 provided the highest total productivity (wet-weight) of 11 ± 1 kg fish/m^3 whereas the densities of 20 and 25 fish/m^3 could only produce a total wet-weight of 6.8 ± 0.3 and 5.3 ± 0.5 kg fish/m^3, respectively. The average wet-weight of Nile tilapia at density of 50 fish/m^3 increased from 2.4 ± 0.6 g/fish to 240 ± 16 g/fish within 112 days of growth which corresponds to the average daily weight gain of 2.1 ± 0.1 g/fish-day. Feed conversion ratio (FCR) and survival rate were 1.5 ± 0.2 and $91\% \pm 1\%$, respectively. These are reasonably good when compared with the results obtained from the other density condition and therefore the density of 50 fish/m^3 was selected for further investigation. The summary of growth characteristics (weight gain, daily weight gain, FCR, and survival rate) is given in Table 1. Note that the growth characteristics obtained from this system is comparable to those from other reported systems, e.g., the daily weight gain from the "recirculating greenwater system" was 1.75 g/fish-day [34] and from the "cage culture system" was 1.15 ± 0.02 g/fish-day [35].

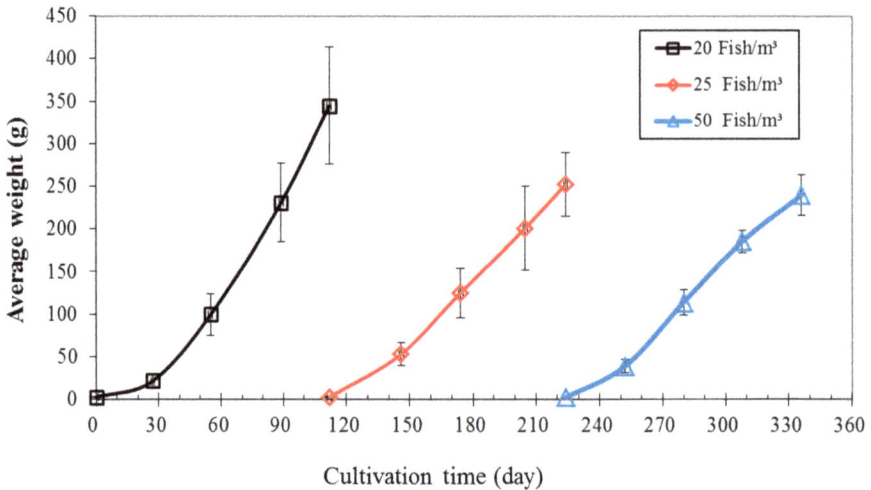

Figure 2. Growth curve of Nile tilapia (note that error bars in this report represent standard deviation of the duplicate results).

Table 1. Growth characteristics of Nile tilapia in the treatment system (mean value).

Growth Parameter	Fish Stocking (Fish/m³)		
	20	25	50
Initial mean weight (g)	2.6 ± 0.5	2.6 ± 0.6	2.4 ± 0.6
Final mean weight (g)	344 ± 69	252 ± 37	240 ± 16
Weight gain (g)	342 ± 15	249 ± 3	237 ± 16
Daily weight gain (g/d)	3.1 ± 0.1	2.23 ± 0.03	2.1 ± 0.1
Feed Conversion Ratio; FCR	1.36 ± 0.06	1.2 ± 0.1	1.5 ± 0.2
Survival rate (%)	95	86 ± 3	91 ± 1
Productivity (kg/m³)	6.8 ± 0.3	5.3 ± 0.5	11 ± 1

It should be mentioned here that the results seem to indicate that there is a trade-off between the higher growth obtained from the low density culture and the total weight gain of the fish in the high density culture, and in this case, the high density resulted in greater productivity. However, the systems employed in this work were continuously aerated, which eliminates the effect of night-time oxygen depletion that might occur from the intensive oxygen consumption through respiration. In large scale systems, this aeration might not be practically feasible, and the nocturnal depletion of oxygen might occur and lead to a different conclusion.

3.2. Growth of Chlorella sp. and Moina macrocopa

Chlorella sp. could grow reasonably well from 0.01 to 0.2 g/L (Figure 3, ◊) within 4 days considering that the system was operated under uncontrolled environmental parameters (light intensity and temperature). Figure 4 illustrates that such growth could be significantly enhanced if the cultivation parameters, e.g., temperature, light intensity and exposure period, could be well controlled [36]. With the outdoor *Chlorella* culture as a feed, *Moina macrocopa* could grow well and the density increased from 0.1 to 0.4 g/L within 4–5 days (Figure 3, Δ) which was slightly better than the value reported in previous literature [37].

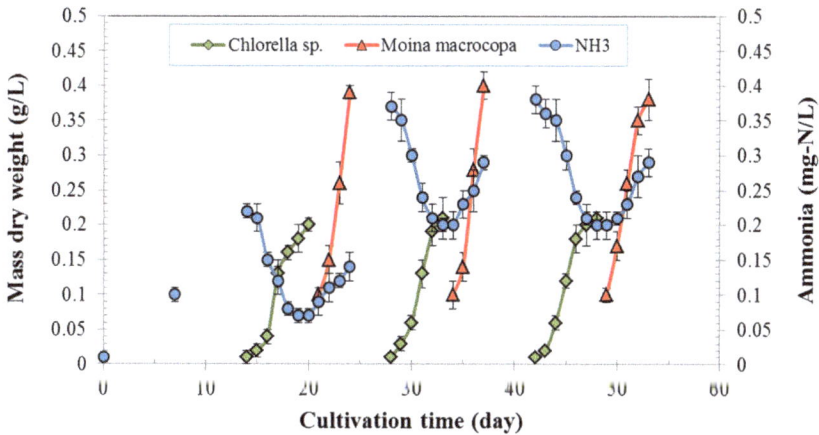

Figure 3. Growth curve of *Chlorella* sp. (◊) and *Moina macrocopa* (Δ) and NH$_3$ concentration profile (○).

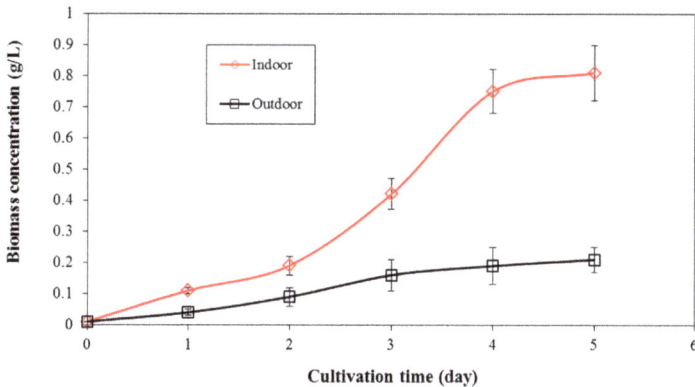

Figure 4. Growth curve of *Chlorella* sp. in Indoor (T = 30 °C, light intensity = 10,000 LUX, light exposure period = 24 h) and Outdoor cultivations (uncontrolled environmental parameters).

249

The results demonstrate that it was possible to enhance the economics of the system by introducing proper bio-components in the food chain. In this case, a high value animal feed *Moina macrocopa* (2–3 $/kg) was introduced to convert and upgrade the low value *Chlorella* biomass which was again fed on NH_3 excreted from the fish culture. It is interesting to observe that, during the growth of *Chlorella* sp., NH_3 was being used for growth and the concentration of NH_3 dropped significantly. The level of NH_3 bounced back again (Figure 3, ○) as *Moina macrocopa* grew and it also excreted NH_3 during its growth stage.

3.3. Growth of Aquatic Plants

Figure 5 illustrates the wet-weights of the five aquatic plants which were harvested every 14 days. The different plants grew at different rates but the productivities of all aquatic plants followed the same pattern. Most plants grew at a relatively slow rate at the beginning which was due to the limited nitrogen source. In other words, the initial concentrations of nitrogen and phosphorus levels from the fish tank that flew into the aquatic plants tank were inadequate for growth (as the fish was still small). The growth rate increased considerably particularly for *Egeria densa* and *Ceratophyllum demersum* during the first 42 days, implying that there were more abundant nitrogen/phosphorus compounds not only due to the accumulation of the uneaten or remaining feed, but also from the acquisition of the plants to the environment of the tank. The other plants, i.e., *Hygrophila difformis, Vallisneria spiralis, and Vallisneria americana*, also grew at a faster rate but the changes in the growth rate were not as obvious when compared to the two species mentioned above. The total fresh weight of all aquatic plants could be ordered from high to low as follows: *Egeria densa* (14.9 ± 0.7 kg), *Ceratophyllum demersum* (13.2 ± 0.5 kg), *Vallisneria americana* (3.87 ± 0.09 kg), *Vallisneria spiralis* (3.67 ± 0.03 kg), and *Hygrophila difformis* (1.74 ± 0.06 kg).

It is noted that aquatic plants directly assimilated nitrogen and phosphorus into the biomass. However, the analysis of N balance in the following section shows that this nitrogen assimilation only accounted for a small fraction of nitrogen input (in animal feed), and most of the nitrogen was lost from another unknown mechanism which, in this case, was believed to be the conversion of NH_3 and NH_4^+ to NO_3^- via nitrifying bacteria and perhaps also via the denitrifying activities as the level of oxygen under the water level in the tank could well exhibit anaerobic condition. These groups of bacteria were generally found in the sediment of the culture tank [38,39].

Aquatic plants were also reported to have beneficial effect as they acted like a filter for the suspended sediment [40]. This helped enhance the level of dissolved oxygen (DO) in the fish tank, as this sediment is usually organic matters which could undergo aerobic decomposition in the fish tank. Preventing this organic decomposition therefore eliminated the unnecessary oxygen uptake in the fish tank resulting in a better control of DO level in the cultivating system [41,42].

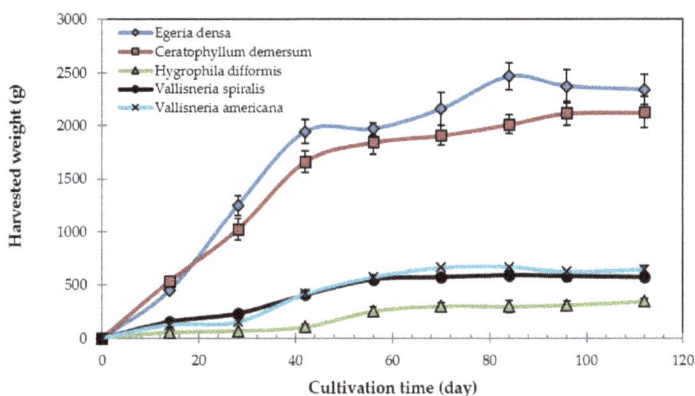

Figure 5. Average harvested weight of aquatic plants (14-day harvesting interval).

The findings from this section suggest that the selection of the aquatic plant species used in the treatment tank should be carefully considered to ensure high treatment efficiency and also a reasonably level of economic benefit. The rapid growth plants must be used to provide a reliable water treatment/filtering capacity whereas the slow growth plants should also be provided as they are usually of high value and could enhance the feasibility of the system.

3.4. Water Quality

Figure 6 illustrates nitrogen and phosphorous compounds profiles in the fish and aquatic plant tanks where nitrogen (NH_3, NH_4^+, NO_2^- and NO_3^-) and phosphorous (PO_4^{3-}) levels in both tanks continuously increased during the first 50 days and remained constant until the end of experiment. Maximum nitrogen and phosphorous concentrations in the fish tank were 0.38 ± 0.02, 0.57 ± 0.02, 55 ± 2 mgN/L and 0.32 ± 0.03 mgP/L, for ammonia, nitrite, nitrate and phosphate, respectively. These levels of nitrogen compounds were still lower than the dangerous level for Nile tilapia (dangerous level indicated by the dash line in Figure 6 as suggested by Hart et al. [43]; Liao and Mayo [44]; Masser et al. [45]), but still higher than those in the aquatic plants tank where the corresponding concentrations were reduced to 0.28 ± 0.02, 0.33 ± 0.02, 38 ± 2 mgN/L and 0.20 ± 0.02 mgP/L, respectively. This indicates that the water in the treatment system could be self-cleaned by the provided concocted ecosystem. It is noted that the levels of ammonia, nitrite, nitrate and phosphate at the end of the control system were 0.52 ± 0.04, 1.20 ± 0.04, 135 ± 11 mgN/L and 2.45 ± 0.04 mgP/L which were relatively high, indicating inadequate treatment capacity in such a system.

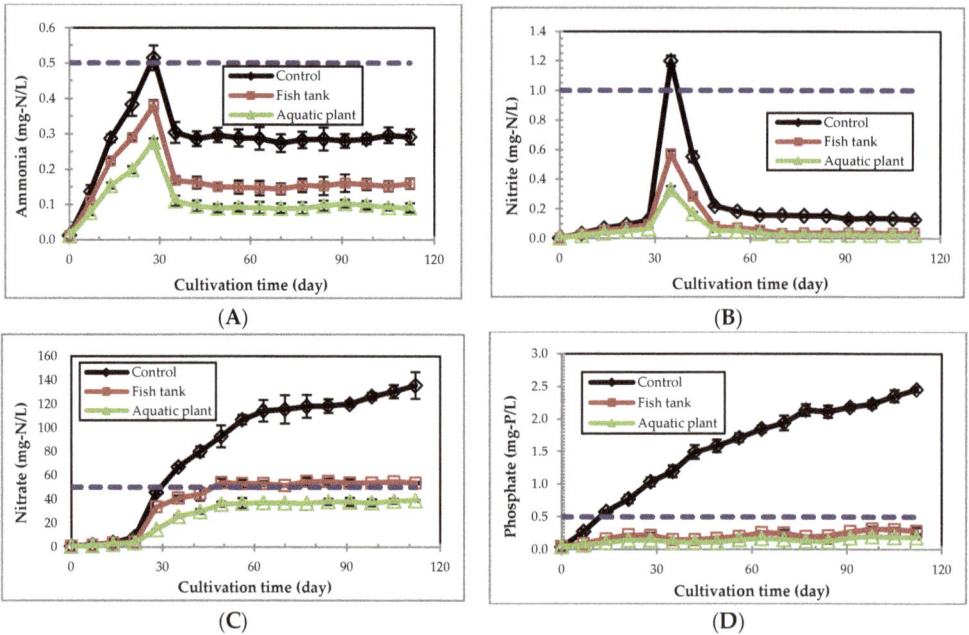

Figure 6. Concentration of ammonia, nitrite, nitrate, and phosphate in fish and aquatic plants tank (mg/L). (**A**) Total ammonia nitrogen (NH_3 and NH_4^+); (**B**) Nitrite nitrogen (NO_2^-); (**C**) Nitrate nitrogen (NO_3^-); (**D**) Phosphate phosphorous (PO_4^{3-}).

Dissolved oxygen (DO) gently decreased both in the fish and aquatic plant tanks. The initial DO concentrations in both tanks were 6.45 and 5.97 mg/L and the final concentrations were 5.55 and 4.65 mg/L, respectively (Figure 7A). In the fish tank, the reduction in DO would be due to a greater need for oxygen from the larger fish [46]. On the other hand, despite oxygen generated from photosynthesis, more oxygen was also required in the aquatic plant tank due primarily to the decomposition of uneaten feed and fish feces and nitrogen compounds through nitrification reaction. DO in the *Chlorella* sp. and *Moina macrocopa* tanks remained mostly unchanged (data not shown) indicating that the activities of the tank could be maintained regardless of the conditions in the other tanks.

Figure 7 also demonstrates the variation in temperature, alkalinity and pH in the system. Due to a large quantity of water, the uncontrolled system temperature was in the range of 27–32 °C with an average of 28–29 °C which was considered within the optimum range (25–30 °C) for Nile tilapia (Figure 7B) [47–49]. Figure 7C demonstrates that alkalinity dropped with time which was potentially due to the activity of nitrifying bacteria and some other algae that might grow in the system. However, Hart, O'Sullivan, Teaching and Aquaculture [43] suggested that the alkalinity for aquatic animals should be maintained above 100 mg-CaCO3/L, therefore, NaHCO$_3$ was added to stabilize the level of alkalinity above this level. The addition of NaHCO$_3$ could also stabilize the pH value in the system, and Figure 7D illustrates that pH (at 2 p.m.) could be naturally controlled within the range of 6–8.5 which was safe for the living organisms involved in the ecology of this system.

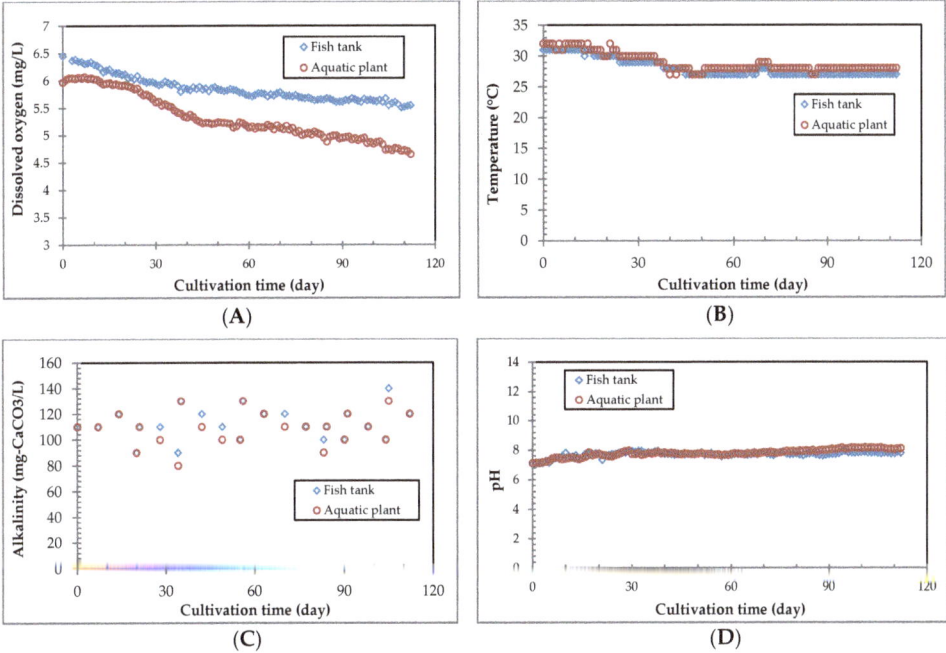

Figure 7. Water qualities in fish and aquatic plant tanks: Dissolved oxygen (**A**); temperature (**B**); alkalinity (**C**) and pH (**D**).

3.5. Nitrogen and Phosphorus Mass Balances

Figure 8 is the summary of the flow of nitrogen compounds within IMRAS. In this experiment, total nitrogen input (790.63 g) came from the use of fish feed throughout the 112 days of each crop. A large quantity of nitrogen (301.13 g·N or 38.09% of the total nitrogen input) could be converted to Nile tilapia. This level is more or less within the range reported elsewhere [16,50]. The remaining nitrogen was converted to: *Chlorella* sp. (7.64 g or 0.97%), *Moina macrocopa* (1.88 g or 0.24%), and aquatic plants (75.29 g or 9.52%). Some nitrogen, e.g., ammonia, nitrite, nitrate (about 91.10 g or 11.52%) was still dissolved in the water at the end of the experiment. Some of nitrogen (48.31 g or 6.11%) might still be adsorbed in the soil while some of nitrogen was not measured directly but calculated as the unaccounted nitrogen. As much as 265.30 g (33.56%) could undergo the decomposition reaction carried out by denitrifying bacteria residing within the eco-system such as in the soil sediment in the aquatic plants tank.

Similarly, Figure 9 displays the flow of phosphorus within IMRAS where the total phosphorus entering the system was 172.01 g (mostly in the fish feed). The amounts of phosphorus converted to Nile tilapia, *Chlorella* sp., *Moina macrocopa* and aquatic plants were 52.75 g (30.67%), 1.20 g (0.70%), 0.22 g (0.13%), and 19.61 g (11.40%), respectively (Figure 9). Again, some phosphorus was still soluble in the water at the end of the experiment and this accounted for about 2.65% (or 4.55 g) of the total phosphorus input. Some of phosphorus (7.07 g or 4.11%) might still be adsorbed in the soil while some of phosphorus was not measured directly but calculated as unaccounted phosphorus. As much as 86.60 g or 50.35% of phosphorus could not be accounted for by the measurement employed in this work. This phosphorus was anticipated to remain partially in the excretion matrix and some could be assimilated to the microorganisms cultivated within the system. It is noted here that the remaining phosphorus in the sediment could pose some concerns on the long-term operation of this system and will need to be extracted at some point.

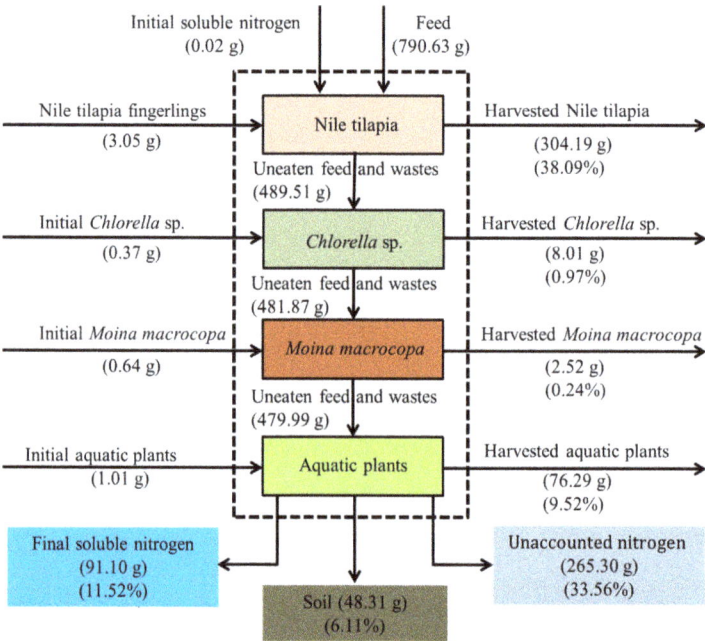

Figure 8. Nitrogen balance of IMRAS.

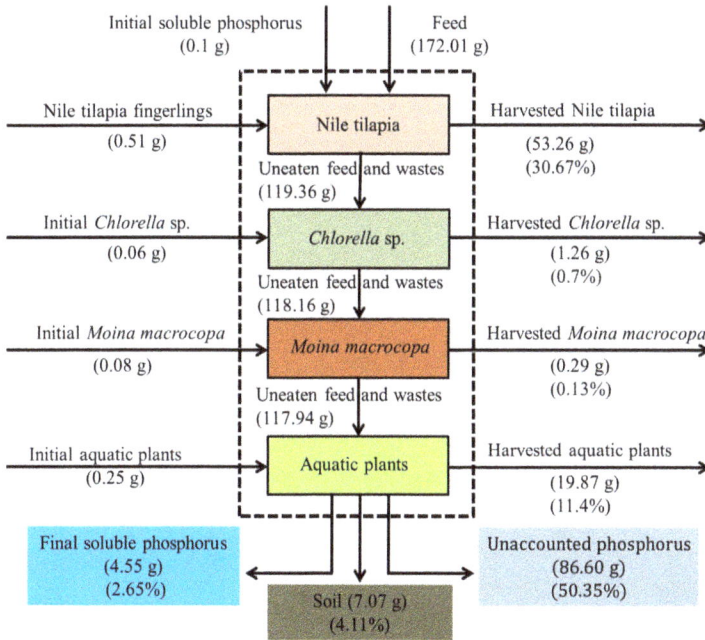

Figure 9. Phosphorous balance of IMRAS.

Figure 10 summarizes that nitrogen and phosphorous of 1311 ± 26 mgN/m^3/d and 230 ± 21 mgP/m^3/d were converted to Nile tilapia mass with an initial fresh weight of 2.4 ± 0.6 g/fish and with a density of 50 fish/m^3. This was from the cultivation period of 112 days where the final fresh weight was 240 ± 16 g/fish. Phytoplankton and zooplankton could only convert a small fraction of nitrogen and phosphorous to biomass, i.e., at about 41 ± 7 mgN/m^3/d and 6 ± 1 mgP/m^3/d. Nitrogen and phosphorous of 328 ± 80 mgN/m^3/d and 85 ± 16 mgP/m^3/d, respectively, were converted into all aquatic plants (*Egeria densa, Ceratophyllum demersum, Vallisneria americana, Vallisneria spiralis,* and *Hygrophila difformis*). Nitrogen and phosphorous amounts of 397 ± 108 mgN/m^3/d and 20 ± 3 mgP/m^3/d, respectively, were dissolved in the water. Nitrogen and phosphorous amounts of 210 ± 31 mgN/m^3/d and 31 ± 6 mgP/m^3/d, respectively, were adsorbed in the soil, while nitrogen of 1155 ± 114 mgN/m^3/d and phosphorous of 377 ± 17 mgP/m^3/d could not be utilized. This finding suggested that the remaining nitrogen and phosphorous could still be utilized by aquatic plants provided that there is enough area for the plants to grow. A rough linear estimate recommended that the area for the aquatic plants should increase 4–5 times to accommodate the amount of the remaining nitrogen and phosphorus.

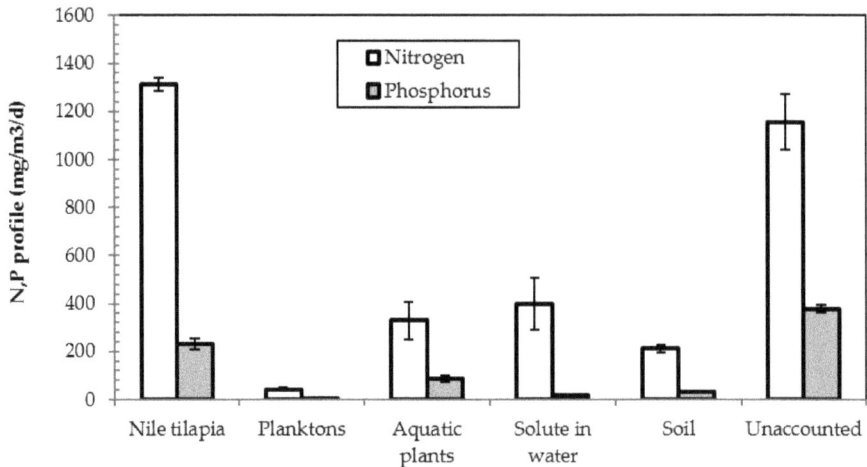

Figure 10. Nitrogen and phosphorous final profile.

4. Conclusions

This work demonstrates the success of the implementation of a closed-loop aquacultural system where a treatment tank is introduced. Although the results revealed that most of the nitrogen and phosphorus were unaccounted for by the fish, planktons and aquatic plants as they were taken up by other metabolisms in the sediment, the introduction of the treatment tank helps to complete the ecology of the system by providing proper conversion of the waste generated by the fish, as some of this waste was converted into valuable products, in the case of phytoplankton, zooplankton, and aquatic plants. Not only does this system benefit from these added-value by-products, but it also enables the recirculation of the culture water, enhancing the reliability of the water management within the system. With the treatment tank, the water quality, ammonia, nitrite nitrate, phosphate, pH and DO could be effectively controlled at safe levels for the cultivation duration of 112 days, and the observed fish productivity was reasonably high.

Acknowledgments: This research was supported by the Dutsadi Phiphat Scholarship of Chulalongkorn University and Thailand National Research Universities (NRU).

Author Contributions: These research was designed and written by Puchong Sri-uam and Prasert Pavasant; Puchong Sri-uam and Seri Donnuea performed the experiments; Puchong Sri-uam, Sorawit Powtongsook and Prasert Pavasant analyzed the data. All authors have read and approved the final manuscript.

Conflicts of Interest: The authors declare no conflict of interest.

References

1. Bahnasawy, M.H.; El-Ghobashy, A.E.; Abdel-Hakim, N.F. Culture of the Nile tilapia (Oreochromis niloticus) in a recirculating water system using different protein levels. *Egypt. J. Aquat. Biol. Fish.* **2009**, *12*, 1–15.

2. Menasveta, P.; Panritdam, T.; Sihanonth, P.; Powtongsook, S.; Chuntapa, B.; Lee, P. Design and function of a closed, recirculating seawater system with denitrification for the culture of black tiger shrimp broodstock. *Aquac. Eng.* **2001**, *25*, 35–49.

3. Silapakul, S.; Powtongsook, S.; Pavasant, P. Nitrogen compounds removal in a packed bed external loop airlift bioreactor. *Korean J. Chem. Eng.* **2005**, *22*, 393–398.

4. Uemoto, H.; Morita, M. Nitrogen removal with a dual bag system capable of simultaneous nitrification and denitrification. *Biochem. Eng. J.* **2010**, *52*, 104–109.

5. Wei, H.; Zhang, Y.; Jiang, Y.; Zhang, C. Influence of carbon source on simultaneous nitrification and denitrification of biomembrane reactor. *J. Beijing Univ. Technol.* **2010**, *36*, 506.

6. Zhang, Y.; Guo, Y.; Bai, Y.; Tan, L.; Wang, Y.; Koyama, T. Simultaneous nitrification and denitrification by catching bed biofilm reactor. *Huan Jing Ke Xue* **2010**, *31*, 134–139.

7. Brito, L.O.; Chagas, A.M.; Silva, E.P.D.; Soares, R.B.; Severi, W.; Gálvez, A.O. Water quality, vibrio density and growth of pacific white shrimp *Litopenaeus vannamei* (Boone) in an integrated biofloc system with red seaweed *Gracilaria birdiae* (Greville). *Aquac. Res.* **2014**, *47*, 940–970.

8. Correia, E.S.; Wilkenfeld, J.S.; Morris, T.C.; Wei, L.; Prangnell, D.I.; Samocha, T.M. Intensive nursery production of the Pacific white shrimp *Litopenaeus vannamei* using two commercial feeds with high and low protein content in a biofloc-dominated system. *Aquac. Eng.* **2014**, *59*, 48–54.

9. De Souza, D.M.; Martins, Á.C.; Jensen, L.; Wasielesky, W., Jr.; Monserrat, J.M.; Garcia, L.d.O. Effect of temperature on antioxidant enzymatic activity in the pacific white shrimp *Litopenaeus vannamei* in a BFT (Biofloc technology) system. *Mar. Freshwat. Behav. Physiol.* **2014**, *47*, 1–10.

10. Emerenciano, M.; Ballester, E.L.; Cavalli, R.O.; Wasielesky, W. Biofloc technology application as a food source in a limited water exchange nursery system for pink shrimp *Farfantepenaeus brasiliensis* (Latreille, 1817). *Aquac. Res.* **2012**, *43*, 447–457.

11. Emerenciano, M.; Cuzon, G.; Arevalo, M.; Gaxiola, G. Biofloc technology in intensive broodstock farming of the pink shrimp *Farfantepenaeus duorarum*: Spawning performance, biochemical composition and fatty acid profile of eggs. *Aquac. Res.* **2014**, *45*, 1713–1726.

12. Lewis, W.M.; Yopp, J.H.; Schramm, H.L., Jr.; Brandenburg, A.M. Use of hydroponics to maintain quality of recirculated water in a fish culture system. *Trans. Am. Fish. Soc.* **1978**, *107*, 92–99.

13. Naegel, L.C. Combined production of fish and plants in recirculating water. *Aquaculture* **1977**, *10*, 17–24.

14. Seawright, D.E.; Stickney, R.R.; Walker, R.B. Nutrient dynamics in integrated aquaculture–hydroponics systems. *Aquaculture* **1998**, *160*, 215–237.

15. Thanakitpairin, A.; Pungrasmi, W.; Powtongsook, S. Nitrogen and phosphorus removal in the recirculating aquaculture system with water treatment tank containing baked clay beads and Chinese cabbage. *EnvironmentAsia* **2014**, *7*, 81–88.

16. Trang, N.T.D.; Brix, H. Use of planted biofilters in integrated recirculating aquaculture-hydroponics systems in the Mekong Delta, Vietnam. *Aquac. Res.* **2014**, *45*, 460–469.

17. Ardjosoediro, I.; Ramnarine, I.W. The influence of turbidity on growth, feed conversion and survivorship of the Jamaica red tilapia strain. *Aquaculture* **2002**, *212*, 159–165.

18. Coward, K.; Bromage, N. Reproductive physiology of female tilapia broodstock. *Rev. Fish Biol. Fish.* **2000**, *10*, 1–25.

19. El-Sayed, A.-F.M. *Tilapia Culture*; CABI: Wallingford, UK, 2006.

20. Hassanien, H.A.; Elnady, M.; Obeida, A.; Itriby, H. Genetic diversity of Nile tilapia populations revealed by randomly amplified polymorphic DNA (RAPD). *Aquac. Res.* **2004**, *35*, 587–593.

21. Toguyeni, A.; Fauconneau, B.t; Fostier, A.; Abucay, J.; Mair, G.; Baroiller, J.-F. Influence of sexual phenotype and genotype, and sex ratio on growth performances in tilapia, Oreochromis niloticus. *Aquaculture* **2002**, *207*, 249–261.

22. El-Sayed, A.-F.M.; Kawanna, M. Effects of photoperiod on the performance of farmed Nile tilapia Oreochromis niloticus: I. Growth, feed utilization efficiency and survival of fry and fingerlings. *Aquaculture* **2004**, *231*, 393–402.

23. Ross, L. Environmental physiology and energetics. In *Tilapias: Biology and Exploitation*; Springer: Heidelberg, Germany, 2000; pp. 89–128.

24. Siddiqui, A.; Howlader, M.; Adam, A. Culture of Nile tilapia, Oreochromis niloticus (l.), at three stocking densities in outdoor concrete tanks using drainage water. *Aquac. Res.* **1989**, *20*, 49–58.

25. Burut-Archanai, S.; Eaton-Rye, J.J.; Incharoensakdi, A.; Powtongsook, S. Phosphorus removal in a closed recirculating aquaculture system using the cyanobacterium *Synechocystis* sp. PCC 6803 strain lacking the SphU regulator of the Pho regulon. *Biochem. Eng. J.* **2013**, *74*, 69–75.

26. Körner, S.; Das, S.K.; Veenstra, S.; Vermaat, J.E. The effect of pH variation at the ammonium/ammonia equilibrium in wastewater and its toxicity to Lemna gibba. *Aquat. Bot.* **2001**, *71*, 71–78.

27. Lin, C.K.; Shrestha, M.K.; Yi, Y.; Diana, J.S. Management to minimize the environmental impacts of pond effluent: Harvest draining techniques and effluent quality. *Aquac. Eng.* **2001**, *25*, 125–135.

28. Thompson, F.L.; Abreu, P.C.; Wasielesky, W. Importance of biofilm for water quality and nourishment in intensive shrimp culture. *Aquaculture* **2002**, *203*, 263–278.

29. APHA. *Standard Methods for the Examination of Water and Wastewater*; APHA, AWWA, WPCF: Washington DC, USA, 2005.

30. Riche, M.; Garling, D. Feeding tilapia in intensive recirculating systems. *North Cent. Reg. Aquac. Center Fact Sheet Ser.* **2003**, *114*, 1–4.

31. Ribeiro, F.B.; Lanna, E.A.T.; Bomfim, M.A.D.; Donzele, J.L.; Freitas, A.S.d.; Sousa, M.P.D.; Quadros, M. Dietary total phosphorus levels for Nile tilapia fingerlings. *Rev. Bras. Zootec.* **2006**, *35*, 1588–1593.

32. Lee, C.; Kim, M.; Sanjay, K.; Kwag, J.; Ra, C. Biomass production potential of Chlorella vulgaris under different CO_2 concentrations and light intensities. *J. Anim. Sci. Technol.* **2011**, *53*, 261–268.

33. Sung, K.-D.; Lee, J.-S.; Shin, C.-S.; Park, S.-C. Effect of iron and EDTA on the growth of Chlorella sp. KR-1. *J. Microbiol. Biotechnol.* **1998**, *8*, 409–411.

34. Al-Hafedh, Y.S.; Alam, A. Operation of a water recirculating greenwater system for the semi-intensive culture of mixed-sex and all-male Nile tilapia, Oreochromis niloticus. *J. Appl. Aquac.* **2005**, *17*, 47–59.

35. Gibtan, A.; Getahun, A.; Mengistou, S. Effect of stocking density on the growth performance and yield of Nile tilapia [Oreochromis niloticus (l., 1758)] in a cage culture system in lake Kuriftu, Ethiopia. *Aquac. Res.* **2008**, *39*, 1450–1460.

36. Sri-uam, P.; Linthong, C.; Powtongsook, S.; Kungvansaichol, K.; Pavasant, P. Manipulation of biochemical compositions of chlorella sp. *Eng. J.* **2015**, *19*, 13–24.

37. Martínez-Jerónimo, F.; Gutierrez-Valdivia, A. Fecundity, reproduction, and growth of Moina macrocopa fed different algae. *Hydrobiologia* **1991**, *222*, 49–55.

38. Cossu, R.; Haarstad, K.; Lavagnolo, M.C.; Littarru, P. Removal of municipal solid waste COD and NH 4–N by phyto-reduction: A laboratory–scale comparison of terrestrial and aquatic species at different organic loads. *Ecol. Eng.* **2001**, *16*, 459–470.

39. Zhao, Y.; Fang, Y.; Jin, Y.; Huang, J.; Bao, S.; Fu, T.; He, Z.; Wang, F.; Zhao, H. Potential of duckweed in the conversion of wastewater nutrients to valuable biomass: A pilot-scale comparison with water hyacinth. *Bioresour. Technol.* **2014**, *163*, 82–91.

40. Ostroumov, S. Biological filtering and ecological machinery for self-purification and bioremediation in aquatic ecosystems: Towards a holistic view. *Riv. Biol.* **1998**, *91*, 221–232.

41. Cheng, J.J.; Stomp, A.-M. Growing duckweed to recover nutrients from wastewaters and for production of fuel ethanol and animal feed. *CLEAN—Soil Air Water* **2009**, *37*, 17–26.

42. Kapuscinski, K.L.; Farrell, J.M.; Stehman, S.V.; Boyer, G.L.; Fernando, D.D.; Teece, M.A.; Tschaplinski, T.J. Selective herbivory by an invasive cyprinid, the rudd Scardinius erythrophthalmus. *Freshw. Biol.* **2014**, *59*, 2315–2327.

43. Hart, P.; O'Sullivan, D.; (University of Tasmania. National Key Centre for Teaching and Research in Aquaculture). *Recirculation Systems: Design, Construction and Management*; Turtle Press: Tasmania, Australia, 1993.

44. Liao, P.B.; Mayo, R.D. Salmonid hatchery water reuse systems. *Aquaculture* **1972**, *1*, 317–335.

45. Masser, M.P.; Rakocy, J.; Losordo, T.M. Recirculating aquaculture tank production systems. In *Management of Recirculating Systems*; SRAC Publication: Stoneville, MA, USA, 1999.

46. Van Dam, A.A.; Pauly, D. Simulation of the effects of oxygen on food consumption and growth of Nile tilapia, Oreochromis niloticus (l.). *Aquac. Res.* **1995**, *26*, 427–440.

47. Balarin, J.D.; Haller, R.D. The intensive culture of tilapia in tanks, raceways and cages. In *Recent Advances in Aquaculture*; Muir, J.F., Roberts, R.J., Eds.; Croom Helm: London, UK, 1982; pp. 266–355.

48. Chervinski, J. Environmental physiology of tilapias. In *The Biology and Culture of Tilapias: Proceedings of the International Conference on the Biology and Culture of Tilapias, Issue 7 of Iclarm Conference Proceedings*; Pullin, R.S.V., Lowe-McConnell, R.H., Eds.; WorldFish: Manila, Philipphines, 1982; pp. 119–128.

49. Philippart, J.-C.; Ruwet, J.-C. Ecology and distribution of tilapias. In *The Biology and Culture of Tilapias: Proceedings of the International Conference on the Biology and Culture of Tilapias, Issue 7 of Iclarm Conference Proceedings*; Pullin, R.S.V., Lowe-McConnell, R.H., Eds.; WorldFish: Manila, Philipphines, 1982; pp. 15–60.

50. Rafiee, G.; Saad, C.R. Nutrient cycle and sludge production during different stages of red tilapia (Oreochromis sp.) growth in a recirculating aquaculture system. *Aquaculture* **2005**, *244*, 109–118.

Urban Cultivation and Its Contributions to Sustainability: Nibbles of Food but Oodles of Social Capital

George Martin, Roland Clift and Ian Christie

Abstract: The contemporary interest in urban cultivation in the global North as a component of sustainable food production warrants assessment of both its quantitative and qualitative roles. This exploratory study weighs the nutritional, ecological, and social sustainability contributions of urban agriculture by examining three cases—a community garden in the core of New York, a community farm on the edge of London, and an agricultural park on the periphery of San Francisco. Our field analysis of these sites, confirmed by generic estimates, shows very low food outputs relative to the populations of their catchment areas; the great share of urban food will continue to come from multiple foodsheds beyond urban peripheries, often far beyond. *Cultivation* is a more appropriate designation than *agriculture* for urban food growing because its sustainability benefits are more social than agronomic or ecological. A major potential benefit lies in enhancing the ecological knowledge of urbanites, including an appreciation of the role that organic food may play in promoting both sustainability and health. This study illustrates how benefits differ according to local conditions, including population density and demographics, operational scale, soil quality, and access to labor and consumers. Recognizing the real benefits, including the promotion of sustainable diets, could enable urban food growing to be developed as a component of regional foodsheds to improve the sustainability and resilience of food supply, and to further the process of public co-production of new forms of urban conviviality and wellbeing.

Reprinted from *Sustainability*. Cite as: Martin, G.; Clift, R.; Christie, I. Urban Cultivation and Its Contributions to Sustainability: Nibbles of Food but Oodles of Social Capital. *Sustainability* **2016**, *8*, 409.

1. Introduction

There is a widespread resurgence in urban food growing. One sign of its popularity is the endorsement of political leaders. For example, Michelle Obama planted a garden in 2009 with the help of schoolchildren—the first White House plot since Eleanor Roosevelt's World War II Victory Garden. The rise in interest is also indicated by a change in the status of urban food growing: increasingly it is referred to as urban agriculture by academics [1] and others. As pointed out in the *New York Times*, it has become a bandwagon phenomenon: "In recent years, chefs, writers, academics, politicians, funders, activists and entrepreneurs have jumped on the hay wagon for urban agriculture" [2] (p. D1).

The designation as *agriculture*, rather than *gardening*, projects a new frame and a larger scale that raise research questions about urban food growing's output and sustainability. We address these questions based on three case studies: a community garden, a community farm, and an agricultural park. The study is exploratory and descriptive, and addresses only cities of the global North; the picture is quite different in the global South [3–5], which is home to 80% of urban croplands [6]. Furthermore, the study does not consider the practice that some have advocated of growing food in underground chambers using artificial light [7], because it is a thermodynamic nonsense. Photosynthesis captures a tiny percentage of incident radiant energy; therefore, the energy used to illuminate plants grown underground is necessarily one to two orders of magnitude greater than the energy content of the plants. Allow for the inefficiencies in converting primary energy to light and the imbalance between the source and the product gets worse by typically a further order of magnitude. Growing food without daylight may have a role for mushrooms, which, traditionally and iconically, can be managed by keeping them in the dark and intermittently covering them in fertilizer, but not more generally and not for any main dietary constituent. Underground growing of photosynthesizing plants has no place in a serious discussion of sustainable food production, whether or not it is organic. As one researcher has commented: "why (does) it make sense to put a lot of intellectual activity and resources into something that negates the direct use of our one and only absolutely renewable resource—the sun—and totally replace it with artificial light?" [8] (p. 5).

There are compelling background reasons for the mounting interest in urban food growing. The world has a rising and increasingly urban population [9]. There will be 2 billion more people to feed by 2050, when around 70% of our population of 9 billion will be urban, compared to 50% today. This is projected to increase food demand from a growing and richer population by 2050. The extent of the increased demand is uncertain, but estimates range up to 70% more crop calories than produced in 2006 ([10] cited in [11]). Urbanization leads to loss of farmland [12]. Between 1970 and 2000 the land equivalent of Denmark was converted from farmland to urban settlement globally. The projection for 2000 to 2030 is the equivalent of Mongolia, 36 times the area of Denmark. Thus, at the same time that more food will be needed, less farmland will be available.

To exacerbate the problem, climate change is projected to result in farm yield loss [13,14]. Although there is debate around how large the loss may be, there is agreement that food security is one of the principal concerns humanity must address in the context of global climate change. For example, the United States, the world's third largest food producer and largest food exporter [15], projects its yield of major crops to decline by mid-century due to soil degradation, rising temperature, and precipitation extremes [14].

As a consequence of these global trajectories, agriculture faces the major challenge of increasing production levels substantially and doing so sustainably. Designating the upsurge in urban food growing as *agriculture* implies that increasing output in cities can contribute significantly to meeting these challenges. To avoid this implicit assumption, we propose the term *cultivation*.

There is also a foreground context for the rising interest in urban food growing: the environmental movement and related campaigns for organic, locally sourced, healthy, and sustainable diets. Community food growing evokes a cultural orientation different from that of traditional urban allotments and domestic gardens. Domestic gardening is a private and individual activity. Allotments are on common land but are allocated to individuals; they were institutionalized as compensation for the land clearances involved in the emergence of industrial agriculture in the late 18th and early 19th centuries in northwestern Europe [16]. In the United Kingdom, statutory allotment sites receive protection under the Allotment Act of 1925, although there are fewer safeguards for private and temporary sites [17]. Current urban food growing arose in the late 20th century, largely as neighborhood mobilizations to reclaim vacant and derelict lots in post-industrial cities of North America and Europe.

An analysis in the United Kingdom found that the "sense of community participation and empowerment . . . links examples of community gardening" [18] (p. 285). Community food growing can enhance the creation of locally specific social capital in urban areas. More recent studies of U.K. community food-growing schemes and networks have reinforced this point, and identified a wide range of associated social processes: for example, grassroots innovation and informal local research and development; demonstration projects; opening up debate with existing actors in food systems; and expanding the range of alternatives to established practices [19,20]. Involvement in community food-growing is associated with opportunities for people "to engage more actively around issues of food, health, waste, community and environment" [19] (p. 31). Such findings underline a key point in our analysis: urban food-growing is primarily about cultivation of social skills, capabilities, and virtues that can contribute to sustainable urban living, rather than about major additions to food production. We describe this set of features as social sustainability services.

2. Methodology

In the context of this paper, *urban* refers to metropolitan areas—cities (core) with their nearby suburbs (edge) and distant exurbs (periphery). In this exploratory qualitative research into social contributions, we visited three field sites, representing a range of scales, social activities, and locations. Two of the sites exemplify common modalities of the new urban food growing: one very small inner city community garden ("The Garden", specifically Manhattan's West Side Community Garden) and one larger suburban farm ("The Farm", Sutton Community Farm on the outskirts of South London). The Garden was selected because an author was a member and had begun to study it in 2011. The Farm was selected because it had been the subject of recent research [21], including a Life Cycle Assessment (LCA); one author had worked on a similarly sized commercial market garden (known locally as a "smallholding") at that location some 50 years previously. The third case ("The Park", Sunol Agricultural Park near San Francisco) represents an emergent modality: a larger agricultural park in a peri-urban location; it was selected because it represented another variation in urban food growing in yet another metropolis. Two of the authors visited the Garden in 2012 and the Park in 2014. All three authors visited the Farm in 2013. Data collected through these field visits comprised observations, documentary and verbal information provided by informants, supplemented by archival data made available by informants and by online research. The principal informants were the President of the Garden, the Manager of the Farm, and the President of Sustainable Agriculture Education, parent of the Park. For the discussion in this paper, the primary qualitative data are supplemented by average estimates of potential production; see Section 4.

Urban cultivation's contributions to social sustainability require this kind of empirical scrutiny. One methodological tool available for this task is Life Cycle Assessment. LCA was originally developed to assess the environmental costs associated with the full supply chains delivering products and services, and has since been expanded to include the economic costs. For example, an LCA study of the suburban Farm included in this study [21] found that it could produce reductions in greenhouse gas (GHG) emissions as compared to supermarket food chains provided that the Farm concentrated on crops usually grown in heated greenhouses or flown in from warmer climates. Such niche crops are also the major sources of its income. However, while GHG emissions are one key indicator of environmental performance, they are only one of a suite of sustainability metrics that includes aspects not addressed by conventional LCA, including labor issues and social dynamics [22]. Another challenge facing sustainability research is identifying not just the dietary and GHG-reducing aspects of a mode of food production but also the social and ethical benefits. Production, distribution and consumption are to be seen not just as a one-way flow of resources from supplier to consumer, leaving impacts in their wake, but as a channel by which benefits can flow from the consumer (of food or land use) to other agents [23]. Adapting LCA to this kind of case represents a methodological challenge. The guidelines on social LCA [24] are still very much in the developmental phase [25] and are in any case directed at detecting social "bads" in international supply chains rather than the local social benefits that are among the drivers for urban cultivation. Further development of social LCA depends on the execution of case-by-case studies, using methods and approaches appropriate to the specific scope and scale.

The Garden, Farm, and Park cases illustrate that: "To be a viable alternative in cities and compete with other land uses, the justification for urban agriculture must include the ecological and cultural function these systems offer, in addition to the direct benefits of food produced" [26]. At the three sites examined here, the viability of the activity depends not just on inputs like imported compost but also on consumers—of vegetable boxes delivered to the doorstep in the case of the Farm, of cultural programs at the Garden, and of farming educations at the Park. The social benefits, as counter-posed to the environmental and economic ones, of urban cultivation accrue mainly at a local level rather than being distributed along a supply chain; they are outside the familiar framework of LCA and can flow laterally to the residents of an activity's catchment area; e.g., in the form of educations. Therefore, the kind of field investigation carried out here will remain essential.

3. Results: The Three Cases

3.1. The Garden

The Garden emerged in the context of the massive 1970's Urban Renewal Program in the slums of post-industrial U.S. cities [27]. The City of New York evicted tenants and razed tenement buildings in much of Manhattan's Upper West Side, leaving brownfield land available for redevelopment and gentrification [28]. A high-rise condominium building was built on a site which included the future Garden and another was awaiting capital investment. In the meantime the site became a dump for abandoned automobiles and other urban detritus.

The dump site was transformed into a verdant garden in a spontaneous response by local residents to clean up a dangerous area in their midst that was also an eyesore. With construction imminent, the neighborhood was assisted in saving this open space by the local Community Board and the national Trust for Public Land. City government officials and real estate developers acquiesced, in part because community gardens enhance property values and thereby add to tax revenues—while also of course adding to value for property owners. In an analysis of community gardens established in New York City between 1977 and 2000, Voicu and Been found that "gardens were located on sites that acted as local disamenities within their communities . . . after opening gardens have a positive impact on surrounding property values, which grows steadily over time" [29] (p. 268). The City administered a "sunshine test" and approved the site as a garden, with two stipulations for becoming untaxed land: that it would be open to the public and would pay for its upkeep.

The Garden is located near the geographic center of New York City's Manhattan Borough. The land, 0.15 hectare, is governed by a board elected annually from its 300+ paid members. Membership is open to the public at a nominal annual fee. Only about 1/3 of the Garden's space is used to grow food. Gardeners work on raised beds of 2.8 m². Of the remaining garden space, 1/3 is devoted to horticulture and 1/3 is dominated by an amphitheater used for cultural productions.

Gardeners reported that they do not grow much food—enough vegetables to add to several meals a week over the harvest period. "I just grow some nibbles," one gardener said. Several informants related that growing food was not the main reason they gardened; rather, it was because they liked to garden. "I enjoy my green thumbs," one reported. Also, many gardeners said that they liked the cooperative aspects of the Garden and enjoyed its features—a quiet, safe, and green retreat amid the Manhattan skyscrapers.

The Garden depends on a steady replenishment of labor to take care of its three large compost bins and to keep public areas tidy, as well as to raise money. It requires about $75,000 annually to operate. The bulk of the money goes to maintain pavements, towards insurance, and to purchase gardening supplies and tools. The Garden has no paid labor. Finding volunteer labor has been a general problem for community gardens. The work required is skilled and this limits the available pool. The largest source of gardeners is women, mainly retirees. New York City's gardens have declined in number since the mid-1980s largely due to a lack of participation—many rely on only a few tireless souls [30].

The Garden provides a range of cultural programs that attract thousands of visitors who are potential sources of finance and labor. The programs include music, theatre, poetry, film, and arts and crafts presentations. The Garden's signature cultural event is its annual spring Tulip Festival, when some 12,000 blooming flowers attract visitors from around the world [31].

3.2. The Farm

The Farm comprises 2.9 ha, 1.4 of which are tilled. It lies in the Borough of Sutton at the southern edge of greater London, in what is termed the Green Belt in the U.K. planning system. It occupies greenfield land but the soil is very poor. Until the 20th Century lavender (*Lavandula*) had been grown on the site as it can thrive in poor soil. The land use was changed as part of the mid-20th century drive to increase food production in the United Kingdom and took advantage of labor from a nearby camp for prisoners of war. Fifty years ago, the smallholding was operated by a family who lived there; it depended on high value glasshouse crops, primarily salad vegetables and cut flowers sold via large wholesale markets in London, with high inputs that included horse manure. There are now 500 m² of polytunnels at the

Farm, providing for year-round production, but it requires large inputs of compost, an expensive appetite for a non-residential farm with no manure-producing animals.

The Farm is London's largest community farm. It was started in 2010 with the blessing of Surrey County Council, the local government authority, which owns the land and collects ground rent. The Farm is a cooperative and plans to offer shares within its local community. Other examples of similar social enterprises engaged in scaling up urban food growing include *Urbivore* in Stoke [32], *Farmscape* in Los Angeles [33] and *Lufa Farms* in Montreal [34].

The Farm is not solvent and there are no plans to make a profit. The goals are to make it pay for itself and to become a platform for food growing activities in the local community; examples include providing gardening experiences to school children and to disabled people. However, because of the Farm's location and the lack of local public transport, a visit must be a planned event.

While salad crops are still the most profitable output, accounting for around 1/3 of income and only 1/7 of acreage, the produce is consumed more locally than 50 years ago. About 3/4 is distributed to retail customers in vegetable boxes; this scheme currently has 142 customers, with a capacity for 350. The remaining 1/4 is sold wholesale, largely to local restaurants and cafés. The demographics of vegetable box customers reflect the local residential area: they are largely middle class. Many are seasonal customers who grow their own vegetables and therefore buy much less in the summer. The Farm's unsold produce is collected by a local charity that makes soup from it. Its two major expenses are compost (purchased from a local municipal site), for which haulage is the principal outlay, and water for irrigation.

The Farm's manager is a university graduate who used to be a chef and became interested in food security issues. He has organized an apprentice scheme at the Farm. His view is that expertise in managing small scale farms is generally lacking in the United Kingdom. The Manager also organizes volunteer gardeners, some of whom are employees of local businesses which pay them as they work on the Farm as part of a Corporate Social Responsibility program.

Most of the Farm's tilled land is devoted to leaf crops but it has not applied for Organic Certification. The Manager said that its production is "based on organic principles" and is open to anyone who wants to come and see for themselves. The Farm uses a small tractor but most of its work is manual. The sole full-time employee is the Manager. A "sustainable farming" apprentice grower is paid for three days per week. One grower is paid for one day per week. The vegetable box scheme has one employee working 3.5 days per week to deal with customers, and two drivers are employed, each for one day a week. The total paid labor is equivalent to 2.7 full-time workers.

3.3. The Park

The Park is located in Sunol, an unincorporated former railway town in Alameda County in the San Francisco metropolitan area. Sunol's annual sunshine days are about twice the U.S. average. The Park is adjacent to a Water Temple, a well-known local feature commemorating the opening of a water supply system. The land is owned by the San Francisco Public Utilities Commission (a water supplier) and is managed by Sustainable Agriculture Education (SAGE). The Park was set up in 2006 with the mission to foster sustainable farming and public education programs while protecting natural resources. It rents the land and is currently home to six small-scale organic farming enterprises on 6.5 of its 7.3 hectares. The Park is an example of an incubator farm. One of its models is the Agriculture and Land-based Training Association in Salinas (a farming town 129 km south of the Park and not in the San Francisco metropolitan area), which pioneered a farmworker-to-farmer program. The Salinas program trains Latino farmworkers to take on farm management and operation. The Park aspires to do something similar but in an urban periphery setting. Its brochure states that: "AgParks are an innovative, scalable model that facilitates land access for beginning and immigrant farmers, local food provision for diverse communities, resource conservation, public education, and job training opportunities."

While nominally in a metropolitan area, the Park is remote: there are no significant sources of laborers nearer than Pleasanton, 8 km to the northeast. The land was a hay field until it was taken over by SAGE, which acts as a non-profit intermediary between the farmers and the water utility. It collects rents and other charges from farmers and arranges a supply of irrigation water. It also maintains Organic Certification and permits for chickens. SAGE promotes mutual learning among the farmers, and implicitly tries to screen out any who are not serious or competent.

The soil of the Park is thin clay and very arid—not unlike the soil at the Farm, but much drier—so plentiful irrigation is essential. Compost must be applied at least annually to condition the soil and to ensure water retention. Aphid infestation is a particular problem. Organic practices are *de rigueur* so there is no input of synthetic fertilizers or pesticides. There is no access to grid electrical power.

About 2000 schoolchildren visit the Park annually, mainly from Oakland, also in Alameda County and about 25 km away. The children are aged 10–12 and are largely from low-income families; 60% are on their school district's Free & Reduced Meals Plan. SAGE tries to charge $2 per head as a contribution towards payments to the water utility but do not always succeed. SAGE also helps schools find grants towards the cost of bussing children.

A young and aspiring farming couple have been at the Park since March 2014, initially on 0.4 ha but now on 0.8 ha—the Happy Acre Farm. One was previously a manager at a farmer's market; the other had worked on an organic farm. The couple used their first season to learn the ropes; *i.e.,* what could successfully be grown and what was to be avoided—for example, not growing crops when "bugs" would seriously attack them. The first task was to remove the Bermuda grass (*Cynodon dactylon*) that covered the site. They did this by an undercutting technique taught by a local specialist who is something of a farming guru. They were greatly helped by another farmer; he taught them husbandry practices, introduced them to customers, and gave them some seeds. The couple emphasized that contact with more experienced farmers is one of the benefits at the Park.

The couple plan to keep themselves by farming but not to employ anyone, at least in the short term; in any case they recognize that finding laborers is problematic and expensive. They plan to augment income from produce by selling processed goods such as jellies and preserves and by running classes on making such goods. The couple plan to grow arugula (or rocket: *Eruca sativa*) on a rotation of three crops per year, hoping that this intensity of production will enable them to live off the land. They also plan to introduce chickens for their eggs and meat. The chicken paddock will be moved around to fertilize the soil, subject to SAGE regulation of how long land must be left fallow between keeping chickens and growing food on it.

Happy Acre Farm has been distributing produce to restaurants. In 2015, following a break of several months when they were not farming, the couple started a box system with the help of the Community Supported Agriculture program. This program, modeled on that developed in Switzerland and Japan in the 1960s, was adopted by the U.S. Department of Agriculture in the mid-1980s as a vehicle to help low income farmers find capital, labor, and dependable markets [35]. Associations of individual consumers pledge to financially support one or more local farms, sharing the risks and benefits of food production with growers. This particular farming duo will do the distribution themselves, using their own truck. They have targeted Oakland as a market and are hoping to distribute to neighborhood hubs from which individual consumers will collect produce. They have already spent about $25,000 on their enterprise for insurance, seeds, tools and transport fuel. They pay SAGE $2,000 per annum per acre for land rent and water. They buy compost from a supplier recommended by SAGE and plan on paying about $150 for 5 tons of compost per acre per annum.

3.4. Comparing the Three Cases

All three sites are nominally within the metropolitan areas of very large cities. The Garden is the most centrally located, followed by the Farm and then the Park, the most outlying. The Garden is an example of an inner city brownfield site co-opted for social benefits; it is completely dependent on volunteer labor and contributions. The Farm is an example of a low-productivity greenfield site that has

269

been transmuted into a social enterprise with some income for a paid work force. The Park is an incubator farm that collects rent from aspiring farmer-entrepreneurs. While the Garden supplies its own compost, the Farm and the Park must haul in large quantities. The Garden's relatively low water needs are supplied by a sufficiently damp microclimate but the Farm and the Park must import large quantities of irrigation water.

All three sites have outreach educational programs for their local communities. While the Garden has several schools a short walk away, children visiting the Farm and the Park must be transported. The Garden gets a large number of walk-in users and provides a sizeable cultural program, while the Farm and the Park produce considerably more food. The Garden can be described as place-based because it is embedded within a neighborhood, while the Farm and the Park are interest-based as their service areas are rather large districts [36].

The three cases illustrate the core-edge-periphery trinity of urban form, with the Garden being 8 km from its urban center (seat of the central city's government), the Farm 16 km, and the Park 32 km. The core encompasses a built-up, high-density center; the edge encompasses the city-to-suburb transition area; the periphery, the suburb-to-exurb transition area. Both population density and land value decline with distance from the center. Thus, the proportion of land available for growing food increases with distance from the core; at the same time fewer people have access to the land and distribution to consumers requires longer delivery journeys. In addition to residential density, cultural diversity declines with distance from city cores in these three metropolitan areas. The diversity includes the largest pool of potential urban gardeners: poor immigrants with farming experience.

4. Results: Food Production

4.1. Food Production Metrics

The food growing potentials of the Garden, the Farm, and the Park are shown in Table 1, estimated using average data for metric tons produced per hectare [37] rather than the actual outputs of the three sites. For these estimates, the sites are assumed to be devoted to fruit and vegetable production, although all three currently grow some flowers for aesthetic, commercial, and pollinating purposes.

Table 1. Annual potential fruit and vegetable production of the Garden, Farm, and Park [1].

	Garden	Farm	Park
Growing area (hectares)	0.05	1.42	6.48
Potential production (kilograms) [2]	535	16,264 [3]	69,336
Persons per day fed [4]	4	111	475

[1]: Assumes one harvest per annum for field crops; [2]: Yield figure is 10.7 metric tons/ha from Garnett [37]; [3]: Includes two annual harvests from polytunnels, which account for 7% of output; [4]: Based on a consumption level of 0.4 kg, the minimum recommended by WHO [38].

The World Health Organization recommends the consumption of at least 0.4 kilograms of fruit and vegetable per day in a healthy diet [38]. This is also a reasonable proxy for a sustainable plate, needed to "provide good nutrition" [39]. The "healthy" and "sustainable" plates both focus on higher portions of fruit and vegetable and lower portions of meat than are common in "Northern" diets. Using that minimal standard, the Garden can provide fruit and vegetable for four persons per day, the Farm for 111 persons, and the Park for 475 persons. This represents just "nibbles" of food for two reasons. Firstly, the recommendation of the U.K. National Health Service is that fruit and vegetables comprise 1/3 of a plate. At least as yet, none of the three sites produces any of the food comprising the remaining 2/3 of the plate, including grains, milk and dairy, meat and fish. Secondly, the food output does not approach serving a substantial portion of even fruit and vegetables for the populations of the sites' catchment zones. These catchment zones comprise the areas from which the Garden, Farm, and Park draw the largest share of their members/customers/tenants/visitors, and in which they provide social sustainability services.

Comparison of the food production of the three sites as compared to their catchment area populations reveals a very low per-person output. The Garden's output would provide fruit and vegetables for just 0.002% of the residents of its Upper West Side district (within a radius of 2.1 km), and just 0.02% of the residents of its immediate neighborhood (within a radius of 0.3 km) consisting of two of the 23 census tracts on the Upper West Side. The Farm's output would provide for only 0.06% of the residents of its district (within a radius of 2.4 km—Sutton borough) and just 0.24% of the residents of its immediate neighborhood (within 0.8 km—the suburb of Carshalton). The Park's production would provide for just 0.03% of its district's population (within 24 km—Alameda County) and only 0.1% of the residents of its neighborhood (within 12 km, consisting of its surrounding six suburban cities). Looking at this data from another perspective, it would take over 4500 gardens to provide the fruit and vegetable components of their diet for the people in its immediate neighborhood; over 400 farms and over 1000 parks to do the same.

Thus the three cases—Garden, Farm, and Park—illustrate the problem of scale facing urban food production. Other studies reflect a similar pattern: very low current output and low potential output, relative to the provision of healthy plates. Within London, current production is estimated as 1–2% of potential output; full potential output would still represent only 18% of current fruit and vegetable consumption [37]. Studies of other cities show comparable results: Cleveland [40], Detroit [41], New York City [42], Oakland [43] and Oxford [44]. These six urban areas grew an average of 2–3% and have an (unweighted) average maximum projected output of 20% of the fruit and vegetables consumed by their inhabitants. However, these figures are based on uniform distributions of consumption across urban populations; if urban produce is consumed by specific groups, the benefits may be more significant (see below).

There is wide variation among all seven urban sites, a finding that underscores the value of a case-by-case analysis of urban food growing. A major source of variation is between successful cities such as London, New York, and San Francisco, and distressed cities such as Detroit and Cleveland. The latter cities have considerably more available land; in Detroit, "abandoned houses, vacant lots and empty factories make up about a third" of the landscape [45] (p. 47) and therefore it demonstrates the highest level of potential production.

In reality, reaching the potential outputs will have to overcome some daunting conditions and issues. Cities struggle today to maintain their current green spaces. In London, the area of domestic gardens, which comprise 25% of the land upon which fruit and vegetables would be grown, is declining: between 1998 and 2008 it fell by 12% while the area of hard surfacing increased by 26%, largely paved over for car parking [46]. Land availability is but one example of an imposing array of structural challenges to scaling up urban food production. Globally, the land area used to produce just the vegetables in a healthy or sustainable diet is equivalent to about 3/4 of all urbanized land [47]. It is evident that there is not nearly enough arable land in urban areas to produce more than a small portion of the fruit and vegetables consumed by their residents. Of course, this leaves the remaining 2/3 of the sustainable plate unaccounted for.

More Food Production?

What are the possibilities of creating more urban food growing land by utilizing brownfield sites and by converting green spaces to food production? Both options have major drawbacks. With regard to brownfield sites, the condition of the soil is questionable. For example, a study of lead contamination in Oakland sites found a high level of variability that must be considered when undertaking food growing [43].

With regard to conversion of green spaces to food growing, there are issues of competition in supporting sustainability: these spaces already provide for carbon sequestration, urban cooling, and biological diversity. Any land use change to food production will have an uneven profile with regard to its environmental costs and benefits [48].

One solution proposed for the lack of arable urban land is vertical farming, or "z-farming" for zero acreage [49,50]. However, there are major sustainability obstacles for high-rise farming, including the inputs of energy for artificial lighting to grow plants away from windows (see earlier comments on underground farming, which are also relevant here) and the industrial fertilizers needed to optimize yields from hydroponic production [51]. These inputs add substantially to the environmental impacts: findings from a recent life cycle study indicate that "vertically grown produce has a carbon footprint that is much higher than conventionally grown produce" [52] (p. 76).

Rooftop gardens are another fashionable initiative that does not stand up to examination as a serious contribution to sustainable food production—as distinct from "green roofs," which can play a role in helping to mitigate urban heat islands [53]. Self-evidently, the area available for rooftop cultivation is strictly limited. Rooftop greenhouses are difficult to integrate into the waste management and recycling systems of their buildings [54]. Furthermore, roofs do not provide pavement level viewing and open access; in the absence of access to people other than the residents of the building, they cannot provide the social sustainability services of cultivation at ground level (see below).

In the end, the hope that urban food production might produce enough food to support the population within its borders is a utopian goal. The greatest opportunity for urban areas to reach a higher level of food security lies in the next tier of available land that is beyond the urban periphery: the broader region that is still largely rural. For example, in 2009–2010, 57% of London's consumption of fruit and vegetables was grown in the rural hinterland beyond its urban and peri-urban zones [55].

However, even assuming a full development of the broader foodshed region, it will still be necessary to bring in food, including cereal grains and exotic foods, frequently internationally traded. These imports are determined on a national case basis. For example, the United Kingdom's imported exotics include bananas, citrus fruits, coffee and tea. While our ancient ancestors ate only what they could find by walking within a hunter-gatherer food system, we humans of the Anthropocene have global-range appetites met by industrial-scale production and transport of agricultural products.

5. Results: Sustainability Services

5.1. Ecological Sustainability

Like all urban green spaces, the Garden, Farm and Park make contributions to ecological sustainability—by providing natural habitats, improving soil quality, reducing soil erosion, and mitigating the city heat island effect [17,53,56]. They may also reduce the runoff loss of rainwater exacerbated by the concreting over of cities and their environs [17]; this is significant where, as in London, a principal aquifer lies below the city.

In some respects, food growing plots may contribute more than other urban green spaces to ecological sustainability. However, a locally sensitive design is crucial in maximizing their potential. For example, Kulak *et al.* [21] carried out an LCA of the same Farm examined here. They found that reductions in greenhouse gas emissions are gained from an appropriate choice of local crops that can substitute for foods with high carbon footprints. Such crop prioritizing for sustainability also can be applied to broader multi-foodshed areas, as has been done for New York State [57].

Another way in which food growing may outperform other urban green spaces is that it shelters more biological diversity through its wide variety of flora—agricultural and horticultural. A key component of this diversity is the presence of bees to pollinate plants. The Farm in Sutton has three hives tended by a volunteer keeper and is close to woodland and commercial hives, while marigolds and other flowers are grown to encourage pollination. The Park in Sunol grows wild flowers and is serviced by bee colonies in its semi-rural locale. Despite being in the middle of a large city, the Garden in New York City has a good supply of bees from hives on nearby roofs and wild colonies in nearby Central Park [58]. Thus, all three sites support bee populations by providing a diversity of flora, paralleling the practice of spacing ribbons of flowers amidst mono-crop fields in rural areas.

5.2. Social Sustainability

Although its contribution to dietary provision will always be slight, urban food growing can contribute to two other components of social sustainability: environmental justice and public health. Both are needed now more than ever—environmental justice because of dramatically widening inequalities [59] and public health because of the new obesity epidemic [60]. There is abundant evidence of the ways in which urban green spaces contribute to physical, psychological, and social health [17,61–68], and growing evidence of their contributions to environmental justice. With regard to public health, gardening provides easily accessible opportunities for physical, mental, and social well-being. Growing food is a physical pursuit. Its physicality ranges from the fine motor involvement of cutting flower stems to the aerobic gross motor tasks of turning compost. While gardening promotes physical health, it also "has been observed to be a way to relax and release stress" [61] (p. 28). Finally, all types of urban green space provide natural locales for people living in densely-populated built environments. Access to nature can be a form of therapy, allowing for solitude, serenity, and reflection. This has been found to be related to mental health by mitigating a psychological nature deficit disorder [65].

With regard to environmental justice, urban food growing appears to produce more than just "nibbles of food" for some socially excluded sub-groups. One study shows that it can make a significant contribution to the tables of low-income immigrants from agricultural backgrounds [69]. Another study shows that it can make a substantial input to improving the diets of low-income persons with high rates of obesity and diabetes and with limited sources of fresh produce [70]. However, these are special cases, in which volunteer and experienced gardeners had convenient access to free plots of arable land, whereas this access is not usually available to low-income residents of cities in the global North.

Even a very small food growing space can contribute to environmental justice. An apt example is the half-acre Brook Park Community Garden in the Mott Haven neighborhood of the Bronx borough, one of the poorest communities in New York City. It employs a dozen teenage boys with criminal records to grow serrano peppers (*Capsicum annuum*), working under court orders as an alternative to incarceration. Their small stipends come from the profits from selling the garden's "Bronx Greenmarket Hot Sauce" [71].

However, it is in education, rather than environmental justice or public health, that the Garden, the Farm and the Park make their most impressive contributions to social sustainability. The inter-generational principle of sustainability relies on ecological education. The Garden reserves six plots for schoolchildren who participate in an ecology learning module during which they grow vegetables. As a follow-up to their experiences, children and their teachers have constructed several raised beds in their schoolyard. In New York City, the number of registered school-based gardens has multiplied six-fold [72]. (It is noteworthy that most of the adult participants in the Garden have had previous gardening experience, many in their childhoods.) The Farm operates a funded school program, the Green Grub Club, in which pupils and staff, after school, grow, cook and eat vegetables. In addition, 16 students and their caretakers from a local school participate in a sponsored Disabled Farming Assistance program.

The Park, like the Garden and the Farm, operates environmental education programs for school pupils. In an increasingly important aspect of education, the Park also provides a rare learning opportunity for urban young people to start farms. Increasing urbanization has progressively reduced the number of persons with farming knowledge. In 2012, the average age of U.S. farm operators was 58.3 years, up 1.2 years from 2007 and, continuing a 30-year trend, the number of beginning farmers was down 20% from 2007 [73]. This loss of farming expertise threatens both food security and climate change resilience. The Park addresses both, by inhibiting the conversion of farmland to settlement and by training a new generation of aspiring farmers. SAGE specializes in conserving peri-urban food growing based on the model of the European Association of Peripheral Parks, *Fedenatur,* used, for example, in Barcelona, Lille and Milano. The Sutton Farm addresses the same issues on a smaller scale by providing a sustainable farming apprenticeship program.

Informal education is also part of urban cultivation. For example, whether the land-limited Garden should grow food or flowers is a subject of continuing debate. There are four parties to the debate. Foodists make an environmental justice argument for converting flower plots to vegetables in order to shorten the queue for beds (now a year's wait) and to provide more opportunity for low-income persons to grow food. Ornamentalists make an aesthetic point about the beauty of flowers and gardening's social psychological rewards, which constitute a public health benefit. Pragmatists make an economic argument that flowers attract people who then contribute money and labor to the Garden. Ecologists make a sustainability case for the biological diversity added by flowers and the bees that depend on them.

A latent result of the Garden's debate is its contribution to the ecological knowledge of its participants. Gardeners hear from each other about some of the complexities of flora production and its relationship to sustainability. This communal learning is an example of the synergies that exist between ecological and social sustainability [74] and supports the argument here that small inner-city plots mainly have social value. The communality is a basis for the development of social capital rooted in its use value.

Growing food is both a physical and a mental activity. It simultaneously involves an active mind and body; thus, it is an embodied experience [75,76]. For this reason, it may have more learning impact in a person's life than do other environmental education activities. Communal learning takes place in urban food growing through a sharing of ecological observation and monitoring by gardeners [77]. There is also experiential learning, which can stimulate change in individual lifestyles. For example, food growers may gravitate to healthier diets (with more vegetables) and they may also take up sustainable practices such as composting. A good number of the Garden's members regularly carry food waste from their apartments to its compost bins, whether or not they have plots to tend.

6. Discussion

6.1. Urban Cultivation and Its Future

The estimates in Table 1 indicate that urban food growing in the global North does not now, and likely will not in the future, make more than a trivial direct contribution to food security. There is significant growth potential, with wide variations in the upper limits of production based on local circumstances, but urban food production can at best supply a limited proportion of some components of the diet of urban populations. There are strong structural factors that constrain production. The fundamental limitation is a lack of arable land. Field cultivation is simply beyond the land capacities of urban agglomerations. Cities can produce a small portion of the fruit and vegetable consumption of their residents but they lack the potential to grow the basic food of humanity—cereal grains, which are the stuff of the "staff of life." Cereals supply over half of the dietary energy of the human population [78] and large rural fields will remain the venue for their efficient production. Research indicates that peak economic efficiency is achieved in production units of 160–325 ha for soybean and 325–490 ha for corn ([79] cited in [80]).

Cities are places where great numbers of people live in small areas; they do not contain the expanses of ground level land fully exposed to the sun needed for field crops. Instead of food growing being the goal, it is more realistic to cast it as a secondary gain. This reasoning underlies our suggestion that cultivation rather than agriculture is the appropriate designation for urban food growing: cultivation better captures the sensibility as well as the output of this activity. Agriculture combines two Greek roots: *agros* for field and *cultura* for cultivation. As compared to agriculture, cultivation covers a range of etymological meanings beyond food production, including education and development, all of which are insufficiently recognized in the contemporary enthusiasm for urban food growing.

As to the future, the differences among the three sites of this study illustrate a dilemma for urban cultivation. The Garden is too small to provide a significant food output but has a high social amenity value due to its location and accessibility. The Farm is large enough to provide marketable food and some jobs, but its location makes its social value educational rather than amenity-based. The Park is too remote to provide for general environmental education (in a sustainable way) but the same remoteness allows for large tracts of land suitable for providing specialized farming educations. Given the demand and price for land in successful cities such as London, New York City and San Francisco, the distinction between urban and peri-urban and the demand for urban land are likely to persist although the pressures are currently mitigated in distressed cities like Detroit. At least for the present, being in the Green Belt provides a measure of protection against encroaching development for the Farm. The same can be said of the Park, although its relative remoteness already provides some measure of protection. Additionally, cultivable land can be generated by integrating it into urban new build and re-build plans. A study in Waterloo, Ontario, found that about half the land in its suburbs had the potential to support cultivation [81]. However, as we have indicated, assuring the land is not contaminated and can be sustainably converted from lawns to gardens is a formidable task.

6.2. A Systematic Scaling of Urban Cultivation

Despite scaling and sustainability obstacles, there is space available to continue the increase of food growing across urban areas—parcels of varying sizes of affordable and arable land. This growth should strive to meet local, national, and global sustainability standards, not only for its food production but also with regard to its ecological and social dimensions. These sustainability assessments will be better served if they take a systems perspective of urban food growing, with links among sites in the core, edge and periphery [80]. Such a system can connect diverse willing and able growers and consumers to accessible sites of arable land. Recognizing that urban and peri-urban cultivation play different roles in foodsheds suggests a different approach to land distribution: planning and regulation on a regional or ecosystem basis.

Foodshed is a concept that was developed by Hedden [82] in his 1929 book: "*How Great Cities Are Fed*". Hedden contrasted foodsheds with watersheds by noting that water flows depend on natural land elevations while food movements are based in economic markets. The term was reintroduced by Getz [83] in a 1991 article on permaculture. More recently it was used to describe "the geographic areas that feed population centers" [84] (p.1). One part of the foodshed is the urban nexus of core, edge, and periphery. The three sites examined here illustrate the wide variations in access to land, to people, and to sustainability benefits that exist in a metropolitan area across these three zones (see Table 2).

Table 2. Schematic for the coordination of a potential urban cultivation system.

Site	Scale	Area	Access to		Sustainability Benefits		
			Land	People	Food	Ecological	Social
Garden	micro	core	very low	very high	very low	very low	very high
Farm	meso	edge	low	high	low	low	mod-erate
Park	macro	periphery	moderate	moderate	moderate	low	low

One type of facility, the food hub, may be well placed to coordinate these variations to best advantage. Food hubs are urban facilities that engage in aggregation (which can include growing), preparation, distribution, and marketing of food. They tend to be social enterprises that make small profits and receive benefits (including tax relief) from local government [85]. Food hubs are proliferating rapidly—over 60% of those in a U.S. national study had begun operation in the last five years [86]. From that survey, the three most common food hub customers are restaurants, small grocery stores and school food services. About 3/4 of their customers live within 160 km. Food hubs are usually sited in the moderately dense urban edge. They are possible lynchpins in local food systems because they are best positioned to reach an entire urban nexus efficiently—inward to the core and outward to the periphery. There is potential for the network of food banks—a response to economic hardship and hunger in towns and cities in the United Kingdom and United States in recent years—to evolve into a system of food hubs that could offer not only access to affordable food but also services concerning nutrition, healthier living and urban cultivation. A U.S. initiative, the Healthy Food Bank Hub [87], is a case in point. We suggest that the food hub innovation is an important field for further research on sustainable urban cultivation.

7. Conclusions

The three cases examined here provide a new perspective on the current widespread enthusiasm for urban food growing. While structural limits will prevent urban food growing from becoming urban agriculture (at least in cities of the global North), there is a strong case to be made for it on the grounds of its contributions to social sustainability. Urban agriculture can produce little more than "nibbles" of food but it can contribute "oodles" of social sustainability services [19,20]. Identifying the real benefits enables some forms of urban cultivation, most obviously underground or "vertical" farming, to be recognized as no more than "magical realism." Realistic assessment leads to a basis for promoting urban cultivation as part of the physical and social structure of urban areas, and highlights a potentially important systemic role for Food Hubs.

To assess the real role of urban cultivation, we should perhaps be looking at food provision differently. If the food system is recognized as involving produce supplied by oligopolistic intermediaries (retailers) from ever more consolidated primary producers (industrial scale farmers), many parallels with *energy generation and distribution* become apparent, suggesting a need for reforms to promote sustainability and, in particular, resilience [88]. Urban cultivation is a local phenomenon, the base level in a *food system approach* ([89] cited in [90]). Its unique grassroots activities contain and incubate adaptive, flexible possibilities for social sustainability services whose effects can extend throughout food systems.

What if we actively promoted food hubs to produce some food but primarily as vehicles to promote the social dimension of sustainability, with sub-stations in residential areas (allotment tillers, keen back garden food growers, and community gardeners) acting as the equivalent of a localized energy grid? As with decentralized or community energy, the aim would not be to achieve self-sufficiency and grid independence, except at the margins, but instead to boost system-wide resilience via redundancy, diversity and storage–and also to generate social benefits, such as local collaboration and trust, healthier lifestyles and grassroots innovative capacities.

As for boosting sustainable food production, enhancing food security depends on surmounting an inventory of difficult challenges: (a) reducing waste, which accounts for up to 1/3 of production through the food chain [91]; (b) shifting crops away from animal feeds and biofuels to human foods, which can increase global calorie availability by up to 70% [92], and at the same time shifting to sustainable plates on the consumption side [93–96]; and (c) adopting sustainable intensification practices in which productivity is raised without increasing environmental impact and without using more land [5]. The looming food security threat will not be tempered by the limited amount of food that can be grown in urban areas. Nevertheless, urban food growing can play a small but significant role in evolving a sustainable food system by contributing both to reducing waste and to the adoption of sustainable plates—through the provision of environmental, dietary, and farming education, for example.

While this study is limited to just three cases, the results indicate some useful areas for further research, particularly in exploring the role of urban food growing in contributing to social sustainability services. Science has provided the basic ecological metrics needed to specify parameters for sustainable food security. The present gap in our knowledge is an understanding of the processes and practices necessary to develop the corresponding parameters within social structures—society has been the neglected child in the sustainability family. Urban cultivation provides a potentially informative vehicle for assessing the value and scope of social sustainability services and their synergies with ecological sustainability and food production. The success of assessment depends on the development of empirical measures of "soft data," positive services such as environmental education. This form of assessment requires a different approach from the "hard data" approach of conventional environmental LCA, focused on negative impacts, and suggests a direction in which social LCA should be developed.

Author Contributions: All authors contributed to designing the study, analyzing the data and drafting the paper.

Conflicts of Interest: The authors declare no conflict of interest.

References

1. McClintock, N. Why farm the city? Theorizing urban agriculture through a lens of metabolic rift. *Camb. J. Reg. Econ. Soc.* **2010**, *3*, 191–207.
2. Tortorello, M. Mother Nature's Daughters. *The New York Times*, 28 August 2014, p. D1.
   ` Available online: https://naturalchild.bandcamp.com/track/mother-natures-daughter (accessed on 25 March 2016).

3. Altieri, M. *The Scaling up of Agroecology: Spreading the Hope for Food Sovereignty and Resiliency, Rio+20 Position Paper*; SOCLA: Bruxelles, Belgium, 2012; Available online: futureoffood.org/pdfs/SOCLA _2012_Scaling_Up_Agroecology_Rio20.pdf (accessed on 25 March 2016).

4. Hill, A. A helping hand and many green thumbs: Local government, citizens and the growth of a community-based food economy. *Local Environ.* **2011**, *16*, 539–553.

5. Zezza, A.; Tasciotti, L. Urban agriculture, poverty, and food security: Empirical evidence from a sample of developing countries. *Food Policy* **2010**, *35*, 255–273.

6. Thedo, A.L. Global assessment of urban and peri-urban agriculture. *Environ. Res. Lett.* **2014**, *9*, 114002.

7. Yuan, L. Could Underground Farms be the Future of Urban Agriculture? 2015. Available online: https://thisismold.com/space/farm-systems/ (accessed on 25 March 2016).

8. Hamm, M.W. Feeding cities-with indoor vertical farms? 2015. Available online: http://www.cityfarmer. info/2016/03/26/feeding-cities-with-indoor-vertical-farms/ (accessed on 25 March 2016).

9. United Nations. *World Population Prospects, The 2010 revision*; Department of Economic and Social Affairs: New York, NY, USA, 2011.

10. Searchinger, T.; Hanson, C.; Ranganathan, J.; Lipinski, B.; Waite, R.; Winterbottom, R. *Creating a Sustainable Food Future: A Menu of Solutions to Sustainably Feed more than 9 Billion People by 2050*; World Resources Institute: Washington, DC, USA, 2013.

11. The New Climate Economy. Commission on the Economy and Climate, 2014. Available online: http://www.newclimateeconomy.report (accessed on 25 March 2016).

12. Seto, K.; Fragkais, M.; Guneralp, B.; Reill, M. A meta-analysis of global urban land expansion. *PLoS ONE* **2011**, *6*.

13. Intergovernmental Panel on Climate Change. Report of Working Group II: Impacts, Adaptation and Vulnerability. Fifth Assessment Report, 2014. Available online: http://www.ipcc.ch/report/ar5/wg2/ (accessed on 25 March 2016).

14. United States Department of Agriculture. *Climate Change and Agriculture in the United States: Effects and Adaptation, Technical Bulletin 1935*; United States Department of Agriculture: Washington, DC, USA, 2013.

15. Food and Agriculture Organization. *Statistical Yearbook*; FAO: Rome, Italy, 2013.

16. Fairlie, S. A Short History of Enclosure in Britain, 2009. Available online: http://www.thelandmagazine.org. uk/articles/short-history-enclosure-britain (accessed on 25 March 2016).

17. Royal Commission on Environmental Pollution. Twenty Sixth Report: The Urban Environment. The Stationery Office, London, 2007. Available online: https://www.google.com/url?sa=t&rct=j&q =&esrc=s&source=web&cd=2&ved=0ah UKEwjCu5aeiPzLAhULDiwKHS9KDZIQFggiMAE&url=http%3a% 2f%2fwww.rcep.org. uk%2freports%2f26-urban%2fdocuments%2furb-env-summary.pdf&usg=AFQjCNHxTZ -52COWOW2jdYM5zp7PTQK-kQ&sig2=MYOwyx-CbErG25ddNit7Uw&bvm=bv.118817 766,d.bGs&cad=rja (accessed on 25 March 2016).

18. Holland, L. Diversity and connections in community gardens: A contribution to local sustainability. *Local Environ.* **2004**, *9*, 285–305.

19. White, R.; Sterling, A. Sustaining trajectories towards sustainability: Dynamic and diversity in UK communal growing activities. *Glob. Environ. Chang.* **2013**, *23*, 838–846.

20. Durrant, R. Civil Society Roles in Transition: Towards Sustainable Food? SLRG Working Paper 02–14. Guildford: Sustainable Lifestyles Research Group, University of Surrey/Science Policy Research Unit, University of Sussex, 2014. Available online: http://sro.sussex.ac.uk/51587/ (accessed on 25 March 2016).

21. Kulak, M.; Graves, A.; Chatterton, J. Reducing greenhouse gas emissions with urban agriculture: A life cycle assessment perspective. *Landsc. Urban Plan.* **2013**, *111*, 68–78.

22. Harland, J.I.; Buttriss, J.; Gibson, S. *Achieving Eatwell Plate Recommendations: Is It a Route to Improving both Sustainability and Healthy Eating?*; National Health Service (NHS) Nutrition Bulletin: London, UK, 2012.

23. Clift, R.; Sim, S.; Sinclair, P. Sustainable consumption and production: Quality, luxury and supply chain equity. In *Treatise in Sustainability Science and Engineering*; Jawahir, I.S., Sikhdar, S., Huang, Y., Eds.; Springer: Heidelberg, Germany, 2013; pp. 291–309.

24. Benoit, C.; Mazijn, B. *Guidelines for Social Life Cycle Assessment of Products*; UN Environment Program: Nairobi, Kenya, 2009.

25. Paragahawewa, U.; Blackett, P.; Small, B. Social Life Cycle Analysis (S-LCA): Some Methodological Issues and Potential Application to Cheese Production in New Zealand. AgResearch, Hamilton, NZ. Available online: https://www.google.com/url?sa=t&rct=j&q=&esrc=s&source=web&cd=1&ved=0ahUKEwig1L_LqfzLAhXI jSwKHbueB5YQFggcMAA&url=http%3a%2f%2fwww.saiplatform.org%2fuploads%2fLibrary%2fSocialLCA-FinalReport_July2009.pdf&usg=AFQjCNEw-uG8bz1bm89o8VrsfvP_MiJgTA&sig2=QAhAZohkZWsVB3KAh ZGIpQ&cad=rja (accessed on 25 March 2016).

26. Lovell, S.T. Multifunctional Urban Agriculture for Sustainable Land Use Planning in the United States. *Sustainability* **2010**, *2*, 2499–2522.

27. Martin, G. Transforming a derelict public property into a vibrant public space: The case of Manhattan's West Side Community Garden. In Proceedings of the Royal Geographic Society Annual Meeting, London, UK, 31 August–2 September 2011.

28. Wilson, D. Urban revitalization on the Upper West Side of Manhattan: An urban managerialist assessment. *Econ. Geogr.* **1987**, *63*, 35–47.

29. Voicu, I.; Been, V. The effect of community gardens on neighboring property values. *Real Estate Econ.* **2008**, *36*, 241–283.

30. Tortorello, M. Growing Everything but Gardeners. *The New York Times*, 1 November 2012, p. D6. Available online: http://www.nytimes.com/2012/11/01/garden/urban-gardens-grow-everything-except-gardeners. html?rref=collection%2Fcolumn%2Fin-the-garden&action=click&contentCollection=garden&region=stream &module=stream_unit& version=search&contentPlacement=1&pgtype=collection&_r=0 (accessed on 25 March 2016).

31. Your Perfect Weekend. Time Out New York, 24 April 2009, 9.

32. Williams, R. Can an Urban Food Growing Project Cure a "sick city"? *The Guardian*, 12 June 2013, p. 34. Available online: http://www.theguardian.com/society/2013/jun/11/urban-food-growing-project (accessed on 25 March 2016).

33. Collins, E. Farmscape Grows Plant Managers. USA Today, 19 March 2013, B7.

34. Rifkin, G. Cash Crops under Glass and up on the Roof. *The New York Times*, 19 May 2011, p. B5. Available online: http://www.nytimes.com/2011/05/19/business/smallbusiness/19sbiz.html (accessed on 25 March 2016).

35. DeMuth, S. *Defining Community Supported Agriculture*; US Department of Agriculture: Washington, DC, USA, 1993.

36. Firth, C.; Maye, D.; Pearson, D. Developing "community" in community gardens. *Local Environ.* **2011**, *16*, 555–568.

37. Garnett, T. City Harvest: The Feasibility of Growing More Food in London. Sustain: The Alliance for Better Food and Farming. Available online: http://www.fcrn.org.uk/sites/default/files/CityHarvest/ (accessed on 25 March 2016).

38. World Health Organisation. *CINDI Dietary Guide*; WHO, Office for Europe: Copenhagen, Denmark, 2000.

39. United States Department of Agriculture. *Dietary Guidelines: The Sustainable Power Plate*; USDA: Washington, DC, USA, 2015. Available online: http://www.pcrm.org/health/diets/pplate/dietary-guidelines-usda-sustainable-power-plate (accessed on 25 March 2016).

40. Grewal, S.; Grewal, P. Can cities become self-reliant in food? *Cities* **2012**, *29*, 1–11.

41. Colasanti, K.; Litjens, C.; Hamm, M. Growing Food in the City: The Production Potential of Detroit's Vacant Land. In *East Lansing: The CS Mott Group for Sustainable Food Systems*; Michigan State University: East Lansing, MI, USA, 2010.

42. Ackerman, K. *The Potential for Urban Agriculture in New York City: GROWING capacity, Food Security, and Green Infrastructure*; Columbia University Urban Design Lab.: New York, NY, USA, 2012; Available online: http://www.urbandesignlab.columbia.edu/?pid=nyc-urban-argiculture (accessed on 25 March 2016).

43. McClintock, N.; Cooper, J.; Khandeshi, S. Assessing the potential contribution of vacant land to urban vegetable production and consumption in Oakland, California. *Landsc. Urban Plan.* **2013**, *111*, 46–58.

44. Food and Climate Research Network. *Food Printing Oxford: How to Feed a City*; Low Carbon Oxford: Oxford, UK, 2012.

45. Harris, P. Detroit gets growing. *The Observer Magazine*, 11 July 2010, pp. 42–49. Available online: http://www.theguardian.com/environment/2010/jul/11/detroit-urban-renewal-city-farms-paul-harris (accessed on 25 March 2016).

46. Smith, C. *London: Garden City?*; London Wildlife Trust: London, UK, 2010.

47. Martellozzo, F.; Landry, J.-S.; Plouffe, D.; Seufert, V.; Rowhani, P.; Ramankutty, N. Urban agriculture: A global analysis of the space constraint to meet urban vegetable demand. Available online: http://iopscience. iop.org/1748–9326/9/6/064025/article (accessed on 25 March 2016).

48. Fisher, S.; Karunanithi, A. Contemporary comparative LCA of commercial farming and urban agriculture for selected fresh vegetables consumed in Denver, Colorado. In Proceedings of the LCA Food Conference, San Francisco, CA, USA, 8–10 October 2014.

49. Despommier, D. The rise of vertical farms. *Sci. Am.* **2009**, *301*, 60–67.

50. Despommier, D. *The Vertical Farm: Feeding the World in the 21st Century*; St. Martin's Press: New York, NY, USA, 2010.

51. Specht, K.; Siebert, R.; Hartmann, I.; Freisinger, U.; Sawica, M.; Werner, A. Urban agriculture of the future: An overview of sustainability aspects of food production in and on buildings. *Agric. Hum. Values* **2013**, *8*, 1–18.

52. Al-Chalabi, M. Vertical farming: Skyscraper sustainability? *Sustain. Cities Soc.* **2015**, *18*, 74–77.

53. Bousse, Y.S. Mitigating the Urban Heat Island Effect with an Intensive Green Roof during Summer in Reading, UK. Master's Thesis, Reading University, Reading, UK, 2009.

54. Sanye-Mengual, E.; Oliver-Sola, J.; Anton, A.; Montero, J.I.; Rieradevail, J. Environmental assessment of urban horticulture structures: Implementing rooftop greenhouses in Mediterranean cities. In Proceedings of the LCA Food Conference, San Francisco, CA, USA, 8–10 October 2014.

55. Growing Communities. *Growing Communities Food Zone: Towards a Sustainable and Resilient Food & Farming System*; Growing Communities: London, UK, 2012; Available online: http://www.growingcommunities.org/ (accessed on 25 March 2016).

56. Gardening Matters: Urban Gardens. Available online: https://www.rhs.org.uk/Science/PDF/Climate-and- sustainability/Urban-greening/Gardening-matters-urban-greening (accessed on 25 March 2016).

57. Peters, C.J.; Bills, N.; Lembo, A.; Wilkins, J.; Fick, G. Mapping potential foodsheds in New York State by food group: An approach for prioritizing which foods to grow locally. *Renew. Agric. Food Syst.* **2012**, *27*, 125–137.

58. Satow, J. Bees high up Help Keep the City Green. The New York Times, 15 September 2013, 5.

59. Coote, A. *A New Social Settlement for People and Planet: Understanding the Links between Social Justice and Sustainability*; New Economics Foundation: London, UK, 2014.

60. Freund, P.; Martin, G. Fast cars/fast food: Hyperconsumption and its health and environmental consequences. *Soc. Theory Health* **2008**, *6*, 309–322.

61. Brown, K.H.; Jameton, A.L. Public health implications of urban agriculture. *J. Public Health Policy* **2000**, *21*, 20–39.

62. Cattell, V.; Dines, N.; Gesler, W.; Curtis, S. Mingling, observing, and lingering: Everyday public spaces and their implications for well-being and social relations. *Health Place* **2008**, *14*, 544–561.

63. Comstock, N.; Dickinson, M.; Marshall, J.; Soobader, M.-J.; Turbin, M.; Buchenau, M. Neighborhood attachment and its correlates: Exploring neighborhood conditions, collective efficacy, and gardening. *J. Environ. Psychol.* **2010**, *30*, 435–442.

64. Ferris, J.; Norman, C.; Sempik, J. People, land and sustainability: Community gardens and the social dimension of sustainable development. *Soc. Policy Adm.* **2001**, *35*, 559–568.

65. Louv, R. *Last Child in the Woods: Saving Our Children from Nature-Deficit Disorder*; Algonquin Books: Chapel Hill, NC, USA, 2008.

66. Pugh, R. How Gardening is Helping People with Dementia. *The Guardian*, 30 July 2013, p. 36. Available online: http://www.theguardian.com/society/2013/jul/30/dementia-gardening-helping-people (accessed on 25 March 2016).

67. Relf, D., Ed.; *The Role of Horticulture in Human Well-Being and Social Development*; Timber Press: Portland, OR, USA, 1992.

68. SDC. *Health, Place and Nature*; Sustainable Development Commission: London, UK, 2008.

69. Mares, T.M.; Pena, D.G. Urban agriculture in the making of insurgent spaces in Los Angeles and Seattle. In *Insurgent Public Space*; Hou, J., Ed.; Guerrilla Urbanism and the Remaking of Contemporary Cities: Routledge, London, UK, 2010; pp. 241–254.

70. McMillan, T. Urban Farmers' Crops Go from Vacant Land lot to Market. *The New York Times*, 7 May 2008, p. F1. Available online: http://www.nytimes.com/2008/05/07/dining/07urban.html (accessed on 25 March 2016).

71. Winnie, H. Hot Peppers Becoming a Cash Crop for Bronx Community Garden. *The New York Times*, 19 June 2015, p. A15.

72. Foderaro, L.W. In the Book Bag, more Garden Tools. *The New York Times*, 24 November 2012, p. A16.

73. United States. *2012 Census of Agriculture: U.S. Farms and Farmers*; Department of Agriculture: Washington, DC, USA, 2014.

74. Martin, G. Urban agriculture's synergies with ecological and social sustainability: Food, nature, and community. In Proceedings of the Brighton European Conference on Sustainability, Energy & the Environment, Brighton, UK, 4–7 July 2013.

75. Freund, P. The expressive body: A common ground for the sociology of emotions and health and illness. *Sociol. Health Illn.* **2008**, *12*, 452–477.

76. Turner, B. Embodied connections: Sustainability, food systems and community gardens. *Local Environ.* **2011**, *16*, 509–522.

77. Irvine, S.; Johnson, L.; Peters, K. Community gardens and sustainable land use planning: A case-study of the Alex Wilson Community Garden. *Local Environ.* **1999**, *4*, 33–46.

78. Global and Regional Food Consumption Patterns and Trends: Diet, Nutrition and the Prevention of Chronic Diseases. Available online: http://www.fao.org/docrep/005/ac911e (accessed on 25 March 2016).

79. Duffy, M. Economies of size in production agriculture. *J. Hunger Environ. Nutr.* **2009**.

80. Michael, H. City Region Food Systems—Part I. Available online: http://www.fcrn.org.uk/fcrn-blogs/ michaelwhamm/city-region-food/ (accessed on 25 March 2016).

81. Port, C.M.; Moos, M. Growing food in the suburbs: Estimating the land potential for sub-urban agriculture in Waterloo, Ontario. *Plan. Pract. Res.* **2014**, *29*, 152–170.

82. Hedden, W.P. *How Great Cities are Fed*; D.C. Heath: Boston, MA, USA, 1929.

83. Getz, A. Urban Foodsheds. *Permac. Act.* **1991**, *24*, 26–27.

84. Peters, C.J.; Bills, N.L.; Wilkins, J.; Fick, G. Foodshed analysis and its relevance to sustainability. *Renew. Agric. Food Syst.* **2008**, *24*, 1–7.

85. O'Hara, S. Food security: the urban food hub solution. *Solutions* **2015**, *6*, 42–52.

86. Center for Regional Food Systems, Michigan State University. Key findings from the 2013 National Food Hub Survey. Available online: http://www.msu.edu/foodsystems/uploads/files/fh-survey-key-findings (accessed on 25 March 2016).

87. Healthy Food Bank Hub. Available online: http://healthyfoodbankhub.feedingamerica.org/role-of-food-banks/ (accessed on 25 March 2016).

88. Smith, A.; Hargreaves, T.; Hielscher, S.; Martiskainen, M.; Seyfang, G. Making the most of community energies: Three perspectives on grassroots innovation. *Environ. Plan. A* **2015**.

89. Dahlberg, K. Regenerative food systems: Broadening the scope and agenda of sustainability. In *Food for the Future*; Allen, P., Ed.; Wiley: New York, NY, USA, 1993; pp. 75–102.

90. DeLind, L. Are local food and the local food movement taking us here we want to go? Or are we hitching our wagons to the wrong stars? *Agric. Hum. Values* **2011**, *28*, 273–283.

91. Kummu, M.; de Moel, H.; Porkka, M.; Siebert, S.; Varis, O.; Ward, P. Lost food, wasted resources: Global food supply chain losses and their impacts on freshwater, cropland, and fertilizer use. *Sci. Total Environ.* **2012**, *438*, 477–489.

92. Cassidy, E.; West, P.; Gerber, J.; Foley, J. Redefining agricultural yields: From tonnes to people nourished per hectare. *Environ. Res. Lett.* **2013**, *8*, 1–8.

93. Macdiarmid, J.; Kyle, J.; Horgan, G. Livewell: A Balance of Healthy and Sustainable Food Choices. Available online: http://assests.wwf.org/uk/downloads/livewell_report (accessed on 25 March 2016).

94. Sage, C. Addressing the Faustian bargain of the modern food system: Connecting sustainable agriculture with sustainable consumption. *Int. J. Agric. Sustain.* **2012**, *10*, 204–207.

95. Thompson, S. *LiveWell for LIFE: A Balance of Healthy and Sustainable Food Choices for France, Spain, and Sweden*; World Wide Fund for Nature: Woking, UK, 2013; Available online: http://livewellforlife.eu/wp-content/uploads/2013/02/A-balance-of-healthy-and-sustainable-food-choices.pdf (accessed on 25 March 2016).

96. Garnett, T.; Godfray, C. *Sustainable Intensification in Agriculture: Navigating a Course through Competing Food System Priorities*; Oxford University: Oxford, UK, 2012.

Agroforestry—The Next Step in Sustainable and Resilient Agriculture

Matthew Heron Wilson and Sarah Taylor Lovell

Abstract: Agriculture faces the unprecedented task of feeding a world population of 9 billion people by 2050 while simultaneously avoiding harmful environmental and social effects. One effort to meet this challenge has been organic farming, with outcomes that are generally positive. However, a number of challenges remain. Organic yields lag behind those in conventional agriculture, and greenhouse gas emissions and nutrient leaching remain somewhat problematic. In this paper, we examine current organic and conventional agriculture systems and suggest that agroforestry, which is the intentional combination of trees and shrubs with crops or livestock, could be the next step in sustainable agriculture. By implementing systems that mimic nature's functions, agroforestry has the potential to remain productive while supporting a range of ecosystem services. In this paper, we outline the common practices and products of agroforestry as well as beneficial environmental and social effects. We address barriers to agroforestry and explore potential options to alter policies and increase adoption by farmers. We conclude that agroforestry is one of the best land use strategies to contribute to food security while simultaneously limiting environmental degradation.

Reprinted from *Sustainability*. Cite as: Wilson, M.H.; Lovell, S.T. Agroforestry—The Next Step in Sustainable and Resilient Agriculture. *Sustainability* **2016**, *8*, 574.

1. Introduction

Agriculture shapes our planet in profound ways. Roughly 38% of the land surface of the earth is used to grow food, making agriculture the largest anthropogenic land use [1]. Expansion in agricultural land is the leading cause of deforestation and native habitat loss [2,3], a situation that has led to declines in wildlife, including birds [4], insects [5], and mammals [6], some of which are now considered endangered species [2]. Nutrient leaching from fertilizer results in the eutrophication of waterways, leading to oxygen deficient "dead zones" in water bodies around the world [7,8]. Agriculture is the largest human-caused contributor to the greenhouse gas emissions implicated in climate change [1,9].

Humans are not exempt from these effects. Pesticides in measurable quantities can be found in many environments, including the human body [10,11]. In the United States alone, the human health cost of pesticide poisoning has been estimated at $1.2 billion per year [12], and excess nitrate in drinking water caused by over-fertilization can cause illness and is costly to clean up [13–15].

In addition to environmental and human impacts, there are disconcerting implications for the resilience of our agricultural systems [16]. Worldwide, just fifteen crops produce 90% of food calories, with wheat, rice, and maize alone supplying 60% [17]. A majority of these crops are grown in vast tracts of annual monocultures which have a high risk for pest and disease outbreaks [18,19]. The Irish potato famine of 1845–1850 contributed to the deaths of over a million people and is a stark reminder of what can happen when disease destroys a single crop that is relied upon too heavily [20]. These monocultures require yearly replanting, high inputs, and weed control [21], and it has been suggested that this cycle of plant-fertilize-spray tends to serve the interests of the large agribusiness companies who supply the inputs for this system more than furthering the goal of feeding the world [22].

The long-term sustainability of any agricultural system requires that soils stay productive and that necessary inputs remain available in the future. However, soil loss occurs more rapidly than soil creation in many agricultural landscapes [23], and the soil that remains tends to decline in quality [24]. Heavy reliance on fossil fuels in the form of liquid fuel and fertilizer makes agriculture subject to fluctuations in fuel costs and supply [25]. One-way fertilizer nutrient flows simultaneously cause pollution and scarcity. Phosphorus is one example: this essential plant nutrient is expected to become increasingly expensive to mine and process, while, at the same time, phosphorus runoff causes eutrophication of water bodies [26,27].

In the near future, our agricultural systems will also have to adapt to a changing climate that is expected to bring more extreme weather events like droughts and floods, in addition to increases in outbreaks of diseases and pests [28]. The changes will be more severe in the developing world, where poverty hinders people's ability to adapt [29,30]. The Dust Bowl of the 1930s is an example of destructive agricultural practices paired with an extreme drought that led to catastrophic consequences [31]. Agricultural overreach along with the inability to adapt to changes in climate has toppled civilizations, from the ancient Mesopotamians to the Mayans [32,33].

2. The Rise of Organic Farming

Organic agriculture arose as an alternative to the conventional farming paradigm, pioneered by early practitioners such as Rudolf Steiner in Europe in the 1920s, Sir Albert Howard and Lady Eve Balfour in the UK and J.I. Rodale in the United States in the 1940s, and Masanobu Fukuoka in Japan in the 1970s and 1980s [34]. Several terms were used in these agricultural movements, including "organic", "biodynamic", "ecological", and "biological" [35]. In 1990, the United States Department of Agriculture (USDA) standardized the definition of organic production in the US, giving consumers and producers alike a common understanding of what "Certified Organic" means [35].

Although differing slightly by country and certifying agency, the main guidelines for organic management prohibit the use of synthetically produced pesticides and fertilizers, genetically modified organisms (GMOs), and the prophylactic use of antibiotics in livestock feed. Soil quality must be maintained through various practices such as crop rotation, cover cropping, or mulching [36]. Animals under organic management must be fed certified organic feed and ruminants must have access to pasture for a prescribed number of days [36]. Fertility is typically maintained by leguminous cover crops, applications of manure and compost, biologically derived inputs such as blood and feather meal, and mined mineral substances [36]. Weeds in organic grain and vegetable systems are usually controlled through tillage, though cover cropping and crop rotation also play an important role in breaking up weed cycles [37]. Pests control entails providing habitat for beneficial predators, selecting resistant plant stock, and using biologically derived pesticides as a last resort when needed [36].

The guidelines of organic production usually lead to more sustainable outcomes on the ground. Organic farms foster higher biodiversity than conventional farms, including insects, plants, soil biota, and even birds and larger animals [38–40]. Often, organic farms are more diverse in their cropping systems due to the inclusion of livestock and longer crop rotations [10]. The use of mechanical and cultural control methods for weeds and other pests can leave low levels of these populations that further contribute to biodiversity [40]. Soil quality tends to improve under organic management based on measurements of soil organic matter [39,41], though no-till conventional agriculture measured highest of all in some studies [9]. Although organic yields typically lag behind conventional yields [42], in drought years the opposite has been shown, which is attributed to the higher water holding capacity of soils under organic management [43,44]. Overall, organic production uses less energy per production unit due to the high energy costs of conventional fertilizer and pesticides [39,44,45].

Worth noting is the fact that, although organic certification makes hard distinctions about the use of pesticides, synthetic fertilizers, and GMO technology, a wide spectrum of practices are available for both conventional and organic producers that have beneficial environmental outcomes. Cover cropping, integrated pest management, application of manure and composts to build soil organic matter, crop rotation, and the integration of livestock and crops are important tools that should not be overlooked when considering impacts. Indeed, in some studies that compared organic vs. conventional crop systems, the authors conclude that improvements under organic management were likely due to practices like manure application and cover cropping that were included in the organic system which could be employed in a conventional system to similar effect [46–48].

3. Challenges in Organic Agriculture

Even with the good intentions of organic certification practices, most organic crop production systems utilize the same basic methodology as conventional farming, and therefore can have some of the same negative consequences. The pattern of cultivating annual monocultures that require yearly replanting, application of fertilizer, intensive weed control, and highly mechanized equipment to accomplish the work remains relatively unchanged, especially at scales larger than small market gardens [49]. The undesirable conventional tools are simply swapped out for those that are more benign: organic seeds for GMO seeds, cultivation or mulch instead of herbicides for weed control, and cover crops and manure for fertilization instead of fossil-fuel derivatives [36]. Although these changes can lessen environmental impacts, they may not eliminate them.

The issue of nitrogen leaching offers a good example of environmental impacts that are not eliminated entirely. Even though some studies show an improvement in nitrate leaching under organic management, the levels may still contribute to groundwater pollution. Pimentel *et al.* compared three rotations with differing sources of nitrogen: an organic rotation with legume cover crops, an organic rotation with animal manures, and a conventional rotation utilizing synthetic fertilizers. They found that leachate samples for all three treatments sometimes exceeded the 10 ppm regulatory limit for nitrate concentration in drinking water. The organic animal, organic legume, and conventional rotations lost 20%, 32%, and 20%, respectively, of the nitrogen applied to the crops in the form of nitrate [44]. In Swedish studies, Bergström *et al.* concluded that organic sources of nitrogen leached more than conventional fertilizers. They attributed this to the fact that the manures and legume cover crops released the most nutrients during fallow periods or at times that did not sync with nitrogen demand of the crop [50].

Even though soil quality can improve under organic management relative to conventional management [39,44,51], soil loss and degradation are still risks due to the fact that tillage is required for weed control and for incorporating biomass from cover crops [37,46]. Tillage has been shown to have adverse effects including compaction, erosion, and lowering of biological activity in the soil [23,52,53]. As reported in Arnhold *et al.*, studies comparing erosion in organic and conventional systems have had variable results that depend upon the crop rotation, crops used, and tillage systems. The authors' study in mountainous regions in Korea concluded that soil loss under either conventional or organic management was too high for sustained productivity [54].

Recognizing the benefits of reducing tillage, there has been interest in adapting no-till techniques for organic farming [37]. The process usually entails growing a cover crop ahead of the main cash crop, then crushing it down mechanically and planting through the residue [55]. If done correctly, weeds are suppressed by the mulch and no cultivation is needed for that crop. However, it can be a challenge to grow the necessary biomass in the cover crop to provide effective weed control, and the technique may not be possible in water-limited environments due to water competition by the cover crop [37]. Perennial weeds pose a particular problem, as they are typically able to grow through the mulch [37,42,55].

Studies exploring the impact of organic agriculture on greenhouse gas emissions have shown mixed results [35,41,56]. When measured on a per area basis, organic systems may fare better than conventional systems, but when the yield gap in organic is taken into effect, the emissions may be higher per unit of output [39,48]. Even when soil carbon increases, other gasses such as nitrous oxide are emitted by annual systems that contribute to climate change, negating potential benefits [9].

Differences in yields between organic and conventional systems may also have indirect environmental implications. Organic systems are generally agreed upon as less productive, with an average decrease in yield of around 20% to 25%, though the literature shows ranges anywhere from 5% to 50% depending upon the crop, soils, intensity of management, and methods by which the study was conducted [38,42,57]. Critics argue that under organic management, more land would need to be put into agricultural production in order to maintain global food security. This would result in deforestation and other habitat loss, leading to an overall negative environmental outcome [42,58].

Given these challenges within the organic/conventional debate, there seems to be an opportunity to evaluate additional tools and techniques that may yield other possible solutions. Instead of an "either-or" approach to thinking about our agricultural landscapes, a "yes-and" mentality might be more useful. Indeed, many have called for a multidisciplinary, multifunctional approach to designing agroecosystems [24,39,59,60]. In terms of feeding the world while sustaining the planet, perhaps Foley enunciates this best: "No single strategy is sufficient to solve all our problems. Think silver buckshot, not a silver bullet" [1] (p. 65).

4. Agroforestry as a Transformative Solution

One multifunctional approach for our food system is agroforestry, the intentional combination of trees and shrubs with crops or livestock. Agroforestry has been recognized for nearly half a century as a sustainable agricultural practice [61], and the concept of integrating trees into the agricultural landscape is as old as the practice of cultivating land. The beneficial outcomes of agroforestry include reductions in nutrient and pesticide runoff, carbon sequestration, increased soil quality, erosion control, improved wildlife habitat, reduced fossil fuel use, and increasing resilience in the face of an uncertain agricultural future [21,62–67]. In short, adding trees and other perennials to a landscape can help mitigate many of the harmful effects of agriculture. The fact that it can simultaneously provide economic, ecological, and cultural benefits gives agroforestry great potential as a land use strategy in both the developing and developed world [68].

4.1. Agroforestry Practices and Products

In addition to the environmental benefits, agroforestry can supply products such as timber, crops, fruits, nuts, mushrooms, forages, livestock, biomass, Christmas trees, and herbal medicine [69]. A diverse portfolio of products would allow revenue streams to be spread out over the short-term (crops, forage, livestock, mushrooms, certain fruits like currants), medium-term (nuts, fruits such as apples or persimmons, biomass, medicinal plants), and long-term (lumber, increased property value). This diversity of products can also reduce risk for farmers, though it may require creative marketing [69].

Different types of agroforestry are practiced across the world. Tropical agroforestry has traditionally enjoyed more focus and has been more widely adopted than temperate agroforestry. Systems like shade-grown coffee and tea are well developed, and the availability of hand labor makes some tropical agroforestry practices more practical than in areas where machine harvesting is more common [31,70]. Culturally, agroforestry has played an important role in both indigenous tropical areas and in temperate places like Europe, though land abandonment and agricultural intensification in northern areas has led to declines in traditional agroforestry practices [71]. This review focuses primarily on temperate agroforestry.

There are five generally recognized agroforestry practices promoted in the temperate zone, especially in North America: alley cropping, silvopasture, riparian buffers, windbreaks and forest farming [67,69]. These practices fit within a variety of cropping systems, topographies, and climatic zones.

4.1.1. Alley Cropping

Alley cropping involves growing field crops between rows of trees [72]. The trees can be grown for timber or fruits and nuts, while the alley crops can include a variety of grains, vegetables, or forages cut for hay. The crops provide short-term income while the trees provide longer-term revenue. The tree and crop species also may interact in ways that allow increased production due to the different niches that the trees and crops occupy [73]. For example, one study in France showed walnuts and winter wheat to be good companions because they grow at different times of the year and have differing rooting depths. The researchers concluded that the system produces 40% more product per given area than if the two crops were grown separately [74].

4.1.2. Silvopasture

Silvopasture incorporates livestock into an intentional mixture of trees and pasture. Silvopasture is different from just "grazing the woods", because the spacing of the trees is carefully planned to allow enough sunlight for the forages below, and the livestock are kept from damaging the trees. The trees offer protection for livestock through shade during the heat of the summer and wind reduction in the cold winter [75,76]. Additionally, the pasture quality in partial shade may increase, although it is usually slightly less productive in terms of biomass [77]. Livestock grazed on silvopasture versus open pasture show equal gains [76]. If the trees are also being grown for timber, the long-term bottom line of the farmer will improve without compromising current production [76].

4.1.3. Riparian Buffers

Riparian buffers are planted areas around waterways that are at risk from erosion, nutrient leaching, or habitat loss [78]. Usually there are two or three "zones" of vegetation that vary in composition based on the proximity to the waterway, slope, and producer needs [69]. Riparian zones tend to be marginal for agricultural production, making them prime candidates for alternative uses. There has been concerted effort by the United States Department of Agriculture (USDA) to implement conservation practices on areas around waterways due to their beneficial impact on water and soil quality. The Environmental Quality Incentive Program (EQIP) through the Natural Resources Conservation Service (NRCS) and Conservation Reserve Program (CRP) through the Farm Service Agency (FSA) are examples of some government funded initiatives [79].

4.1.4. Windbreaks

Windbreaks, also known as shelterbelts, were recognized early on as a useful agroforestry practice. Windbreaks prevent wind erosion, provide habitat for wildlife, and can increase water availability to nearby crops due to lower evapotranspiration and the effects of catching snow [75]. More water can mean higher production, leading to important economic benefits to farmers [80]. On a farmstead, windbreaks can decrease the heating and cooling needs for living and working spaces by reducing indoor air exchange caused by wind [81].

The Dust Bowl years in North America led to the U.S. government initiating the Prairie States Forestry Project, a massive shelterbelt stretching from Canada to Texas [75]. Another notable example is China's Three-North Shelter Forest Program, the world's largest afforestation effort [82]. Started in 1978 and expected to be completed in 2050, it is known as "China's Great Green Wall" [82]. Similar strategies have been employed in Russia, northern Europe, Australia, New Zealand and other countries [75,80,83].

4.1.5. Forest farming

Forest farming includes practices such as raising mushrooms, harvesting medicinal herbs like ginseng and goldenseal, and marketing woody ornamental material [69]. This agroforestry approach usually occurs in established forests that are grown for timber and allows for income generation without major disturbance [84]. Management of forest farming systems can range from intensive to minimal, depending on the product and desired market. For example, woods-grown ginseng may involve extensive site preparation, fertilizer, tillage, and fungicides that can increase yields but are costlier and therefore riskier. Alternately, wild-simulated ginseng may involve simply raking leaves back, planting seeds, and letting the ginseng grow for several years until it is ready to harvest [69].

It is noteworthy that, of the five practices, only alley cropping and silvopasture are typically practiced on land that is suitable for conventional agriculture. Even then, conventional cropping is often continued for several years before the trees are fully grown [69]. Riparian buffers, windbreaks, and forest farming usually occur on field margins or on land not suitable for farming, although, in some cases, may require setting aside some cropland to obtain the required width to be effective [85]. These practices therefore tend to complement, rather than compete with, existing production systems and may provide ways to contribute to food security by using resources that are otherwise underutilized.

In practice, agroforestry can contribute to either conventional or organic systems. In either case, the beneficial effects of agroforestry can improve environmental outcomes beyond what is already possible within each system. In this way, agroforestry may be able to address some of the challenges outlined earlier for organic agriculture, including soil loss, greenhouse gas emissions, and nutrient leaching. The next section summarizes these benefits, as promoted in the agroforestry literature.

4.2. Benefits of Agroforestry

Agroforestry has positive effects on soil and water quality. Soil quality is improved by increased levels of organic matter, more diverse microbial populations, and improved nutrient cycling, which may increase crop productivity and the ability to cope with drought [65,86,87]. The water quality benefits occur as non-point source pollution from row crops is reduced by incorporating agroforestry vegetative buffer strips [88–90]. On a "paired" watershed study in Missouri, agroforestry and grass buffer strips reduced phosphorus and nitrogen loss from a corn-soybean rotation [88]. The perennial vegetation increases above-ground biomass that slows runoff and can trap as much as 95% of the sediment at risk of being lost [91], while the below-ground roots can take up 80% or more of excess nutrients as well as hosting microbial populations that can break down pesticides [68,90,92].

The increase in soil organic matter in the form of carbon not only improves the health of the soil, but it can also help reduce atmospheric carbon dioxide that is implicated in climate change [23]. Compared to a monoculture of crops or pasture, adding trees and shrubs to an agricultural landscape increases the level of carbon sequestration [65,93]. Kim *et al.* did a meta-analysis on greenhouse gas emissions in agroforestry and showed an overall mitigation of 27 ± 14 tons CO_2 per hectare per year. Biomass accounted for 70% of sequestered carbon, with the remaining 30% sequestered in the soil [94]. A North American analysis performed by Udawatta and Jose showed that agroforestry practices implemented on a modest scale could potentially sequester 548.4 Tg carbon per year, enough to offset 34% of US emissions from coal, oil, and gas [85].

The mechanisms for increased carbon sequestration include better erosion control, more carbon being stored in woody perennials, reduced organic matter decomposition, and the fact that crop biomass is not harvested in agroforestry to the degree that it is in conventional systems [94].

The link between perennial systems and climate change may be an important one. Robertson *et al.* studied the global warming potential of several annual and perennial systems. They found that none of the annual cropping systems reduced global warming potential, whether conventional, no-till, reduced input, or organic. Although the cropping systems did accumulate carbon in the soil, the gains were offset by nitrous oxide emissions. However, the perennial and early successional forest treatments including alfalfa, hybrid poplar, and abandoned early successional sites all reduced global warming potential. Mid-successional and late successional systems stored less carbon per year as they matured. The authors concluded that the best option for mitigation was the early successional forest system [9]. Many agroforestry practices effectively mimic these early successional forests.

Reducing fossil fuel use is another important strategy for climate change mitigation [95]. Bioenergy is one avenue to reduce fossil fuel dependence, but there are concerns about using valuable cropland to grow crops for energy instead of food [16]. Currently, 40% of the U.S. corn harvest goes to producing ethanol, which seems counterproductive to the goal of reducing world hunger [16]. By producing biomass from trees in combination with food on the same land, agroforestry may be one way to contribute to a secure energy future without compromising food production capabilities [96,97].

When comparing mixes of species (*i.e.*, polycultures) with individual crops, a useful measure is the LER, or Land Equivalent Ratio [98]. This metric considers the yield of the polyculture and calculates the amount of land that would be required if the crops were grown separately. For example, when comparing loblolly pine and switchgrass mixes with pure stands of each crop, Haile *et al.* noted that, although each crop yielded less in the mix, the system produced an overall LER of 1.47 [99]. This means that if switchgrass and loblolly pine were grown separately, it would require 47% more land than the agroforestry system to grow the same amount of biomass.

Modeling of agroforestry systems in Europe using the Yield-SAFE (YIeld Estimator for Long-term Design of Silvoarable AgroForestry in Europe) model predicted LER values between 1–1.4 for scenarios in Spain, France, and the Netherlands, indicating higher productivity when integrating trees and crops than when grown separately [100]. In another study in Switzerland, agroforestry models focusing on walnut (*Juglans* hybrid) and wild cherry (*Prunus avium*) showed that in 12 out of 14 scenarios, mixing crops led to LER measurements higher than one. In addition, 68% of the Swiss financial scenarios were found to be more profitable than current practices [101].

When compared with conventional and organic monocultures, agroforestry contributes to the conservation of biodiversity. Adding trees, shrubs, and other perennial vegetation to an agricultural landscape provides habitat for greater numbers and more diverse populations of wildlife [68,90]. In addition to intrinsic value, biodiversity can provide useful services. More birds and predatory insects can help keep pests under control [19,102]. Habitat for pollinator species can mean better pollination of horticultural crops [103]. Even incidences of disease generally decrease in more diverse populations, for both plants and wildlife [104,105].

Livestock can benefit from agroforestry as well. Windbreaks protect animals from harsh winds, while shade provided by trees can increase comfort in the heat of the summer and may encourage more even grazing over a paddock [71]. Forest-based foraging systems for poultry and hogs can decrease the need for grain and provide surroundings closer to these species' natural habitat [106]. The cork oak *dehesas* of the Mediterranean are an example of a multifunctional landscape that has endured for hundreds of years, providing grass and acorns for grazing livestock and a valuable cash crop in the form of bark for making traditional corks [107].

Compared to annual monocultures, perennial polycultures like agroforestry are inherently more stable in the face of global market volatility and extreme climatic events [16]. In the event of fossil fuel scarcity, mature fruit and nut trees would continue to produce their products with relatively little interruption, though labor may have to be substituted for other inputs. Not only do agroforests sequester greenhouse gasses that are driving global climate change, they are also more resilient to its likely effects. Deeper rooting systems and improved infiltration and water storage lessen the impact of drought, while trees' abilities to pump excess water out of the soil as well as withstanding inundation better than field crops means they are also more resilient to floods [30].

Though often overlooked, there are additional cultural benefits to agroforestry. Many landowners value the preservation of nature, both for its beauty and for perceived benefits including a sense of improved health and the peace and quiet of a rural life [108]. Research shows that aesthetics provided by practices such as vegetative buffers are preferred by rural residents [109]. There are also opportunities for recreation, including bird watching, nature hikes, and hunting [110].

5. Challenges to Agroforestry Adoption

The opportunities for agroforestry are exciting, but not without challenges. Agroforestry adoption has been surprisingly low, considering the well-documented benefits [111–113]. Barriers have included the expense of establishment [114], landowner's lack of experience with trees [108,113], and the time and knowledge required for management [115].

Many farmers learn about new agricultural practices through extension personnel or agricultural product dealers, and these professionals typically do not have training or experience with agroforestry [116]. In addition, lack of established demonstration plots makes it hard for landowners to see these systems in action [3]. Since many of the useful outcomes from agroforestry are less tangible or longer-term, it may be difficult for landowners to envision them [117].

For agroforestry systems that produce edible products such as fruits and nuts, the logistics of harvest can be challenging. For agroforestry systems to be economically competitive, mechanization may be required for larger plantings [118]. This can be complicated if multiple fruit or nut species are grown.

Non-traditional markets and delayed profits may be another deterrent [108]. The economic feasibility of some agroforestry systems such as silvopasture have been shown to be profitable, whereas other practices such as biomass plantings or riparian buffers may need the development of markets that offer compensation for the ecosystem services provided in order to make financial sense [62,97,119]. Social change and networking will also play a role as mindsets evolve to include alternatives to the norm [112,114].

Moving Forward—Policy and Research Needs

Given these challenges, a number of strategies have been proposed to move agroforestry forward. Policy changes could include increased funding for government cost-share programs for installing practices and credits for environmental services rendered, such as pollination and carbon sequestration [68,97,113,116]. Current USDA programs through the NRCS and FSA often stipulate that land set aside for conservation may not be harvested, but agroforestry systems could provide a harvestable product without compromising conservation potential. A policy change to allow non-destructive harvest of consumable products from such systems might encourage more farmers to adopt agroforestry practices, leading to better conservation outcomes [68].

Although it is reasonable that the majority of government funding goes toward major cropping systems such as corn and soybean, the fact that agroforestry has the capability to remediate the negative effects of these very systems suggests it should be given more attention [68]. Some of this support could be used for education through extension and university programs [120]. In fact, education may be the most important factor for adoption, as many studies on the adoption of conservation practices cite lack of access to information and technical assistance as one of the primary barriers [3,108,116,120].

The opportunity to expand the production potential of agroforestry systems is underdeveloped. More research is needed to study the use of trees and shrubs to provide marketable products [121]. Recently, interest has grown in the development of multifunctional, edible polycultures that mimic natural ecosystems such as the native oak savannas of the Midwest [122]. These polycultures include multiple crops stacked together to take advantage of different ecological niches as well as to provide multiple streams of income [123]. For example, field trials at the University of Illinois at Champaign-Urbana were established to study a mixture of chestnuts, hazelnuts, apples, currants and raspberries. Control plots of a conventionally managed corn and soy rotation will allow for comparative analysis of a variety of environmental, ecological, and economic metrics. A large-scale, replicated study established in 2015 will look at different spatial layouts of these polycultures compared to monocultures of each species as they might be grown in a commercial orchard, in addition to being able to compare them to a corn/soybean rotation. Included in the treatments are plantings of native trees and shrubs that also have edible products, including aronia, elderberry, pecan, pawpaw, persimmon, plum, and serviceberry. This native edible plot explores what is possible within the confines of conservation easements that mandate the use of native species [123].

6. Conclusions

Various pathways have been proposed to safely and sustainably feed a growing population. Organic farming shows promise for lowering the use of agrichemicals and improving certain environmental and human health metrics, while proponents of conventional systems point out the advantages of using genetic engineering, fertilizers, and pest control in improving yields.

Broader strategies include limiting the expansion of farmland via deforestation, minimizing food waste, eating less meat, closing the yield gaps for underperforming cropland in the developing world, and more efficient use of resources like water, fertilizer, and fuel [1,48]. These efforts, and others, will be needed as part of a multi-faceted approach if we are going to successfully and sustainably feed the world.

Nature produces its bounty while requiring no plowing, no fertilizer, and no pest control—in fact, no inputs of any kind. It runs entirely on solar energy and generates no harmful waste products. Its biological diversity allows dynamic adaptation in the face of external change. If our agricultural systems can more closely mimic the functionality of nature, they can become more stable and resilient. Building such a system is without a doubt a challenging task, requiring a variety of tools. Agroforestry can provide the next step in sustainable agriculture by promoting and implementing integrated, biodiverse processes to increase yields, decrease harmful effects, and advance our understanding of the complex interactions involved in increasing food production while minimizing damage.

Acknowledgments: This work was made possible by the Jonathan Baldwin Turner Fellowship through the Department of Crop Sciences at the University of Illinois, Champaign-Urbana.

Author Contributions: Matthew Heron Wilson conducted the initial literature review and contributed a majority of the writing, and Sarah Taylor Lovell provided direction, editing, and proofreading, including writing revisions of some sections.

Conflicts of Interest: The authors declare no conflict of interest.

Abbreviations

The following abbreviations are used in this manuscript.

USDA	United States Department of Agriculture
NRCS	National Resources Conservation Service
FSA	Farm Service Agency
EQIP	Environmental Quality Incentive Program
CRP	Conservation Reserve Program
LER	Land equivalency ratio
GMO	Genetically modified organism
Yield-SAFE	Yield Estimator for Long- term Design of Silvoarable AgroForestry in Europe

References

1. Foley, J.A. Can We Feed the World and Sustain the Planet? *Sci. Am.* **2011**, *305*, 60–65.
2. Norris, K. Agriculture and biodiversity conservation: Opportunity knocks. *Conserv. Lett.* **2008**, *1*, 2–11.
3. Jacobson, M.; Kar, S. Extent of Agroforestry Extension Programs in the United States. *J. Ext.* **2013**, *51*, Article 4.
4. Johnson, R.J.; Jedlicka, J.A.; Quinn, J.E.; Brandle, J.R. Global Perspectives on Birds in Agricultural Landscapes. In *Integrating Agriculture, Conservation and Ecotourism: Examples from the Field*; Campbell, W.B., Ortiz, S.L., Eds.; Springer: Dordrecht, The Netherlands, 2011; pp. 55–140.

5. Thomas, J.A.; Telfer, M.G.; Roy, D.B.; Preston, C.D.; Greenwood, J.J.D.; Asher, J.; Fox, R.; Clarke, R.T.; Lawton, J.H. Comparative Losses of British Butterflies, Birds, and Plants and the Global Extinction Crisis. *Science* **2004**, *303*, 1879–1881.

6. Daleszczyk, K.; Eycott, A.E.; Tillmann, J.E. Mammal Species Extinction and Decline: Some Current and Past Case Studies of the Detrimental Influence of Man. In *Problematic Wildlife*; Angelici, F.M., Ed.; Springer International Publishing: New York, NY, USA, 2016; pp. 21–44.

7. Boesch, D.; Brinsfield, R. Coastal Eutrophication and Agriculture: Contributions and Solutions. In *Biological Resource Management Connecting Science and Policy*; Balázs, E., Galante, E., Lynch, J.M., Schepers, J.S., Toutant, J.-P., Werner, D., Werry, P.A.T.J., Eds.; Springer: Berlin, Germany; Heidelberg, Germany, 2000; pp. 93–115.

8. McIsaac, G.F.; David, M.B.; Gertner, G.Z.; Goolsby, D.A. Eutrophication: Nitrate flux in the Mississippi River. *Nature* **2001**, *414*, 166–167.

9. Robertson, G.P.; Paul, E.A.; Harwood, R.R. Greenhouse Gases in Intensive Agriculture: Contributions of Individual Gases to the Radiative Forcing of the Atmosphere. *Science* **2000**, *289*, 1922–1925.

10. Reganold, J.P.; Wachter, J.M. Organic agriculture in the twenty-first century. *Nat. Plants* **2016**, *2*, 15221.

11. Bonny, S. Genetically Modified Herbicide-Tolerant Crops, Weeds, and Herbicides: Overview and Impact. *Environ. Manag.* **2015**, *57*, 31–48.

12. Pimentel, D.; Burgess, M. Environmental and Economic Costs of the Application of Pesticides Primarily in the United States. In *Integrated Pest Management*; Pimentel, D., Peshin, R., Eds.; Springer: Dordrecht, The Netherlands, 2014; pp. 47–71.

13. Di, H.J.; Cameron, K.C. Nitrate leaching in temperate agroecosystems: sources, factors and mitigating strategies. *Nutr. Cycl. Agroecosystems* **2002**, *64*, 237–256.

14. Neider, R.; Benbi, D.K. Leaching Losses and Groundwater Pollution. In *Carbon and Nitrogen in the Terrestrial Environment*; Springer: Dordrecht, The Netherlands, 2008; pp. 219–233.

15. US EPA Clean Water Rule Litigation Statement. Available online: http://www.epa.gov/cleanwaterrule/clean-water-rule-litigation-statement (accessed on 11 March 2016).

16. Foley, J. It's Time to Rethink America's Corn System. Available online: http://www.scientificamerican.com/article/time-to-rethink-corn/ (accessed on 24 April 2016).

17. FAO Dimensions of Need—Staple Foods: What do People Eat? Available online: http://www.fao.org/docrep/u8480e/u8480e07.htm (accessed on 12 April 2016).

18. Altieri, M.A. The ecological role of biodiversity in agroecosystems. *Agric. Ecosyst. Environ.* **1999**, *74*, 19–31.

19. Malézieux, E.; Crozat, Y.; Dupraz, C.; Laurans, M.; Makowski, D.; Ozier-Lafontaine, H.; Rapidel, B.; de Tourdonnet, S.; Valantin-Morison, M. Mixing plant species in cropping systems: Concepts, tools and models—A review. *Agron. Sustain. Dev.* **2009**, *29*, 43–62.

20. O'Boyle, E.J. Classical economics and the Great Irish Famine: A study in limits. *Forum Soc. Econ.* **2006**, *35*, 21–53.

21. Davis, A.S.; Hill, J.D.; Chase, C.A.; Johanns, A.M.; Liebman, M. Increasing Cropping System Diversity Balances Productivity, Profitability and Environmental Health. *PLoS ONE* **2012**, *7*, e47149.

22. Jacobsen, S.-E.; Sørensen, M.; Pedersen, S.M.; Weiner, J. Feeding the world: Genetically modified crops versus agricultural biodiversity. *Agron. Sustain. Dev.* **2013**, *33*, 651–662.

23. Amundson, R.; Berhe, A.A.; Hopmans, J.W.; Olson, C.; Sztein, A.E.; Sparks, D.L. Soil and human security in the 21st century. *Science* **2015**, *348*, 1261071.

24. Hertel, T.W. The challenges of sustainably feeding a growing planet. *Food Secur.* **2015**, *7*, 185–198.

25. FAO. Energy-Smart Food at FAO: An Overview. Available online: http://www.fao.org/docrep/015/an913e/an913e00.htm (accessed on 27 April 2016).

26. Elser, J.; Bennett, E. Phosphorus cycle: A broken biogeochemical cycle. *Nature* **2011**, *478*, 29–31.

27. Cordell, D.; White, S. Peak Phosphorus: Clarifying the Key Issues of a Vigorous Debate about Long-Term Phosphorus Security. *Sustainability* **2011**, *3*, 2027–2049.

28. Frison, E.A.; Cherfas, J.; Hodgkin, T. Agricultural Biodiversity Is Essential for a Sustainable Improvement in Food and Nutrition Security. *Sustainability* **2011**, *3*, 238–253.

29. Beddington, J.; Asaduzzaman, M.; Clark, M.; (Eds.) Achieving Food Security in the Face of Climate Change: Final report from the Commission on Sustainable Agriculture and Climate Change. AGRIS: Rome, Italy, 2012.

30. Verchot, L.V.; Noordwijk, M.V.; Kandji, S.; Tomich, T.; Ong, C.; Albrecht, A.; Mackensen, J.; Bantilan, C.; Anupama, K.V.; Palm, C. Climate change: Linking adaptation and mitigation through agroforestry. *Mitig. Adapt. Strateg. Glob. Change* **2007**, *12*, 901–918.

31. Muschler, R.G. Agroforestry: Essential for Sustainable and Climate-Smart Land Use? In *Tropical Forestry Handbook*; Springer: Berlin, Germany; Heidelberg, Germany, 2015.

32. Brewbaker, J.L. Diseases of maize in the wet lowland tropics and the collapse of the Classic Maya civilization. *Econ. Bot.* **1979**, *33*, 101–118.

33. Wilkinson, T.J. Environmental Fluctuations, Agricultural Production and Collapse: A View from Bronze Age Upper Mesopotamia. In *Third Millennium BC Climate Change and Old World Collapse*; Dalfes, H.N., Kukla, G., Weiss, H., Eds.; NATO ASI Series; Springer: Berlin, Germany; Heidelberg, Germany, 1997; pp. 67–106.

34. Nandwani, D.; Nwosisi, S. Global Trends in Organic Agriculture. In *Organic Farming for Sustainable Agriculture*; Nandwani, D., Ed.; Sustainable Development and Biodiversity; Springer International Publishing: New York, NY, USA, 2016; pp. 1–35.

35. McGee, J.A. Does certified organic farming reduce greenhouse gas emissions from agricultural production? *Agric. Hum. Values* **2014**, *32*, 255–263.

36. USDA. Introduction to Organic Practices 2015. Available online: https://www.ams.usda.gov/publications/content/introduction-organic-practices (accessed on 29 April 2016).

37. Gallandt, E. Weed Management in Organic Farming. In *Recent Advances in Weed Management*; Chauhan, B.S., Mahajan, G., Eds.; Springer: New York, NY, USA, 2014; pp. 63–85.

38. Maeder, P.; Fliessbach, A.; Dubois, D.; Gunst, L.; Fried, P.; Niggli, U. Soil Fertility and Biodiversity in Organic Farming. *Science* **2002**, *296*, 1694–1697.

39. Tuomisto, H.L.; Hodge, I.D.; Riordan, P.; Macdonald, D.W. Does organic farming reduce environmental impacts?—A meta-analysis of European research. *J. Environ. Manag.* **2012**, *112*, 309–320.

40. Hole, D.G.; Perkins, A.J.; Wilson, J.D.; Alexander, I.H.; Grice, P.V.; Evans, A.D. Does organic farming benefit biodiversity? *Biol. Conserv.* **2005**, *122*, 113–130.

41. Mondelaers, K.; Aertsens, J.; van Huylenbroeck, G. A meta-analysis of the differences in environmental impacts between organic and conventional farming. *Br. Food J.* **2009**, *111*, 1098–1119.

42. Kirchmann, H.; Bergström, L.; Kätterer, T.; Andrén, O.; Andersson, R. Can Organic Crop Production Feed the World? In *Organic Crop Production—Ambitions and Limitations*; Kirchmann, H., Bergström, L., Eds.; Springer: Dordrecht, The Netherlands, 2009; pp. 39–72.

43. Letter, D.W.; Seidel, R.; Liebhardt, W. The performance of organic and conventional cropping systems in an extreme climate year. *Am. J. Altern. Agric.* **2003**, *18*, 146–154.

44. Pimentel, D.; Hepperly, P.; Hanson, J.; Douds, D.; Seidel, R. Environmental, Energetic, and Economic Comparisons of Organic and Conventional Farming Systems. *BioScience* **2005**, *55*, 573–582.

45. Clark, S.; Khoshnevisan, B.; Sefeedpari, P. Energy efficiency and greenhouse gas emissions during transition to organic and reduced-input practices: Student farm case study. *Ecol. Eng.* **2016**, *88*, 186–194.

46. Teasdale, J.R.; Coffman, C.B.; Mangum, R.W. Potential Long-Term Benefits of No-Tillage and Organic Cropping Systems for Grain Production and Soil Improvement. *Agron. J.* **2007**, *99*, 1297–1305.

47. Reganold, J.P.; Elliott, L.F.; Unger, Y.L. Long-term effects of organic and conventional farming on soil erosion. *Nature* **1987**, *330*, 370–372.

48. Muller, A.; Aubert, C. The Potential of Organic Agriculture to Mitigate the Influence of Agriculture on Global Warming—A Review. In *Organic Farming, Prototype for Sustainable Agricultures*; Bellon, S., Penvern, S., Eds.; Springer: Dordrecht, The Netherlands, 2014; pp. 239–259.

49. Dimitri, C. Organic Agriculture: An Agrarian or Industrial Revolution? *Agric. Resour. Econ. Rev.* **2010**, *39*, 384–395.

50. Bergström, L.; Kirchmann, H.; Aronsson, H.; Torstensson, G.; Mattsson, L. Use Efficiency and Leaching of Nutrients in Organic and Conventional Cropping Systems in Sweden. In *Organic Crop Production—Ambitions and Limitations*; Kirchmann, H., Bergström, L., Eds.; Springer: Dordrecht, The Netherlands, 2009; pp. 143–159.

51. Marriott, E.E.; Wander, M.M. Total and Labile Soil Organic Matter in Organic and Conventional Farming Systems. *Soil Sci. Soc. Am. J.* **2006**, *70*, 950–959.

52. Ismail, I.; Blevins, R.L.; Frye, W.W. Long-Term No-tillage Effects on Soil Properties and Continuous Corn Yields. *Soil Sci. Soc. Am. J.* **1994**, *58*, 193–198.

53. Zuber, S.M.; Villamil, M.B. Meta-analysis approach to assess effect of tillage on microbial biomass and enzyme activities. *Soil Biol. Biochem.* **2016**, *97*, 176–187.

54. Arnhold, S.; Lindner, S.; Lee, B.; Martin, E.; Kettering, J.; Nguyen, T.T.; Koellner, T.; Ok, Y.S.; Huwe, B. Conventional and organic farming: Soil erosion and conservation potential for row crop cultivation. *Geoderma* **2014**, *219–220*, 89–105.

55. Schonbeck, M. What Is "Organic No-Till", and Is It Practical? Available online: http://articles.extension.org/pages/18526/what-is-organic-no-till-and-is-it-practical (accessed on 28 April 2016).

56. Bos, J.F.F.P.; de Haan, J.; Sukkel, W.; Schils, R.L.M. Energy use and greenhouse gas emissions in organic and conventional farming systems in The Netherlands. *NJAS Wagening J. Life Sci.* **2014**, *68*, 61–70.

57. Seufert, V.; Ramankutty, N.; Foley, J.A. Comparing the yields of organic and conventional agriculture. *Nature* **2012**, *485*, 229–232.

58. Trewavas, A. Urban myths of organic farming. *Nature* **2001**, *410*, 409–410.

59. Lovell, S.T.; DeSantis, S.; Nathan, C.A.; Olson, M.B.; Ernesto Méndez, V.; Kominami, H.C.; Erickson, D.L.; Morris, K.S.; Morris, W.B. Integrating agroecology and landscape multifunctionality in Vermont: An evolving framework to evaluate the design of agroecosystems. *Agric. Syst.* **2010**, *103*, 327–341.

60. Tilman, D. Global environmental impacts of agricultural expansion: The need for sustainable and efficient practices. *Proc. Natl. Acad. Sci. USA* **1999**, *96*, 5995–6000.

61. Nair, P.K.R.; Garrity, D. *Agroforestry—The Future of Global Land Use, Advances in Agroforestry*; Springer: Dordrecht, The Netherlands, 2012; Volume 9.

62. Winans, K.S.; Tardif, A.-S.; Lteif, A.E.; Whalen, J.K. Carbon sequestration potential and cost-benefit analysis of hybrid poplar, grain corn and hay cultivation in southern Quebec, Canada. *Agrofor. Syst.* **2015**, *89*, 421–433.

63. Dixon, R.K.; Winjum, J.K.; Andrasko, K.J.; Lee, J.J.; Schroeder, P.E. Integrated land-use systems: Assessment of promising agroforest and alternative land-use practices to enhance carbon conservation and sequestration. *Clim. Change* **1994**, *27*, 71–92.

64. Jordan, N.R.; Davis, A.S. Middle-way strategies for sustainable intensification of agriculture. *BioScience* **2015**, *65*, 513–519.

65. Baah-Acheamfour, M.; Carlyle, C.N.; Bork, E.W.; Chang, S.X. Trees increase soil carbon and its stability in three agroforestry systems in central Alberta, Canada. *For. Ecol. Manag.* **2014**, *328*, 131–139.

66. Caudill, S.A.; DeClerck, F.J.A.; Husband, T.P. Connecting sustainable agriculture and wildlife conservation: Does shade coffee provide habitat for mammals? *Agric. Ecosyst. Environ.* **2015**, *199*, 85–93.

67. Brandle, J.R.; Schoeneberger, M.M. Working Trees: Supporting Agriculture and Healthy Landscapes. *J. Trop. For. Sci.* **2014**, *26*, 305–308.

68. Jose, S.; Gold, M.A.; Garrett, H.E. The Future of Temperate Agroforestry in the United States. In *Agroforestry—The Future of Global Land Use*; Nair, P.K.R., Garrity, D., Eds.; Springer: Dordrecht, The Netherlands, 2012; pp. 217–245.

69. *Training Manual for Applied Agroforestry Practices*, 2015 ed.; Gold, M., Cernusca, M., Hall, M., Eds.; University of Missouri Center for Agroforestry: New Franklin, MO, USA, 2006.

70. Jha, S.; Bacon, C.M.; Philpott, S.M.; Rice, R.A.; Méndez, V.E.; Läderach, P. A Review of Ecosystem Services, Farmer Livelihoods, and Value Chains in Shade Coffee Agroecosystems. In *Integrating Agriculture, Conservation and Ecotourism: Examples from the Field*; Campbell, W.B., Ortiz, S.L., Eds.; Springer: Dordrecht, The Netherlands, 2011; pp. 141–208.

71. Fagerholm, N.; Torralba, M.; Burgess, P.J.; Plieninger, T. A systematic map of ecosystem services assessments around European agroforestry. *Ecol. Indic.* **2016**, *62*, 47–65.

72. Campbell, G.E.; Lottes, G.J.; Dawson, J.O. Design and development of agroforestry systems for Illinois, USA: Silvicultural and economic considerations. *Agrofor. Syst.* **1991**, *13*, 203–224.

73. Cardinael, R.; Mao, Z.; Prieto, I.; Stokes, A.; Dupraz, C.; Kim, J.H.; Jourdan, C. Competition with winter crops induces deeper rooting of walnut trees in a Mediterranean alley cropping agroforestry system. *Plant Soil* **2015**, *391*, 219–235.

74. Dupraz, C.; Talbot, G.; Marrou, H.; Wery, J.; Roux, S.; Liagre, F.; Ferard, Y.; Nogier, A. To Mix or Not to Mix: Evidences for the Unexpected High Productivity of New Complex Agrivoltaic and Agroforestry Systems. 2011, pp. 202–203. Available online: https://www.researchgate.net/publication/230675951_To_mix_or_not_to_mix_ _evidences_for_the_unexpected_high_productivity_of_new_complex_agrivoltaic_ and_agroforestry_systems (accessed on 9 December 2015).

75. Brandle, J.R.; Hodges, L.; Zhou, X.H. Windbreaks in North American agricultural systems. *Agrofor. Syst.* **2004**, *61–62*, 65–78.

76. Kallenbach, R.L.; Kerley, M.S.; Bishop-Hurley, G.J. Cumulative Forage Production, Forage Quality and Livestock Performance from an Annual Ryegrass and Cereal Rye Mixture in a Pine Walnut Silvopasture. *Agrofor. Syst.* **2006**, *66*, 43–53.

77. Buergler, A.L.; Fike, J.H.; Burger, J.A.; Feldhake, C.M.; McKenna, J.R.; Teutsch, C.D. Forage Nutritive Value in an Emulated Silvopasture. *Agron. J.* **2006**, *98*, 1265–1273.

78. Blanco-Canqui, H.; Lal, R. Buffer Strips. In *Principles of Soil Conservation and Management*; Springer: Dordrecht, The Netherlands, 2010; pp. 223–257.

79. Skelton, P.; Josiah, S.J.; King, J.W.; Brandle, J.R.; Helmers, G.A.; Francis, C.A. Adoption of riparian forest buffers on private lands in Nebraska, USA. *Small-Scale For. Econ. Manag. Policy* **2005**, *4*, 185–203.

80. Kort, J. Benefits of windbreaks to field and forage crops. *Agric. Ecosyst. Environ.* **1988**, *22*, 165–190.

81. Mize, C.W.; Brandle, J.R.; Schoeneberger, M.M.; Bentrup, G. Ecological Development and function of Shelterbelts in Temperate North America. In *Toward Agroforestry Design*; Jose, S., Gordon, A.M., Eds.; Springer: Dordrecht, The Netherlands, 2008; pp. 27–54.

82. Zheng, X.; Zhu, J.; Xing, Z. Assessment of the effects of shelterbelts on crop yields at the regional scale in Northeast China. *Agric. Syst.* **2016**, *143*, 49–60.

83. Nuberg, I.K. Effect of shelter on temperate crops: A review to define research for Australian conditions. *Agrofor. Syst.* **1998**, *41*, 3–34.

84. Valdivia, C.; Poulos, C. Factors affecting farm operators' interest in incorporating riparian buffers and forest farming practices in northeast and southeast Missouri. *Agrofor. Syst.* **2008**, *75*, 61–71.

85. Udawatta, R.P.; Jose, S. Carbon Sequestration Potential of Agroforestry Practices in Temperate North America. In *Carbon Sequestration Potential of Agroforestry Systems*; Kumar, B.M., Nair, P.K.R., Eds.; Springer: Dordrecht, The Netherlands, 2011; Volume 8, pp. 17–42.

86. Udawatta, R.P.; Kremer, R.J.; Nelson, K.A.; Jose, S.; Bardhan, S. Soil Quality of a Mature Alley Cropping Agroforestry System in Temperate North America. *Commun. Soil Sci. Plant Anal.* **2014**, *45*, 2531–2551.

87. Rivest, D.; Lorente, M.; Olivier, A.; Messier, C. Soil biochemical properties and microbial resilience in agroforestry systems: Effects on wheat growth under controlled drought and flooding conditions. *Sci. Total Environ.* **2013**, *463–464*, 51–60.

88. Udawatta, R.P.; Krstansky, J.J.; Henderson, G.S.; Garrett, H.E. Agroforestry practices, runoff, and nutrient loss: A paired watershed comparison. *J. Environ. Qual.* **2002**, *31*, 1214–1225.

89. Jose, S. Agroforestry for ecosystem services and environmental benefits: An overview. *Agrofor. Syst.* **2009**, *76*, 1–10.

90. Garrett, H.E.; McGraw, R.L.; Walter, W.D. Alley Cropping Practices. *North Am. Agrofor. Integr. Sci. Pract. 2nd Ed.* **2009**, 133–162.

91. Schultz, R.C.; Isenhart, T.M.; Colletti, J.P.; Simpkins, W.W.; Udawatta, R.P.; Schultz, P.L.; Garrett, H.E.G. Riparian and Upland Buffer Practices. In *ACSESS Publications*; American Society of Agronomy: Madison, WI, USA, 2009.

92. Udawatta, R.P.; Garrett, H.E.; Kallenbach, R.L. Agroforestry and grass buffer effects on water quality in grazed pastures. In *Agroforestry Systems*; Springer: Dordrecht, The Netherlands, 2010; Volume 79, pp. 81–87.

93. Jose, S.; Holzmueller, E.J.; Gillespie, A.R.; Garrett, H.E.G. Tree–Crop Interactions in Temperate Agroforestry. In *ACSESS Publications*; American Society of Agronomy: Madison, WI, USA, 2009.

94. Kim, D.-G.; Kirschbaum, M.U.F.; Beedy, T.L. Carbon sequestration and net emissions of CH_4 and N_2O under agroforestry: Synthesizing available data and suggestions for future studies. *Agric. Ecosyst. Environ.* **2016**, *226*, 65–78.

95. Reddy, P.P. Impacts of Climate Change on Agriculture. In *Climate Resilient Agriculture for Ensuring Food Security*; Springer: New Delhi, India, 2015; pp. 43–90.

96. Jose, S.; Bardhan, S. Agroforestry for biomass production and carbon sequestration: An overview. *Agrofor. Syst.* **2012**, *86*, 105–111.

97. Holzmueller, E.J.; Jose, S. Biomass production for biofuels using agroforestry: Potential for the North Central Region of the United States. *Agrofor. Syst.* **2012**, *85*, 305–314.

98. Mead, R.; Willey, R.W. The Concept of a "Land Equivalent Ratio" and Advantages in Yields from Intercropping. *Exp. Agric.* **1980**, *16*, 217–228.

99. Haile, S.; Palmer, M.; Otey, A. Potential of loblolly pine: Switchgrass alley cropping for provision of biofuel feedstock. *Agrofor. Syst.* **2016**.

100. Graves, A.R.; Burgess, P.J.; Palma, J.H.N.; Herzog, F.; Moreno, G.; Bertomeu, M.; Dupraz, C.; Liagre, F.; Keesman, K.; van der Werf, W.; *et al.* Development and application of bio-economic modelling to compare silvoarable, arable, and forestry systems in three European countries. *Ecol. Eng.* **2007**, *29*, 434–449.

101. Sereke, F.; Graves, A.R.; Dux, D.; Palma, J.H.N.; Herzog, F. Innovative agroecosystem goods and services: Key profitability drivers in Swiss agroforestry. *Agron. Sustain. Dev.* **2014**, *35*, 759–770.

102. Bianchi, F.J.J.; Booij, C.J.; Tscharntke, T. Sustainable pest regulation in agricultural landscapes: A review on landscape composition, biodiversity and natural pest control. *Proc. R. Soc. B* **2006**, *273*, 1715–1727.

103. Hanley, N.; Breeze, T.D.; Ellis, C.; Goulson, D. Measuring the economic value of pollination services: Principles, evidence and knowledge gaps. *Ecosyst. Serv.* **2015**, *14*, 124–132.

104. Civitello, D.J.; Cohen, J.; Fatima, H.; Halstead, N.T.; Liriano, J.; McMahon, T.A.; Ortega, C.N.; Sauer, E.L.; Sehgal, T.; Young, S.; *et al.* Biodiversity inhibits parasites: Broad evidence for the dilution effect. *Proc. Natl. Acad. Sci. USA* **2015**, *112*, 8667–8671.

105. Keesing, F.; Ostfeld, R.S. Is biodiversity good for your health? *Science* **2015**, *349*, 235–236.

106. Smith, J.R. *Tree Crops: A Permanent Agriculture*; Island Press: Washington, DC, USA, 1950.

107. Acha, A.; Newing, H.S. Cork Oak Landscapes, Promised or Compromised Lands? A Case Study of a Traditional Cultural Landscape in Southern Spain. *Hum. Ecol.* **2015**, *43*, 601–611.

108. Strong, N.; Jacobson, M.G. A case for consumer-driven extension programming: Agroforestry adoption potential in Pennsylvania. *Agrofor. Syst.* **2006**, *68*, 43–52.

109. Sullivan, W.C.; Anderson, O.M.; Lovell, S.T. Agricultural buffers at the rural–urban fringe: An examination of approval by farmers, residents, and academics in the Midwestern United States. *Landsc. Urban Plan.* **2004**, *69*, 299–313.

110. Valdivia, C.; Gold, M.; Zabek, L.; Arbuckle, J.; Flora, C. Human and Institutional Dimensions of Agroforestry. *North Am. Agrofor. Integr. Sci. Pract.* **2009**.

111. Trozzo, K.E.; Munsell, J.F.; Chamberlain, J.L. Landowner interest in multifunctional agroforestry Riparian buffers. In *Agroforestry Systems*; Springer: Dordrecht, The Netherlands, 2014; Volume 88, pp. 619–629.

112. Sereke, F.; Dobricki, M.; Wilkes, J.; Kaeser, A.; Graves, A.R.; Szerencsits, E.; Herzog, F. Swiss farmers don't adopt agroforestry because they fear for their reputation. *Agrofor. Syst.* **2015**.

113. Faulkner, P.E.; Owooh, B.; Idassi, J. Assessment of the Adoption of Agroforestry Technologies by Limited-Resource Farmers in North Carolina. *J. Ext.* **2014**, *52*, Article 5. Available online: http://www.joe.org/joe/2014october/rb7.php (accessed on 16 October 2015).

114. Thevathasan, N.V.; Gordon, A.M.; Bradley, R.; Cogliastro, A.; Folkard, P.; Grant, R.; Kort, J.; Liggins, L.; Njenga, F.; Olivier, A.; *et al.* Agroforestry Research and Development in Canada: The Way Forward. In *Agroforestry—The Future of Global Land Use*; Nair, P.K.R., Garrity, D., Eds.; Springer: Dordrecht, The Netherlands, 2012; pp. 247–283.

115. Valdivia, C.; Barbieri, C.; Gold, M.A. Between Forestry and Farming: Policy and Environmental Implications of the Barriers to Agroforestry Adoption. *Can. J. Agric. Econ.* **2012**, *60*, 155–175.

116. Current, D.A.; Brooks, K.N.; Ffolliott, P.F.; Keefe, M. Moving agroforestry into the mainstream. *Agrofor. Syst.* **2008**, *75*, 1–3.

117. Quinn, C.E.; Quinn, J.E.; Halfacre, A.C. Digging Deeper: A Case Study of Farmer Conceptualization of Ecosystem Services in the American South. *Environ. Manag.* **2015**, *56*, 802–813.

118. Chestnuts—Harvesting. Available online: http://msue.anr.msu.edu/topic/chestnuts/harvest_storage/harvesting (accessed on 6 June 2016).

119. Grado, S.C.; Husak, A.L. Economic Analyses of a Sustainable Agroforestry System in the Southeastern United States. In *Valuing Agroforestry Systems*; Advances in Agroforestry; Springer: Dordrecht, The Netherlands, 2004; pp. 39–57.

120. Prokopy, L.S.; Floress, K.; Klotthor-weinkauf, D.; Baumgart-getz, A. Determinants of agricultural best management practice adoption: Evidence from the Literature. *J. Soil Water Conserv.* **2008**, *63*, 300–311.

121. Gold, M.A.; Godsey, L.D.; Josiah, S.J. Markets and marketing strategies for agroforestry specialty products in North America. In *New Vistas in Agroforestry*; Nair, P.K.R., Rao, M.R., Buck, L.E., Eds.; Springer: Dordrecht, The Netherlands, 2004; pp. 371–382.

122. The Savanna Institute. What Is Savannah-Based Restoration Agriculture? Available online: http://www.savannainstitute.org/about (accessed on 9 December 2015).

123. Multifunctional Landscape Analysis and Design. Available online: http://multifunctionallandscape.com/Home_Page.html (accessed on 29 April 2016).

Genetic Engineering and Sustainable Crop Disease Management: Opportunities for Case-by-Case Decision-Making

Paul Vincelli

Abstract: Genetic engineering (GE) offers an expanding array of strategies for enhancing disease resistance of crop plants in sustainable ways, including the potential for reduced pesticide usage. Certain GE applications involve transgenesis, in some cases creating a metabolic pathway novel to the GE crop. In other cases, only cisgenesis is employed. In yet other cases, engineered genetic changes can be so minimal as to be indistinguishable from natural mutations. Thus, GE crops vary substantially and should be evaluated for risks, benefits, and social considerations on a case-by-case basis. Deployment of GE traits should be with an eye towards long-term sustainability; several options are discussed. Selected risks and concerns of GE are also considered, along with genome editing, a technology that greatly expands the capacity of molecular biologists to make more precise and targeted genetic edits. While GE is merely a suite of tools to supplement other breeding techniques, if wisely used, certain GE tools and applications can contribute to sustainability goals.

Reprinted from *Sustainability*. Cite as: Vincelli, P. Genetic Engineering and Sustainable Crop Disease Management: Opportunities for Case-by-Case Decision-Making. *Sustainability* **2016**, *8*, 495.

1. Introduction and Background

Disease management practices can contribute to sustainability by protecting crop yields, maintaining and improving profitability for crop producers, reducing losses along the distribution chain, and reducing the negative environmental impacts of diseases and their management. Crop disease management supports sustainability goals through contributions to food security, food safety, and food sovereignty for producers and consumers alike [1].

While pesticides have done much to contribute to food security and food sovereignty for many millions of people worldwide, pest and disease control through the regular use of pesticides is neither desirable nor sustainable over the long term. Pesticide use raises significant concerns over impacts on health [2–7] and the environment [8–12]. Furthermore, we cannot address the challenges to sustainability posed by synthetic pesticides by simply switching to the application of natural pesticides, because the same concerns apply to them [12–19].

308

Practices for managing crop diseases fall into four general categories: host plant resistance, cultural practices, biological control, and chemical control. If pesticide use is to be reduced, it will be necessary to depend more on the remaining three approaches. Cultural practices (examples include crop rotation, polyculture, manipulation of planting date, *etc.*) certainly play a central role in disease management [20]. However, control achieved via cultural practices is sometimes inadequate, impractical, or economically nonviable. Natural biological control of plant pathogens is a fact of life, as it undoubtedly occurs at some level in all agricultural soils. However, there are many destructive diseases for which years of research have failed to lead to practical, commercially viable biocontrol options. Thus, in order to reduce the need for pesticides while still attaining acceptable yields, it will be critical to judiciously take full advantage of plant genetics. After all, if farmers are to reduce pesticide use, they must have viable alternatives for controlling diseases.

Host plant resistance is an ecologically sound way to manage crop diseases. Approaches to genetic crop improvement span an ever-expanding range of techniques, from simple phenotypic selection through techniques of genome editing (discussed below). Conventional breeding can often produce adequate levels of disease control, and we can expect that all breeding techniques will continue to play important roles indefinitely. However, when conventional breeding and other management options are inadequate, or when linkage drag limits the usefulness of conventionally derived traits, GE offers alternatives. As examples of diseases for which GE presently appears to represent the only acceptable disease management in culturally or economically important crops, consider papaya ring spot in Hawaii [21], cassava brown streak disease in Africa [22], and citrus greening in Florida [23–25]. The loss of important crops to infectious diseases is completely contrary to the principles of sustainability. Thus GE can make it possible to save crops in the face of virulent disease epidemics, crops that may be integral to food security, sources of farmer income, or culturally important dietary components. In addition, GE can make it possible to reduce farmers' dependence on pest-control products, with undeniable benefits for sustainability. The deployment of Cry proteins for insect control serves as an excellent example of how a GE approach can contribute to sustainability through reduced application of pesticides, resulting in fewer pesticide poisonings, increased biodiversity, and increased biocontrol services [26–35]. Certainly other examples seem eminently possible as we employ additional GE traits for pest and disease control. One must also acknowledge that Cry proteins serve as an example of how overreliance on single genetic traits can allow for pest evolution to overcome such a trait [36–39]. In fact, this presents another justification for taking full advantage of the opportunities offered by GE. Selection pressure towards virulence is a "given" whenever managing pests and diseases. Breeders therefore need a wide array of genetic options in order to diversify the arsenal of resistance traits deployed

309

in crops, thereby reducing this selection pressure. As will be apparent in this review, GE is greatly expanding the genetic options for disease control available to breeders.

2. Strategies for Engineering Resistance

There is a wide variety of published GE strategies for engineering disease resistance, and ongoing research and expanding genetic resources [40] are likely to lead to additional strategies. Furthermore, within most of those strategies, diverse applications are conceivable. Taken together, these suggest that GE presents a vast pool of genetic possibilities for future generations. This will allow breeding for disease resistance to remain highly dynamic in the face of pathogen adaptation towards virulence on resistant cultivars.

In contrast to typical pesticides, GE mechanisms are often designed to have selective efficacy against particular target pathogens. High target selectivity is advantageous, in that it minimizes health concerns for consumers as well as risks to non-target biota in and around agroecosystems. However, the drawback is that one GE trait is unlikely to protect against the full spectrum of damaging pathogens on a given crop—which is also true of many conventional genes for disease resistance.

While it is difficult to foretell which GE strategies will have the greatest impact on crop disease control in the coming decades, all those described below hold promise and, in the author's opinion, merit continued research attention. Some have demonstrated proof-of-concept, while others have been evaluated in the field and, in certain cases, introgressed into commercially viable varieties. All strategies described below take advantage of—and in most cases, mimic—processes that occur in Nature.

2.1. Boosting Plant Recognition of Infection

Plants have evolved to trigger basal defenses upon recognition of certain conserved molecules of an invading pathogen. These molecules, which are highly conserved evolutionarily and are metabolically important for the pathogen, are referred to as pathogen-associated molecular patterns (PAMPs) [41–43]. Receptor molecules in the host membrane recognize PAMPs and elicit a natural defense response called PAMP-tiggered immunity (PTI). PAMP receptor molecules differ among plant species. Thus, genes encoding PAMP receptors from crops and other plants can be transformed into other crops, expanding the range of pathogen molecules that trigger PTI in the latter [43]. A gene encoding a PAMP receptor does not introduce a novel defense mechanism into the plant. The transferred PAMP receptor merely allows the receiving plant to recognize infection, so it can respond with its own, natural immune system. Increased resistance has been obtained using this strategy against a range of bacterial diseases in both monocots and dicots [44–47].

An important question is whether the transfer of PAMP receptors among plant species would increase the risk of selection towards wider pathogen host ranges. Since PAMPs are highly conserved molecules that are metabolically important for the pathogen [44], rapid evolution of these molecules is unlikely. Rational deployment strategies, such as those described in Section 3, can also reduce this risk.

2.2. Mining R Genes

PTI places strong selection pressure on pathogens to restore a virulent host-parasite interaction. According to the prevailing model of disease resistance, pathogens produce one or more effector molecules which enhance virulence, resulting in effector-triggered susceptibility (ETS) [42,48,49]. Over evolutionary time scales, plants respond to ETS by producing an intracellular receptor (R protein) which detects the presence or activity of particular pathogen effectors, restoring a resistance response called effector-triggered immunity or effector-triggered defense [41,42,48,50]. In the face of a renewed defense response in the host, a pathogen may eventually evolve to produce a new effector to restore compatibility. In turn, the plant may evolve a new R protein. This coevolutionary, gene-for-gene, "molecular arms race" [48,50] between pathogen effectors and their corresponding R proteins has yielded pools of R genes (resistance genes) useful in breeding crops for disease resistance [42].

One way GE can contribute to resistance breeding is through cisgenics: engineering only with genetica obtained from a crop's sexually compatible gene pool [51]. Conventional breeding techniques are often suitable for introgressing cisgenes into new varieties, in which case GE is unnecessary. However, in some crops, such as potato, grape, banana, apple, and strawberry, conventional breeding is exceptionally difficult or time-consuming. For crops such as these, cisgenes can be transferred via GE [43,51,52], resulting in a genetic outcome that would be conceivable—although perhaps impractical—by conventional means. A major advantage of cisgenics over conventional breeding is that it circumvents linkage drag [43,51].

For some plants, hybridization is difficult or impossible using current techniques. In such cases, GE offers an alternative for introgressing R genes, even from plants that are not part of a crop's normal breeding pool. For example, in tomato, bacterial leaf spot, a highly destructive disease, was controlled in the field with a single R gene obtained from pepper [53,54]. Indeed, the level of control obtained was higher than that obtained by any conventional breeding approach. This R gene is expected to provide an alternative to the repeated use of foliar copper applications, benefiting both field workers and the environment [54]. Other examples of "mining" of R genes from related as well as unrelated plant species have been published for both monocots and dicots [47,55,56]. Recent research has also shown that it is possible to enhance disease resistance by modifying the target of a pathogen effector so that it recognizes other pathogen effectors [57]. For example, the target molecule of pathogen effector "A" can be modified (with modest edits) so that its product is activated (and thereby triggers a defense reaction) by another pathogen's effector "B." This creative approach provides new disease resistance traits while avoiding any transfer of genetic material. Durability of R genes could be enhanced by engineering resistance based on recognition of effectors critical to pathogenicity [57].

It is worth recalling that R genes do not code for new biochemical pathways; they merely code for receptor molecules. This allows the plant to recognize the presence of an invading pathogen, thereby taking advantage of their native, natural mechanisms of disease resistance.

Resistance conferred by individual R genes is often not durable, because widespread deployment of an R genes selects for pathogen strains capable of overcoming it [42,58–60]. The ability to "mine" R genes from plants outside of a crop's breeding pool may be especially important for sustainability, in that it opens a vast pool of R genes potentially useful for breeding.

2.3. Upregulating Defense Pathways

Molecules involved in defense signaling, defense regulation, or other processes can be upregulated, boosting general defense responses. Such defenses include generation of reactive oxygen species, callose deposition, synthesis of pathogenesis-related (PR) proteins, and increased activation of systemic acquired resistance (SAR) [23,61]. As with the previously described strategies, this strategy takes advantage of the plant's own natural immune system and does not introduce new metabolic pathways. This approach has been successful against bacterial pathogens attacking several host species [62–64], and it offers promising results for enhancing resistance to citrus greening [23], a disease of urgency for the citrus industry. Upregulation of defense pathways was also successful against destructive fungal pathogens, including *Rhizoctonia solani* (the cause of many diseases) and *Magnaporthe oryzae* (the cause of rice blast) [61,65]. In both cases, resistance was achieved by expressing a native rice gene under the control of a constitutive promotor from maize, introducing neither a novel pathway nor a non-crop gene. It may eventually be possible to upregulate defense responses using native cisgenic promotors, avoiding the use of any DNA outside of the crop's breeding pool.

2.4. Disarming Host Susceptibility Genes

Plants possess genes whose products are important in its normal physiology, but in some way also function to facilitate pathogen infection and colonization. These can be considered susceptibility genes [66]. (See the Supplemental Table 1 in [66] for a long list of examples.) Changes in such genes by natural means can result in increased disease resistance [47,67]. The same is true for GE-induced changes [66,68–70].

While we must remain aware that susceptibility genes may have pleiotropic effects, disarming susceptibility genes may hold promise for durable resistance for two reasons: first, in some pathosystems, many host factors contribute to host-parasite compatibility, offering many potential targets to disarm through very modest changes in DNA sequence; and second, overcoming a disarmed susceptibility gene requires the pathogen to gain a new function to replace the lost host factor it was exploiting. Gaining a new function is not likely to be easily accomplished [66]. Disarming susceptibility genes can be achieved without introducing a novel metabolic pathway or leaving exogenous DNA in the final product.

2.5. Producing Antimicrobial Compounds

Genes encoding antimicrobial compounds can be expressed in crop plants, resulting in restricted pathogen activity and, consequently, increased disease resistance. As a result of citrus greening, a highly destructive bacterial disease, the economic health and even survival of the Florida orange juice industry is uncertain [25,71]. Thus far, the only potentially viable, environmentally acceptable solution may be citrus trees that express antimicrobial peptides called defensins, produced by genes obtained from spinach [25,72,73].

Resistance to diverse fungal diseases was obtained in grape and cotton when plants were transformed to constitutively produce chitin-degrading enzymes [74,75]. All of the diseases controlled in these studies were caused by fungi that contain chitin as an important component of their cell walls. The sources of the chitinase genes were *Trichoderma* species, fungal parasites of other fungi. Plants may be engineered to deliver pest-control substances that act in particular tissues or organs of multicellular, anatomically complex pathogens [76], which may have particular relevance to nematode control.

One advantage of transforming crops with genes for natural antimicrobial substances is that one can employ *in-vitro* techniques of molecular evolution to broaden the range of molecular targets of such antimicrobials [77]. Such techniques potentially can be employed to reverse the buildup of pathogen resistance to the antimicrobial.

Microorganisms could potentially serve as a source of many antimicrobial compounds, though public acceptance of transgenes from microorganisms is mixed [78]. In contrast to several strategies described in this review, this strategy does not take advantage of existing defense mechanisms; rather, it creates a new one.

2.6. Silencing Essential Pathogen Genes

The presence of double-stranded RNA (dsRNA) in the cytoplasm of eukaryotic cells triggers the natural and targeted process of post-transcriptional gene silencing (RNA silencing, RNA interference, or RNAi) [79]. Through the use of genetic constructs with sequence identity to important pathogen genes (and, ideally, with little to no identity to mammalian genes), RNAi can be elicited in plants to silence such genes, resulting in reduced disease. In RNAi, no novel protein or biochemical pathway is created in the crop; the natural process of RNAi is invoked in order to silence a particular target gene in the pathogen.

The papaya industry in Hawaii was saved by transforming papaya with the coat protein gene of papaya ringspot virus. This gene elicits RNAi against this highly destructive virus [21,80]. Such a GE application mimics cross-protection, a phenomenon in which symptoms due to severe strains of a virus can be reduced by prior infection by a mild strain. Cross-protection is a perfectly natural phenomenon. Unfortunately, implementing it for disease management has practical drawbacks [81], which is why transgenic coat-protein-mediated resistance was utilized against this devastating virus disease. While consumers may be hesitant to eat transgenic papaya containing a viral coat-protein gene, they may be surprised to know that they are eating complete virus particles in fruit harvested from non-transgenic, infected trees, including fruit from cross-protected trees. RNAi provides control of other destructive viruses of crops, including the viral complexes that attack cassava in East Africa [22,82], soybean [83], and summer squash [84], and others [47,85].

Recent research clearly highlights the substantial potential which RNA silencing offers for management of diseases caused by biotrophic fungi, necrotrophic fungi, and oomycetes [86–91]. These studies report partial to complete control of diseases caused by several of the most important pathogens worldwide. Likewise, gene silencing holds much promise for pesticide-free nematode management [92–94]. Diverse pathogenicity genes in nematodes present many molecular targets [95], highlighting the promise RNA silencing holds for sustainable, long-term nematode management.

Some success in RNAi-based insect control has been obtained by feeding insects dsRNA constructs that trigger RNAi [96,97]. Commercial products based on this technology are being pursued. Since foliar applications of small RNAs require no genetic changes in the plant, this technology may appeal to consumers for its "non-GMO" status. Of course, compared to genetic changes, there are sustainability costs (both economic and environmental) to the use of products that must be applied repeatedly and indefinitely.

2.7. Modifying Host Targets of Pathogenicity/Virulence Factors

Certain plant pathogens produce molecules (virulence factors) that play a role in virulence by binding to host target molecules [98]. The molecular targets of these in the crop can be engineered so as to result in reduced binding, thereby increasing disease resistance [99]. Genetic modification of targets of pathogen virulence factors increases host resistance without introducing an exogenous biochemical pathway into the plant, and also can be achieved without transgene insertion.

2.8. Detoxifying Pathogen Toxins

Pathogen-produced toxins can disrupt important biochemical processes of their hosts, thereby facilitating disease development [100]. In turn, plant resistance may be conferred by a host enzyme that inactivates a pathogen toxin, whether that enzyme is native [101] or the result of GE. As an example of the latter, the phytotoxin oxalic acid is central to pathogenicity of *Cryphonectria parasitica*, the cause of catastrophic epidemics of chestnut blight [102]. Significantly less disease development was observed in American chestnut trees transformed with a wheat gene coding for the production of the degradative enzyme, oxalate oxidase [103]. As another example, a toxin-degrading enzyme encoded by a barley gene was transformed into wheat, resulting in resistance in the wheat to the highly destructive disease, Fusarium head blight [104]. In both examples, the gene constructs used included a viral promotor and a bacterial selectable marker, so in their present configuration, these GE crops clearly qualify as transgenic. However, these potential concerns may be addressed by employing native promotors derived from the engineered crop and marker-free transformation [85].

2.9. Engineering CRISPR/Cas Immune System

CRISPR (Clustered Regularly Interspaced Short Palindromic Repeats) is a prokaryotic defense system that targets the DNA of invading viruses and plasmids [105,106]. In this system, an endonuclease (commonly CRISPR associated protein 9, abbreviated Cas9) is directed to cut the invading DNA at a particular target, where the DNA sequence matches the sequence of an RNA guide strand (gRNA) associated with Cas9. Plants can be transformed to produce both Cas9 and a target-specific gRNA, in order to cleave a specified target of invading DNA. For example, a Cas9/gRNA complex can be engineered to target the replicating DNA of Geminiviruses, which are highly destructive to crops in tropical and subtropical climates [106–109]. Such an engineered Cas9/gRNA complex produces a sequence-specific, targeted immune response which can result in significant host resistance against a DNA virus. These laboratory-based results are exciting if they are reproduced in the field, since conventional breeding has not been universally successful against Geminiviruses [108,110]. A variety of viral genetic elements can be successfully targeted [106,108], which would confer long-term utility to this strategy. Crops engineered to express a CRISPR/Cas immune system are transgenic, containing DNA sequences which are bacterial and viral in origin (coding for Cas9 and gRNA, respectively), which may hamper public acceptance.

2.10. Reducing Infection Courts

Transgenic crops expressing δ-endotoxins (Cry proteins) from *Bacillus thuringiensis* (Bt) have been used successfully to control certain insects. Another benefit from the use of Bt corn has been the well-documented reductions in mycotoxin contamination that sometimes occur. Reductions in both fumonisins and aflatoxins have been reported in field studies on several continents [111–115]. These reductions have been associated with reduced insect wounding on kernels expressing a Cry endotoxin, resulting in fewer openings for infection by mycotoxin-producing fungi [114,116]. The Bt trait is not a "silver bullet", eliminating all mycotoxin risk. However, reductions occur often enough that the Bt trait is commonly thought to contribute to food safety and livestock health. It is interesting to note that the application of synthetic insecticides to control kernel-feeding insects on non-Bt plants also sometimes reduces insect feeding and fumonisin contamination. However, to this observer, genetic approaches to reducing mycotoxin contamination are preferred for considerations of both environmental protection and consumer health.

3. Deployment of GE Traits

Just as pathogen populations adapt to conventionally bred resistance, prudence dictates that we anticipate the same in response to the deployment of engineered resistance mechanisms. Reducing disease pressure through integrated disease management remains an essential strategy for reducing selection pressure towards overcoming resistance traits [117]. Thus, GE traits should be deployed in conjunction with appropriate management practices for disease control. This will help to promote sustainability by extending the useful life of resistance traits. In addition, GE traits must be deployed with attention to genetic diversity. Widespread deployment of a single gene conferring high levels of disease resistance imposes substantial selection pressure for virulence, often resulting in pathogen strains highly virulent on plants possessing that resistance gene [59,118,119]. Thus, the widespread deployment of solo resistance genes—whether conventional or GE-derived—should not be expected to provide sustainable disease control.

One way to introduce genetic diversity is via "stacking" multiple, distinct resistance traits. Creating *in planta* diversity by gene stacking would be expected to increase the durability of resistance traits, since the target pathogen must overcome all genes in the stack to be fully virulent [37,43,52,58,62,66,120–122]. Stacking conventional *R* genes with diverse biological effects on the pathogen has been shown to increase resistance durability [58]. This suggests that traits based on distinct GE strategies could also be stacked in order to disrupt the evolution of virulence. Depending on the crop, stacking via GE may often be more practical than by other breeding techniques [43]. Molecular tools also permit us to identify *R* genes that correspond to "core," conserved pathogen effectors [47], which may impose a high fitness cost on virulent pathogen strains. High-throughput techniques for cloning R-genes [123,124] are expected to greatly expand the libraries of R genes available to breeders. Plant artificial chromosomes [125] will likely facilitate stacking of numerous genes performing diverse functions, thus increasing the durability of deployed genes. Rotation of R genes may also contribute to durability [126]. A long-term, sustainable approach to disease management may involve coordination of breeding programs to systematically substitute or rotate stacked genes at periodic intervals. Deploying diverse genetics through time in this way would further disrupt pathogen adaptation.

4. Selected Concerns

A variety of concerns are raised with respect to GE crops. In-depth consideration of all of these lies beyond the scope of this paper. However, several selected concerns are discussed below.

4.1. Flow of Recombinant DNA

Perhaps the most significant biological risk in the cultivation of GE crops is the possibility of transgene flow to non-GE crops or to wild or weedy relatives [127]. Flow of genes into wild or weedy relatives can occur also from conventionally bred, non-GE crops [128–130], even causing negative environmental consequences [131]. However, transgenesis may create a higher level of uncertainty in terms of environmental risk. In wild or weedy relatives, transgenes may have no net impact on fitness [132], or they may confer fitness costs [133], likely leading to a decline in frequency of the transgene over time. Alternatively, a transgene may confer a fitness advantage on the recipient species or population [134], which creates the possibility of long-term ecological impacts. Of course, if engineered genetics are cisgenes [43,51,52], ecological risks are no different than those that would result from conventional breeding.

Crop residues are sources of environmental DNA [135], including transgenic DNA. DNA from residues—transgenic or not—is potentially available for transformation of soil microorganisms via natural processes [136]. Such events are rare but possible [136–141]. For prevalent transgenes currently in use, the importance of such a risk is unclear, for several reasons: these genes are already widespread in the environment in their source organisms [137]; crop transgenes sometimes are designed with eukaryotic promotors rather than prokaryotic ones; and they may not be codon-optimized for a recipient microorganism. Certainly, assessing such risks would be facilitated by a better understanding of the potential selective impact of transgenes transformed into soil microorganisms [136,142].

Given the long-term uncertainties of uncontrolled dispersal of transgenes, particularly via pollen, mitigation of risk is critical. Several basic precautions can be taken, including spatial separation of GE and non-GE crops [143] and avoiding transgenic crops in areas where wild relatives occur. For plant species whose pollen is almost completely free of chloroplasts, transgenes in chloroplasts would reduce the risk of transmission [144], although the high transgene copy number in chloroplasts may be physiologically taxing on the plant. Breeders can take advantage of gametic incompatibility, which can block fertilization of non-GE corn kernels by GE pollen [145]. Additional options exist for mitigating the risk of transgene flow [130,144,146,147]. Since in some crops, the risk of transgene flow may be greatly minimized but remain non-zero, some may argue for GE to be limited to non-transgenic applications. Numerous non-transgenic approaches are described above in Section 2, and more can be expected with continuing research.

4.2. Consumption of GE Crops

All share a concern for producing safe, wholesome food. This is a fundamental requirement of the social pillar of sustainability. The weight of the evidence in favor of safety of crop improvement using GE is overwhelming, as reflected in the position statements of diverse, prestigious scientific societies [148–163]; scientific review papers [164–171], and hundreds of peer-reviewed research papers. This is not to suggest that there will never be a GE plant produced that has some unintended, negative effect on a consuming animal or human. No one can assure against such risk, whether the crop is conventional or GE. Rather, these facts support the notion, commonly held among scientists, that what matters to food safety is not the process used to create a plant, but the properties of the resulting plant [157,172–175]. In fact, instead of posing a routine food-safety risk, the reverse is true: GE traits can actually increase food safety as compared to conventional crops (see [176] and citations in [177]).

In pondering questions of food safety, two additional facts seem important:

(1) Recombinant DNA is a completely normal part of our diet. Our crops contain much natural recombinant DNA. Naturally produced recombinant DNA can result from: meiotic recombination; the action of diverse and often abundant mobile genetic elements; gene duplication; chromosomal inversions and translocations; novel gene assemblies; shuffling of exons and other gene fragments; chromosomal duplication; horizontal gene transfer; and incorporation of viral genes. In fact, all land plants appear to be "natural GMOs," as all contain genes apparently acquired horizontally [178–194]. To my knowledge, there is no published, validated research showing any fundamental biochemical or biophysical difference between DNA recombined in a test tube *vs.* that recombined in a living cell.

(2) Compared to other breeding techniques, targeted DNA manipulations achieved during transgenesis, cisgenesis, intragenesis, or genome editing are no more disruptive—and are commonly less disruptive—to a plant's genome,

transcriptome, proteome, and composition than other methods of crop improvement [170,171,195–200]. If unanticipated health consequences from GE manipulations merit concern, so do the unanticipated health consequences of each new conventionally bred crop variety [201]. It does not matter that breeding through phenotypic selection is a technique that is thousands of years old—every plant is a unique genetic and epigenetic creation. Therefore, every new plant presents unknown risks as a result of its unique genetic and epigenetic heritage.

4.3. Corporate Influence

Consolidation of the seed industry has been substantial since the introduction of GE crops [202,203]. This consolidation raises concerns as to whether food-system challenges will be decided more based on corporate interests than by broader considerations of sustainability. Related to this are concerns over patenting of GE traits and the restrictions patenting places on farmers with respect to seed saving and sharing [204–206]. In the developing world, where the food security of many smallholders depends on saved or shared seed [207], such restrictions are an acute concern (Figure 1). However, such restrictions are not universal, and royalty-free distribution is certainly compatible with GE [22,208,209].

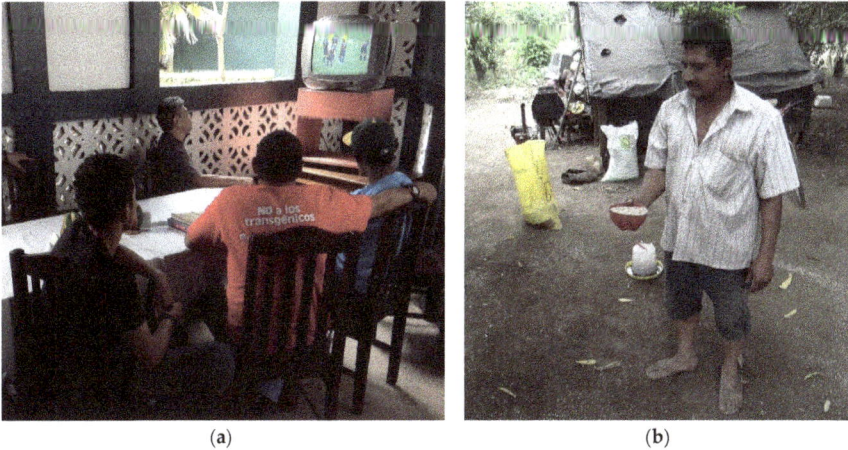

(a)   (b)

Figure 1. Selected perspectives on genetically engineered (GE) crops. (**a**) Some consider transgenic crops to be in conflict with their local culture; note the red t-shirt: "NO to transgenics." It is worth noting that some applications of GE do not employ transgenesis and therefore may be more widely acceptable. (**b**) Many smallholders are understandably concerned about the restrictions patents place on seed saving and sharing. In developing economies, GE traits in the public domain are likely to be the most accepted.

In the USA, GE traits and crops are protected under both the Plant Variety Protection Act and utility patents, the latter providing 20 years of protection against unauthorized use or distribution of the trait and/or genotype by farmers, plant breeders, and researchers [205,210–212]. Such patents protect the substantial investment in their development, and provide benefits that accrue through crop innovation. However, patent restrictions also pose risks to sustainability. Restrictions on the free movement of germplasm among breeders may create risks to the long-term diversity of crop genetics, and they can impede public crop improvement programs [210,213]. In addition, patent restrictions have at least the potential to constrain independent public research [211,214]. It is not surprising that utility patents are sought for GE traits because of their very high development costs, which can be well in excess of $100,000,000 [215]. Seed companies have little incentive to develop and market crops that their competitors can also sell. However, it is significant that patent applications by major seed companies have not been limited to GE traits, as utility patent protection has also been sought for conventionally bred traits [213] and genotypes. The current legal landscape allowing protection of plant traits under utility patents was not created exclusively for GE crops. Rather, it is the result of a series of federal legislative acts and Supreme Court decisions beginning as early as 1930 [204,205]. Thus, even if a less aggressive patent landscape were desirable from a sustainability standpoint, a move away from such protections of intellectual property would likely require either a tectonic shift in market pressure from consumers in support of initiatives such as the Open Source Seed Initiative [216] or, literally, an act of Congress.

4.4. Other Concerns

Some express concern that GE crops promote large-scale agriculture, with an associated loss of agrobiodiversity. It is true that present-day GE crops are often suitable for large-scale farming. However, large-scale monoculture exists even in non-GE crops, as it is driven by economies of scale and not by GE. Furthermore, when GE traits are deployed in diverse germplasms, there appears to be little loss in the genetic base [210]. GE is simply a suite of crop-improvement tools, which can be applied to agriculture at any scale, depending on the particular circumstances. GE traits can be introgressed into an unlimited number of local varieties, thus preserving agrobiodiversity while allowing the benefits of GE traits to accrue to farmers, consumers, rural communities, and/or the environment [22,209,217]. Furthermore, large-scale crop production very likely will continue to play an essential role in food and fiber production for the billions of us who depend on farmers. Food-system challenges should not be framed as "either-or" situations. It is critical to work towards increasing the sustainability of all farming systems, including large-scale systems, and the selective and wise use of GE crops can be expected to contribute towards that goal at all scales of farming.

In some instances, populations may conclude that GE is incompatible with their local food culture. While this concern lies outside the realm of scientific data, scientists have an obligation to respect such concerns.

5. Genome Editing: More Precise, Dynamic Tools for GE

Biology is being revolutionized by genome editing based on CRISPR/Cas9 technologies. These technologies not only provide powerful tools for research and therapeutics; they provide new methods for engineering crops to address genuine human needs and environmental impacts of crop production. Until recently, most applications of GE in crops involved insertion of DNA from an evolutionarily distant organism via either the bacterium *Agrobacterium tumefaciens* (a "natural genetic engineer") or the "gene gun." In contrast to plant transformation, genome editing can produce defined genetic changes in targeted genes much like a word processor, and with high efficiency and limited off-target changes [69,218–220]. Furthermore, it can be done in ways that leave no trace in the plant of foreign DNA (such as antibiotic resistance genes, plasmid fragments, *etc.*) [221,222]. Several examples were cited in Section 2 of the successful application of CRISPR/Cas9 technologies in developing crop disease resistance [69,70,106–109], and many others are expected. Indeed, CRISPR/Cas9 technologies may facilitate the development of entirely novel GE strategies not presented in this review.

Genome editing permits a more dynamic range of possibilities for genetic changes beyond those provided by plant transformation. It has commonly used for targeted mutagenesis and targeted modification: making very modest changes in existing genes in live cells [69,105,219,223,224]. Genetic changes can be as limited as a single nucleotide change, with no trace of introduced DNA. Thus, through targeted mutagenesis via CRISPR-Cas9, it is possible to create a nontransgenic gene edit that cannot be distinguished from a mutation that was naturally occurring or that was introgressed by conventional breeding [221]. Many experts consider such gene edits as excludable from GMO regulation [225] and argue that they should be clearly distinguished from GMOs, referring to them as "genetically edited crops" (GECs) [224]. Through homology-directed repair (HDR), genome editing can be used to edit a crop's genome so as to contain a novel functional DNA string identical to that of any source organism, including unrelated ones [223,226,227]. One application of HDR-based genome editing could be to edit a gene so as to match a gene from a crop's natural gene pool (=cisgenesis). Genome editing may thus be able to expedite genetic outcomes achievable through conventional breeding. For example, cisgenic applications of genome editing may present a particularly important path to increased disease resistance in crops that are difficult to hybridize [43,51].

Alternatively, HDR-based genome editing could conceivably be used to add a gene sequence from some evolutionarily distant organism, which is the equivalent of transgenesis [223,227] and therefore warranting regulatory scrutiny similar to that of transgenic crops [225].

Experienced molecular biologists commonly report that implementing CRISPR/Cas9 techniques is relatively straightforward, efficient, and low-cost, as compared to other techniques of genome editing. Depending on the regulatory environment, some applications of genome editing could help to "democratize" GE making it more accessible to small seed companies, nonprofit organizations, and governments in developing countries. In addition, it may facilitate beneficial applications of GE beyond large-scale agronomic crops (corn, soy, cotton, *etc*.), which currently dominate GE acreage globally.

6. Conclusions

The sustainability of food systems is a unifying interest that lies beyond particular crop production approaches or philosophies. The critical question is not, "Do certain GE strategies fit within a given production philosophy?" but rather, "Can a given practice or technology take us further down the path towards sustainability?" With respect to disease control, GE technologies, used wisely, certainly will permit the expeditious introduction into crops of targeted, diverse resistance mechanisms that mimic natural processes. While recognizing the important benefits GE technologies offer, larger considerations merit attention, especially questions of public acceptability and of whether there are any long-term ecological risks different from those posed by conventional breeding. In considering such issues, it is important to remember that, not only do diverse GE strategies exist, but diverse GE manipulations are possible, ranging from very modest, targeted mutagenesis, through cisgenics and intragenics, to insertion of transgenes from other crops, from other (non-crop) plants, and from evolutionarily distant organisms. Thus, in considering socioeconomic and cultural perspectives of GE, it is important to bear in mind this diversity of strategies and applications: GE crops can differ markedly from one another.

All of the varied conventional breeding techniques in existence today will remain the keystone of sustainable crop improvement, for several reasons. First, GE is commonly not the best breeding approach. If conventional breeding techniques permit breeders to meet their breeding goals, then these will often be preferred. Second, even when a useful GE trait has been created, conventional breeding remains necessary in order to introgress the trait into elite breeding lines. Finally, in any given crop, a useful GE construct may target one or a few pathogens of particular importance, but other breeding techniques still may be important for tackling disease problems not targeted by available GE traits. Thus, GE should be understood, not as the best approach to addressing sustainability challenges, but merely as a suite of tools that capitalizes on the knowledge that biologists gain through our ongoing study of Nature. GE simply expands the breeding "toolbox," providing options to consider on a case-by-case basis for enhancing the sustainability of crop disease management.

Acknowledgments: Thanks to Aardra Kachroo (University of Kentucky), Wayne Parrott (University of Georgia), and the anonymous reviewers for providing helpful suggestions on a previous draft of this paper.

Conflicts of Interest: The author declares no conflict of interest in the topic of genetic engineering. For a detailed disclosure statement, see "Disclosure Statement on Industry Relations" at http://out-of-the- box-vincelli.blogspot.com/2015/11/disclosure-statement-on-industry.html. All costs relating to the development of this review—routine academic expenses such as professorial salary, internet access, library services, *etc.*—were paid by the University of Kentucky. No external funding was solicited nor received for this project. Publication costs were waived by the journal

References

1. Pinstrup-Andersen, P. The future world food situation and the role of plant diseases. *Plant Health Instr.* **2001**.

2. Anway, M.D.; Leathers, C.; Skinner, M.K. Endocrine disruptor vinclozolin induced epigenetic transgenerational adult-onset disease. *Endocrinology* **2006**, *147*, 5515–5523.

3. Taxvig, C.; Hass, U.; Axelstad, M.; Dalgaard, M.; Boberg, J.; Andeasen, H.R.; Vinggaard, A.M. Endocrine- disrupting activities in vivo of the fungicides tebuconazole and epoxiconazole. *Toxicol. Sci.* **2007**, *100*, 464–73.

4. Pezzoli, G.; Cereda, E. Exposure to pesticides or solvents and risk of Parkinson disease. *Neurology* **2013**, *80*, 2035–2041.

5. Beard, J.D.; Umbach, D.M.; Hoppin, J.A.; Richards, M.; Alavanja, M.C.; Blair, A.; Sandler, D.P.; Kamel, F. Pesticide exposure and depression among male private pesticide applicators in the agricultural health study. *Environ. Health Perspect.* **2014**, *122*, 984–991.

6. Costello, S.; Cockburn, M.; Bronstein, J.; Zhang, X.; Ritz, B. Parkinson's disease and residential exposure to maneb and paraquat from agricultural applications in the Central Valley of California. *Am. J. Epidemiol.* **2009**, *169*, 919–926.

7. Collotta, M.; Bertazzi, P.A.; Bollati, V. Epigenetics and pesticides. *Toxicology* **2013**, *307*, 35–41.

8. McMahon, T.A.; Halstead, N.T.; Johnson, S.; Raffel, T.R.; Romansic, J.M.; Crumrine, P.W.; Rohr, J.R. Fungicide-induced declines of freshwater biodiversity modify ecosystem functions and services. *Ecol. Lett.* **2012**, *15*, 714–722.

9. Greitens, T.J.; Day, E. An alternative way to evaluate the environmental effects of integrated pest management: Pesticide risk indicators. *Renew. Agric. Food Syst.* **2007**, *22*, 213.

10. Schafer, R.B.; Gerner, N.; Kefford, B.J.; Rasmussen, J.J.; Beketov, M.A.; de Zwart, D.; Liess, M.; von der Ohe, P.C. How to characterize chemical exposure to predict ecologic effects on aquatic communities? *Environ. Sci. Technol.* **2013**, *47*, 7996–8004.

11. Malaj, E.; von der Ohe, P.C.; Grote, M.; Kuhne, R.; Mondy, C.P.; Usseglio-Polatera, P.; Brack, W.; Schafer, R.B. Organic chemicals jeopardize the health of freshwater ecosystems on the continental scale. *Proc. Natl. Acad. Sci. USA* **2014**, *111*, 9549–9554.

12. Zhu, Y.C.; Adamczyk, J.; Rinderer, T.; Yao, J.; Danka, R.; Luttrell, R.; Gore, J. Spray toxicity and risk potential of 42 commonly used formulations of row crop pesticides to adult honey bees (Hymenoptera: Apidae). *J. Econ. Entomol.* **2015**, *108*, 1–8.

13. Mader, E.; Adamson, N.L. Organic-Approved Pesticides: Minimizing Risks to Bees. Available online: http://www.xerces.org/wp-content/uploads/2009/12/xerces-organic-approved-pesticides-factsheet.pdf (accessed on 12 May 2016).

14. Barbosa, W.F.; De Meyer, L.; Guedes, R.N.C.; Smagghe, G. Lethal and sublethal effects of azadirachtin on the bumblebee *Bombus terrestris* (Hymenoptera: Apidae). *Ecotoxicology* **2015**, *24*, 130–142.

15. Michaud, J.P.; Grant, A.K. Sub-lethal effects of a copper sulfate fungicide on development and reproduction in three coccinellid species. *J. Insect. Sci.* **2003**, *3*, 1–6.

16. Hernández-Colorado, R.R.; Alvarado, A.L.; Romero, R.M. Acumulación de cobre en plantas silvestres de zonas agrícolas contaminadas con el metal. *Cienc. Tecnol.* **2012**, *28*, 55–61. (In Spanish)

17. Dutka, A.; McNulty, A.; Williamson, S.M. A new threat to bees? Entomopathogenic nematodes used in biological pest control cause rapid mortality in *Bombus terrestris*. *PeerJ* **2015**, *3*, e1413.

18. Cannon, J.R.; Tapias, V.M.; Na, H.M.; Honick, A.S.; Drolet, R.E.; Greenamyre, J.T. A highly reproducible rotenone model of Parkinson's disease. *Neurobiol. Dis.* **2009**, *34*, 279–290.

19. Coats, J.R. Risks from natural versus synthetic insecticides. *Annu. Rev. Entomol.* **1994**, *39*, 489–515.

20. Palti, J. *Cultural Practices and Infectious Crop Diseases*; Springer: Berlin, Germany, 1981; p. 246.

21. Ferreira, S.A.; Pitz, K.Y.; Manshardt, R.; Zee, F.; Fitch, M.; Gonsalves, D. Virus coat protein transgenic papaya provides practical control of papaya ringspot virus in Hawaii. *Plant Dis.* **2002**, *86*, 101–105.

22. Taylor, N.J.; Halsey, M.; Gaitán-Solís, E.; Anderson, P.; Gichuki, S.; Miano, D.; Bua, A.; Alicai, T.; Fauquet, C.M. The VIRCA project: Virus resistant cassava for Africa. *GM Crops Food* **2012**, *3*, 93–103.

23. Dutt, M.; Barthe, G.; Irey, M.; Grosser, J. Transgenic citrus expressing an *Arabidopsis* NPR1 gene exhibit enhanced resistance against huanglongbing (HLB; citrus greening). *PLoS ONE* **2015**, *10*, e0137134.

24. Kress, R. Citrus Plants Resistant to *Citrus Huanglongbing* (ex *greening*) Caused by Candidatus Liberibacter Asiaticus (las) and Bacterial Canker Caused by (*Xanthomonas axonopodis* pv. *citri*) (xac). US20150067918 A1, 5 March 2015.

25. National Research Council. *Strategic Planning for the Florida Citrus Industry: Addressing Citrus Greening Disease*; The National Academies Press: Washington, DC, USA, 2010; p. 326.

26. Carpenter, J.E. Impact of GM crops on biodiversity. *GM Crops* **2011**, *2*, 7–23.

27. Huang, J.; Rozelle, S.; Pray, C.; Wang, Q. Plant biotechnology in China. *Science* **2002**, *295*, 674–677.

28. Huang, J.; Hu, R.; Qiao, F.; Yin, Y.; Liu, H.; Huang, Z. Impact of insect-resistant GM rice on pesticide use and farmers' health in China. *Sci. China Life Sci.* **2015**, *58*, 466–471.

29. National Research Council. *Impact of Genetically Engineered Crops on Farm Sustainability in the United States*; The National Academies Press: Washington, DC, USA, 2010; p. 270.

30. Klumper, W.; Qaim, M. A meta-analysis of the impacts of genetically modified crops. *PLoS ONE* **2014**, *9*, e111629.

31. Kouser, S.; Qaim, M. Impact of Bt cotton on pesticide poisoning in smallholder agriculture: A panel data analysis. *Ecol. Econ.* **2011**, *70*, 2105–2113.

32. Fernandez-Cornejo, J.; Wechsler, S.J.; Livingston, M.; Mitchell, L. Genetically Engineered Crops in the United States. Available online: www.ers.usda.gov/publications/err-economic-research-report/err162.aspx (accessed on 12 May 2016).

33. Huesing, J.; English, L. The impact of Bt crops on the developing world. *AgBioForum* **2004**, *7*, 84–95.

34. Zambrano, P.; Smale, M.; Maldonado, J.H.; Mendoza, S.L. Unweaving the threads: The experiences of female farmers with biotech cotton in Colombia. *AgBioForum* **2012**, *15*, 125–137.

35. Smyth, S.J.; Kerr, W.A.; Phillips, P.W.B. Global economic, environmental and health benefits from GM crop adoption. *Glob. Food Secur.* **2015**, *7*, 24–29.

36. Gassmann, A.J.; Petzold-Maxwell, J.L.; Clifton, E.H.; Dunbar, M.W.; Hoffmann, A.M.; Ingber, D.A.; Keweshan, R.S. Field-evolved resistance by western corn rootworm to multiple *Bacillus thuringiensis* toxins in transgenic maize. *Proc. Natl. Acad. Sci. USA* **2014**, *111*, 5141–5146.

37. Tabashnik, B.E.; Brevault, T.; Carriere, Y. Insect resistance to Bt crops: lessons from the first billion acres. *Nat. Biotechnol.* **2013**, *31*, 510–521.

38. Huang, F.; Qureshi, J.A.; Meagher, R.L.Jr.; Reisig, D.D.; Head, G.P.; Andow, D.A.; Ni, X.; Kerns, D.; Buntin, G.D.; Niu, Y.; Yang, F.; Dangal, V. Cry1F resistance in fall armyworm *Spodoptera frugiperda*: Single gene versus pyramided Bt maize. *PLoS ONE* **2014**, *9*, e112958.

39. Campagne, P.; Kruger, M.; Pasquet, R.; Le Ru, B.; Van den Berg, J. Dominant inheritance of field-evolved resistance to Bt corn in *Busseola fusca*. *PLoS ONE* **2013**, *8*, e69675.

40. Cochrane, G.; Karsch-Mizrachi, I.; Nakamura, Y. The international nucleotide sequence database collaboration. *Nucleic Acids Res.* **2011**, *39*, D15–D18.

41. Chisholm, S.T.; Coaker, G.; Day, B.; Staskawicz, B.J. Host-microbe interactions: shaping the evolution of the plant immune response. *Cell* **2006**, *124*, 803–814.

42. Jones, J.D.; Dangl, J.L. The plant immune system. *Nature* **2006**, *444*, 323–329.

43. Jones, J.D.; Witek, K.; Verweij, W.; Jupe, F.; Cooke, D.; Dorling, S.; Tomlinson, L.; Smoker, M.; Perkins, S.; Foster, S. Elevating crop disease resistance with cloned genes. *Philos. Trans. R. Soc. Lond. B. Biol. Sci.* **2014**, *369*, 20130087.

44. Lacombe, S.; Rougon-Cardoso, A.; Sherwood, E.; Peeters, N.; Dahlbeck, D.; van Esse, H.P.; Smoker, M.; Rallapalli, G.; Thomma, B.P.; Staskawicz, B.; *et al.* Interfamily transfer of a plant pattern-recognition receptor confers broad-spectrum bacterial resistance. *Nat. Biotechnol.* **2010**, *28*, 365–369.

45. Tripathi, J.N.; Lorenzen, J.; Bahar, O.; Ronald, P.; Tripathi, L. Transgenic expression of the rice Xa21 pattern-recognition receptor in banana (*Musa* sp.) confers resistance to *Xanthomonas campestris* pv. *musacearum*. *Plant Biotechnol. J.* **2014**, *12*, 663–673.

46. Schwessinger, B.; Bahar, O.; Thomas, N.; Holton, N.; Nekrasov, V.; Ruan, D.; Canlas, P.E.; Daudi, A.; Petzold, C.J.; Singan, V.R.; *et al.* Transgenic expression of the dicotyledonous pattern recognition receptor EFR in rice leads to ligand-dependent activation of defense responses. *PLoS Pathog.* **2015**, *11*, e1004809.

47. Dangl, J.L.; Horvath, D.M.; Staskawicz, B.J. Pivoting the plant immune system from dissection to deployment. *Science* **2013**, *341*, 746–751.

48. Gill, U.S.; Lee, S.; Mysore, K.S. Host versus nonhost resistance: Distinct wars with similar arsenals. *Phytopathology* **2015**, *105*, 580–587.

49. Stotz, H.U.; Mitrousia, G.K.; de Wit, P.J.; Fitt, B.D. Effector-triggered defence against apoplastic fungal pathogens. *Trends Plant Sci.* **2014**, *19*, 491–500.

50. Saunders, D.G.O. Hitchhiker's guide to multi-dimensional plant pathology. *N. Phytol.* **2015**, *205*, 1028–1033.

51. Holme, I.B.; Wendt, T.; Holm, P.B. Intragenesis and cisgenesis as alternatives to transgenic crop development. *Plant Biotechnol. J.* **2013**, *11*, 395–407.

52. Jo, K.R.; Kim, C.J.; Kim, S.J.; Kim, T.Y.; Bergervoet, M.; Jongsma, M.A.; Visser, R.G.; Jacobsen, E.; Vossen, J.H. Development of Late Blight Resistant Potatoes by Cisgene Stacking. Available online: http://www.biomedcentral.com/1472-6750/14/50 (accessed on 12 May 2016).

53. Tai, T.H.; Dahlbeck, D.; Clark, E.T.; Gajiwala, P.; Pasion, R.; Whalen, M.C.; Stall, R.E.; Staskawicz, B.J. Expression of the Bs2 pepper gene confers resistance to bacterial spot disease in tomato. *Proc. Natl. Acad. Sci. USA* **1999**, *96*, 14153–14158.

54. Horvath, D.M.; Stall, R.E.; Jones, J.B.; Pauly, M.H.; Vallad, G.E.; Dahlbeck, D.; Staskawicz, B.J.; Scott, J.W. Transgenic resistance confers effective field level control of bacterial spot disease in tomato. *PLoS ONE* **2012**, *7*, e42036.

55. Yang, S.; Li, J.; Zhang, X.; Zhang, Q.; Huanga, J.; Chen, J.-Q.; Hart, D.L.; Tiana, D. Rapidly evolving *R* genes in diverse grass species confer resistance to rice blast disease. *Proc. Natl. Acad. Sci. USA* **2013**, *110*, 18572–18577.

56. Kawashima, C.G.; Guimaraes, G.A.; Nogueira, S.R.; MacLean, D.; Cook, D.R.; Steuernagel, B.; Baek, J.; Bouyioukos, C.; Melo, B.D.; Tristao, G.; *et al.* A pigeonpea gene confers resistance to Asian soybean rust in soybean. *Nat. Biotechnol.* **2016**.

57. Kim, S.H.; Qi, D.; Ashfield, T.; Helm, M.; Innes, R.W. Using decoys to expand the recognition specificity of a plant disease resistance protein. *Science* **2016**, *351*, 684–687.

58. Fukuoka, S.; Saka, N.; Mizukami, Y.; Koga, H.; Yamanouchi, U.; Yoshioka, Y.; Hayashi, N.; Ebana, K.; Mizobuchi, R.; Yano, M. Gene pyramiding enhances durable blast disease resistance in rice. *Sci. Rep.* **2015**.

59. Fry, W.E. *Principles of Plant Disease Management*; Academic Press: New York, NY, USA, 1982; p. 378.

60. Pel, M.A.; Foster, S.J.; Park, T.H.; Rietman, H.; van Arkel, G.; Jones, J.D.; Van Eck, H.J.; Jacobsen, E.; Visser, R.G.; Van der Vossen, E.A. Mapping and cloning of late blight resistance genes from *Solanum venturii* using an interspecific candidate gene approach. *Mol. Plant Microbe Interact.* **2009**, *22*, 601–615.

61. Bundo, M.; Coca, M. Enhancing blast disease resistance by overexpression of the calcium-dependent protein kinase OsCPK4 in rice. *Plant Biotechnol. J.* **2015**.

62. Tripathi, L.; Tripathi, J.N.; Kiggundu, A.; Korie, S.; Shotkoski, F.; Tushemereirwe, W.K. Field trial of Xanthomonas wilt disease-resistant bananas in East Africa. *Nat. Biotechnol.* **2014**, *32*, 868–870.

63. Ger, M.; Chen, C.; Hwang, S.; Huang, H.; Podile, A.R.; Dayakar, B.V.; Feng, T. Constitutive expression of *hrap* gene in transgenic tobacco plant enhances resistance against virulent bacterial pathogens by induction of a hypersensitive response. *Mol. Plant-Microbe Interact.* **2002**, *15*, 764–773.

64. Huang, H.E.; Ger, M.J.; Yip, M.K.; Chen, C.Y.; Pandey, A.K.; Feng, T.Y. A hypersensitive response was induced by virulent bacteria in transgenic tobacco plants overexpressing a plant ferredoxin-like protein (PFLP). *Physiol. Mol. Plant Pathol.* **2004**, *64*, 103–110.

65. Chen, X.J.; Chen, Y.; Zhang, L.N.; Xu, B.; Zhang, J.H.; Chen, Z.X.; Tong, Y.H.; Zuo, S.M.; Xu, J.Y. Overexpression of OsPGIP1 enhances rice resistance to sheath blight. *Plant Dis.* **2016**, *100*, 388–395.

66. van Schie, C.C.; Takken, F.L. Susceptibility genes 101: How to be a good host. *Annu. Rev. Phytopathol.* **2014**, *52*, 551–581.

67. Berg, J.A.; Appiano, M.; Santillan Martinez, M.; Hermans, F.W.; Vriezen, W.H.; Visser, R.G.; Bai, Y.; Schouten, H.J. A transposable element insertion in the susceptibility gene CsaMLO8 results in hypocotyl resistance to powdery mildew in cucumber. *BMC Plant Biol.* **2015**, *15*, 243.

68. De Almeida Engler, J.; Favery, B.; Engler, G.; Abad, P. Loss of susceptibility as an alternative for nematode resistance. *Curr. Opin. Biotechnol.* **2005**, *16*, 112–117.

69. Wang, Y.; Cheng, X.; Shan, Q.; Zhang, Y.; Liu, J.; Gao, C.; Qiu, J.L. Simultaneous editing of three homoeoalleles in hexaploid bread wheat confers heritable resistance to powdery mildew. *Nat. Biotechnol.* **2014**, *32*, 947–951.

70. Jia, H.; Orbovic, V.; Jones, J.B.; Wang, N. Modification of the PthA4 effector binding elements in Type I CsLOB1 promoter using Cas9/sgRNA to produce transgenic Duncan grapefruit alleviating XccΔpthA4:dCsLOB1.3 infection. *Plant Biotechnol. J.* **2016**, *14*, 1291–1301.

71. Perez, M.G. Florida's Orange Industry Is in Its Worst Slump in 100 Years. Available online: http://www.bloomberg.com/news/articles/2015-11-24/in-florida-the-oj-crop-is-getting-wiped-out-by-an-asian-invader (accessed on 12 May 2016).

72. Harmon, A. A Race to Save the Orange by Altering Its DNA. Available online: http://www.nytimes.com/2013/07/28/science/a-race-to-save-the-orange-by-altering-its-dna.html?pagewanted=all&_r=1 (accessed on 12 May 2016).

73. Ohlemeier, D. GMO Targets Citrus Greening. Available online: http://www.thepacker. com/news/florida-grower-growing-citrus-greening-resistant-gmo-fruit (accessed on 12 May 2016).

74. Rubio, J.; Montes, C.; Castro, Á.; Álvarez, C.; Olmedo, B.; Muñoz, M.; Tapia, E.; Reyes, F.; Ortega, M.; Sánchez, E.; *et al.* Genetically engineered Thompson Seedless grapevine plants designed for fungal tolerance: Selection and characterization of the best performing individuals in a field trial. *Transgenic Res.* **2015**, *24*, 43–60.

75. Emani, C.; Garcia, J.M.; Lopata-Finch, E.; Pozo, M.J.; Uribe, P.; Kim, D.J.; Sunilkumar, G.; Cook, D.R.; Kenerley, C.M.; Rathore, K.S. Enhanced fungal resistance in transgenic cotton expressing an endochitinase gene from *Trichoderma virens*. *Plant Biotechnol. J.* **2003**, *1*, 321–336.

76. Bonning, B.C.; Pal, N.; Liu, S.; Wang, Z.; Sivakumar, S.; Dixon, P.M.; King, G.F.; Miller, W.A. Toxin delivery by the coat protein of an aphid-vectored plant virus provides plant resistance to aphids. *Nat. Biotechnol.* **2014**, *32*, 102–105.

77. Badran, A.H.; Guzov, V.M.; Huai, Q.; Kemp, M.M.; Vishwanath, P.; Kain, W.; Nance, A.M.; Evdokimov, A.; Moshiri, F.; Turner, K.H.; *et al.* Continuous evolution of *Bacillus thuringiensis* toxins overcomes insect resistance. *Nature* **2016**, *533*, 58–63.

78. Lusk, J.L.; Rozan, A. Consumer acceptance of ingenic foods. *Biotechnol. J.* **2006**, *1*, 1433–1434.

79. Carthew, R.W.; Sontheimer, E.J. Origins and mechanisms of miRNAs and siRNAs. *Cell* **2009**, *136*, 642–655.

80. Gonsalves, D.; Ferreira, S. Transgenic papaya: a case for managing risks of papaya ring spot virus in Hawaii. *Plant Health Prog.* **2003**.

81. Fitch, M.M.M.; Manschardt, R.M.; Gonsalves, D.; Slightom, J.L.; Sanford, J.C. Virus resistant papaya derived from tissues bombarded with the coat protein gene of papaya ringspot virus. *Nat. Biotechnol.* **1992**, *10*, 1466–1472.

82. Odipio, J.; Ogwok, E.; Taylor, N.J.; Halsey, M.; Bua, A.; Fauquet, C.M.; Alicai, T. RNAi-derived field resistance to cassava brown streak disease persists across the vegetative cropping cycle. *GM Crops Food* **2014**, *5*, 16–19.

83. Zhang, X.; Sato, S.; Ye, X.; Dorrance, A.E.; Morris, T.J.; Clemente, T.E.; Qu, F. Robust RNAi-based resistance to mixed infection of three viruses in soybean plants expressing separate short hairpins from a single transgene. *Phytopathology* **2011**, *101*, 1264–1269.

84. Klas, F.E.; Fuchs, M.; Gonsalves, D. Fruit yield of virus-resistant transgenic summer squash in simulated commercial plantings under conditions of high disease pressure. *J. Hortic. For.* **2011**, *3*, 46–52.

85. Yang, C.F.; Chen, K.C.; Cheng, Y.H.; Raja, J.A.; Huang, Y.L.; Chien, W.C.; Yeh, S.D. Generation of marker-free transgenic plants concurrently resistant to a DNA geminivirus and a RNA tospovirus. *Sci. Rep.* **2014**, *4*, 5717.

86. Andrade, C.M.; Tinoco, M.L.P.; Rieth, A.F.; Maia, F.C.O.; Aragão, F.J.L. Host-induced gene silencing in the necrotrophic fungal pathogen *Sclerotinia sclerotiorum*. *Plant Pathol.* **2015**, *65*, 626–632.

87. Nowara, D.; Gay, A.; Lacomme, C.; Shaw, J.; Ridout, C.; Douchkov, D.; Hensel, G.; Kumlehn, J.; Schweizer, P. HIGS: Host-induced gene silencing in the obligate biotrophic fungal pathogen *Blumeria graminis*. *Plant Cell* **2010**, *22*, 3130–3141.

88. J.R. Simplot Company. Petition for Determination of Nonregulated Status for Innate™ Potatoes with Late Blight Resistance, Low Acrylamide Potential, Reduced Black Spot, and Lowered Reducing Sugars: Russet Burbank Event W8. Available online: https://www.aphis.usda.gov/brs/aphisdocs/14_09301p.pdf (accessed on 12 May 2016).

89. Jahan, S.N.; Asman, A.K.; Corcoran, P.; Fogelqvist, J.; Vetukuri, R.R.; Dixelius, C. Plant-mediated gene silencing restricts growth of the potato late blight pathogen *Phytophthora infestans*. *J Exp. Bot.* **2015**, *66*, 2785–2794.

90. Krijger, J.J.; Oliveira-Garcia, E.; Astolfi, P.; Sommerfeld, K.; Gase, I.; Kastner, C.; Kumlehn, J.; Deising, H.B. Discovery of candidate genes for defeating fungal pathogens by host-induced gene silencing (HIGS). In *Modern Fungicides and Antifungal Compounds VII*; Dehne, H.W., Deising, B., Fraaije, U., Gisi, D., Hermann, D., Mehl, A., Oerke, E.C., Russell, P.E., Stammler, G., Kuck, K.H., Lyr, H., Eds.; Deutsche Phytomedizinische Gesellschaft: Braunschweig, Germany, 2014; pp. 35–44.

91. Govindarajulu, M.; Epstein, L.; Wroblewski, T.; Michelmore, R.W. Host-induced gene silencing inhibits the biotrophic pathogen causing downy mildew of lettuce. *Plant Biotechnol. J.* **2015**, *13*, 875–883.

92. Fan, W.; Wei, Z.; Zhang, M.; Ma, P.; Liu, G.; Zheng, J.; Guo, X.; Zhang, P. Resistance to *Ditylenchus destructor* infection in sweet potato by the expression of small interfering RNAs targeting *unc-15*, a movement-related gene. *Phytopathology* **2015**, *105*, 1458–1465.

93. Lourenço-Tessutti, I.T.; Souza Junior, J.D.; Martins-de-Sa, D.; Viana, A.A.; Carneiro, R.M.; Togawa, R.C.; de Almeida-Engler, J.; Batista, J.A.; Silva, M.C.; Fragoso, R.R.; *et al.* Knock-down of heat-shock protein 90 and isocitrate lyase gene expression reduced root-knot nematode reproduction. *Phytopathology* **2015**, *105*, 628–637.

94. Huang, G.; Allen, R.; Davis, E.L.; Baum, T.J.; Hussey, R.S. Engineering broad root-knot resistance in transgenic plants by RNAi silencing of a conserved and essential root-knot nematode parasitism gene. *Proc. Natl. Acad. Sci. USA* **2006**, *103*, 14302–14306.

95. Noon, J.B.; Hewezi, T.; Maier, T.R.; Simmons, C.; Wei, J.Z.; Wu, G.; Llaca, V.; Deschamps, S.; Davis, E.L.; Mitchum, M.G.; Hussey, R.S.; Baum, T.J. Eighteen new candidate effectors of the phytonematode *Heterodera glycines* produced specifically in the secretory esophageal gland cells during parasitism. *Phytopathology* **2015**, *105*, 1362–1372.

96. Baum, J.A.; Bogaert, T.; Clinton, W.; Heck, G.R.; Feldmann, P.; Ilagan, O.; Johnson, S.; Plaetinck, G.; Munyikwa, T.; Pleau, M.; *et al.* Control of coleopteran insect pests through RNA interference. *Nat. Biotechnol.* **2007**, *25*, 1322–1326.

97. San Miguel, K.; Scott, J.G. The next generation of insecticides: dsRNA is stable as a foliar-applied insecticide. *Pest Manag. Sci.* **2015**.

98. Friesen, T.L.; Stukenbrock, E.H.; Liu, Z.; Meinhardt, S.; Ling, H.; Faris, J.D.; Rasmussen, J.B.; Solomon, P.S.; McDonald, B.A.; Oliver, R.P. Emergence of a new disease as a result of interspecific virulence gene transfer. *Nat. Genet.* **2006**, *38*, 953–956.

99. Zhang, L.; Yao, J.; Withers, J.; Xin, X.F.; Banerjee, R.; Fariduddin, Q.; Nakamura, Y.; Nomura, K.; Howe, G.A.; Boland, W.; Yan, H.; He, S.Y. Host target modification as a strategy to counter pathogen hijacking of the jasmonate hormone receptor. *Proc. Natl. Acad. Sci. USA* **2015**, *112*, 14354–14359.

100. Mobius, N.; Hertweck, C. Fungal phytotoxins as mediators of virulence. *Curr. Opin. Plant Biol.* **2009**, *12*, 390–398.

101. Johal, G.S.; Briggs, S.P. Reductase activity encoded by the HM1 disease resistance gene in maize. *Science* **1992**, *258*, 985–987.

102. Chen, C.; Sun, Q.; Narayanan, B.; Nuss, D.L.; Herzberg, O. Structure of oxalacetate acetylhydrolase, a virulence factor of the chestnut blight fungus. *J. Biol. Chem.* **2010**, *285*, 26685–26696.

103. Zhang, B.; Oakes, A.D.; Newhouse, A.E.; Baier, K.M.; Maynard, C.A.; Powell, W.A. A threshold level of oxalate oxidase transgene expression reduces *Cryphonectria parasitica*-induced necrosis in a transgenic American chestnut (*Castanea dentata*) leaf bioassay. *Transgenic Res.* **2013**, *22*, 973–982.

104. Li, X.; Shin, S.; Heinen, S.; Dill-Macky, R.; Berthiller, F.; Nersesian, N.; Clemente, T.; McCormick, S.; Muehlbauer, G.J. Transgenic wheat expressing a barley UDP-glucosyltransferase detoxifies deoxynivalenol and provides high levels of resistance to *Fusarium graminearum*. *Mol. Plant Microbe Interact.* **2015**, *28*, 1237–1246.

105. Shan, Q.; Wang, Y.; Li, J.; Zhang, Y.; Chen, K.; Liang, Z.; Zhang, K.; Liu, J.; Xi, J.J.; Qiu, J.L.; Gao, C. Targeted genome modification of crop plants using a CRISPR-Cas system. *Nat. Biotechnol.* **2013**, *31*, 686–688.

106. Ali, Z.; Abulfaraj, A.; Idris, A.; Ali, S.; Tashkandi, M.; Mahfouz, M.M. CRISPR/Cas9-mediated viral interference in plants. *Genome Biol.* **2015**, *16*, 238.

107. Ji, X.; Zhang, H.; Zhang, Y.; Wang, Y.; Gao, C. Establishing a CRISPR–Cas-like immune system conferring DNA virus resistance in plants. *Nat. Plants* **2015**, *1*, 15144.

108. Chaparro-Garcia, A.; Kamoun, S.; Nekrasov, V. Boosting plant immunity with CRISPR/Cas. *Genome Biol.* **2015**, *16*, 254.

109. Baltes, N.J.; Hummel, A.W.; Konecna, E.; Cegan, R.; Bruns, A.N.; Bisaro, D.M.; Voytas, D.F. Conferring resistance to geminiviruses with the CRISPR–Cas prokaryotic immune system. *Nat. Plants* **2015**, *1*, 15145.

110. Rojas, M.R.; Hagen, C.; Lucas, W.J.; Gilbertson, R.L. Exploiting chinks in the plant's armor: evolution and emergence of geminiviruses. *Annu. Rev. Phytopathol.* **2005**, *43*, 361–394.

111. Clements, M.J.; Campbell, K.W.; Maragos, C.M.; Pilcher, C.; Headrick, J.M.; Pataky, J.K.; White, D.G. Influence of Cry1Ab protein and hybrid genotype on fumonisin contamination and Fusarium ear rot of corn. *Crop Sci.* **2003**, *43*, 1283–1293.

112. Folcher, L.; Delos, M.; Marengue, E.; Jarry, M.; Weissenberger, A.; Eychenne, N.; Regnault-Roger, C. Lower mycotoxin levels in Bt maize grain. *Agron. Sustain. Dev.* **2010**, *30*, 711–719.

113. Munkvold, G.P.; Hellmich, R.L.; Rice, L.G. Comparison of fumonisin concentrations in kernels of transgenic Bt maize hybrids and nontransgenic hybrids. *Plant Dis.* **1999**, *83*, 130–138.

114. Williams, W.P.; Windham, G.L.; Buckley, P.M.; Perkins, J.M. Southwestern corn borer damage and aflatoxin accumulation in conventional and transgenic corn hybrids. *Field Crops Res.* **2005**, *91*, 329–336.

115. Williams, W.P.; Windham, G.L.; Buckley, P.M.; Daves, C.A. Aflatoxin accumulation in conventional and transgenic corn hybrids infested with southwestern corn borer (Lepidoptera: Crambidae). *J. Agric. Urban Entomology* **2002**, *19*, 227–236.

116. Munkvold, G.P.; Hellmich, R.L.; Showers, W.B. Reduced Fusarium ear rot and symptomless infection in kernels of maize genetically engineered for European corn borer resistance. *Phytopathology* **1997**, *87*, 1071–1077.

117. Lamichhane, J.R.; Dachbrodt-Saaydeh, S.; Kudsk, P.; Messéan, A. Toward a reduced reliance on conventional pesticides in European agriculture. *Plant Dis.* **2016**, *100*, 10–24.

118. Scheffer, R.P. *The Nature of Disease in Plants*; Cambridge University Press: New York, NY, USA, 1997; p. 325.

119. Burdon, J.J. *Diseases and Plant Population Biology*; Cambridge University Press: Cambridge, United Kingdom, 1997; p. 208.

120. Carriere, Y.; Crickmore, N.; Tabashnik, B.E. Optimizing pyramided transgenic Bt crops for sustainable pest management. *Nat. Biotechnol.* **2015**, *33*, 161–168.

121. Zhu, S.; Li, Y.; Vossen, J.H.; Visser, R.G.; Jacobsen, E. Functional stacking of three resistance genes against *Phytophthora infestans* in potato. *Transgenic Res.* **2012**, *21*, 89–99.

122. Singh, R.P.; Hodson, D.P.; Jin, Y.; Lagudah, E.S.; Ayliffe, M.A.; Bhavani, S.; Rouse, M.N.; Pretorius, Z.A.; Szabo, L.J.; Huerta-Espino, J.; Basnet, B.R.; Lan, C.; Hovmøller, M.S. Emergence and spread of new races of wheat stem rust fungus: Continued threat to food security and prospects of genetic control. *Phytopathology* **2015**, *105*, 872–884.

123. Witek, K.; Jupe, F.; Witek, A.I.; Baker, D.; Clark, M.D.; Jones, J.D. Accelerated cloning of a potato late blight-resistance gene using RenSeq and SMRT sequencing. *Nat. Biotechnol.* **2016**.

124. Steuernagel, B.; Periyannan, S.K.; Hernandez-Pinzon, I.; Witek, K.; Rouse, M.N.; Yu, G.; Hatta, A.; Ayliffe, M.; Bariana, H.; Jones, J.D.; Lagudah, E.S.; Wulff, B.B. Rapid cloning of disease-resistance genes in plants using mutagenesis and sequence capture. *Nat. Biotechnol.* **2016**.

125. Yu, W.; Yau, Y.Y.; Birchler, J.A. Plant artificial chromosome technology and its potential application in genetic engineering. *Plant Biotechnol. J.* **2016**, *14*, 1175–1182.

126. Marcroft, S.J.; Van de Wouw, A.P.; Salisbury, P.A.; Potter, T.D.; Howlett, B.J. Effect of rotation of canola (*Brassica napus*) cultivars with different complements of blackleg resistance genes on disease resistance. *Plant Pathol.* **2012**, *61*, 934–944.

127. Zapiola, M.L.; Campbell, C.K.; Butler, M.D.; Mallory-Smith, C.A. Escape and establishment of transgenic glyphosate-resistant creeping bentgrass *Agrostis stolonifera* in Oregon, USA: A 4-year study. *J. Appl. Ecol.* **2008**, *45*, 486–494.

128. Shivrain, V.K.; Burgos, N.R.; Anders, M.M.; Rajguru, S.N.; Moore, J.; Sales, M.A. Gene flow between Clearfield™ rice and red rice. *Crop Prot.* **2007**, *26*, 349–356.

129. Fuchs, E.J.; Martínez, A.M.; Calvo, A.; Muñoz, M.; Arrieta-Espinoza, G. Genetic structure of *Oryza glumaepatula* wild rice populations and evidence of introgression from *O. sativa* in Costa Rica. *PeerJ PrePrints* **2015**.

130. Kwit, C.; Moon, H.S.; Warwick, S.I.; Stewart, C.N. Jr. Transgene introgression in crop relatives: molecular evidence and mitigation strategies. *Trends Biotechnol.* **2011**, *29*, 284–293.

131. Kiang, Y.T.; Anotonovis, J.; Wu, L. The extinction of wild rice (*Oryza perennis formosa*) in Taiwan. *J. Asian Ecol.* **1979**, *1*, 1–9.

132. Sasu, M.A.; Ferrari, M.J.; Du, D.; Winsor, J.A.; Stephenson, A.G. Indirect costs of a nontarget pathogen mitigate the direct benefits of a virus-resistant transgene in wild *Cucurbita*. *Proc. Natl. Acad. Sci. USA* **2009**, *106*, 19067–19071.

133. Van Etten, L.; Kuester, A.; Chang, S.M.; Baucom, R. Reduced Seed Viability and Reductions in Plant Size Provide Evidence for Costs of Glyphosate Resistance in An Agricultural Weed. Available online: http://biorxiv.org/content/early/2015/11/06/030833 (accessed on 19 May 2016).

134. Wang, W.; Xia, H.; Yang, X.; Xu, T.; Si, H.J.; Cai, X.X.; Wang, F.; Su, J.; Snow, A.A.; Lu, B.R. A novel 5-enolpyruvoylshikimate-3-phosphate (EPSP) synthase transgene for glyphosate resistance stimulates growth and fecundity in weedy rice (*Oryza sativa*) without herbicide. *N. Phytol.* **2014**, *202*, 679–688.

135. Nielsen, K.M.; Johnsen, P.J.; Bensasson, D.; Daffonchio, D. Release and persistence of extracellular DNA in the environment. *Environ. Biosaf. Res.* **2007**, *6*, 37–53.

136. Nielsen, K.M.; Bones, A.M.; Smalla, K.; van Elsas, J.D. Horizontal gene transfer from transgenic plants to terrestrial bacteria—a rare event? *FEMS Microbiol. Rev.* **1998**, *22*, 79–103.

137. de Vries, J.; Wackernagel, W. Microbial horizontal gene transfer and the DNA release from transgenic crop plants. *Plant Soil* **2004**, *266*, 91–104.

138. Syvanen, M. Search for horizontal gene transfer from transgenic crops to microbes. In *Horizontal Gene Transfer*; Syvanen, M., Kado, Cc.i., Eds.; Academic Press: London, UK, 2002; p. 445.

139. Demaneche, S.; Monier, J.M.; Dugat-Bony, E.; Simonet, P. Exploration of horizontal gene transfer between transplastomic tobacco and plant-associated bacteria. *FEMS Microbiol. Ecol.* **2011**, *78*, 129–136.

140. Pontiroli, A.; Rizzi, A.; Simonet, P.; Daffonchio, D.; Vogel, T.M.; Monier, J.M. Visual evidence of horizontal gene transfer between plants and bacteria in the phytosphere of transplastomic tobacco. *Appl. Environ Microbiol.* **2009**, *75*, 3314–3322.

141. Heinemann, J.A.; Traavik, T. Problems in monitoring horizontal gene transfer in field trials of transgenic plants. *Nat. Biotechnol.* **2004**, *22*, 1105–1109.

142. Lu, B.R. Transgene escape from GM crops and potential biosafety consequences: An environmental perspective. *Collect. Biosaf. Rev.* **2008**, *4*, 66–141.

143. Baltazar, B.M.; Castro Espinoza, L.; Espinoza Banda, A.; de la Fuente Martinez, J.M.; Garzon Tiznado, J.A.; Gonzalez Garcia, J.; Gutierrez, M.A.; Guzman Rodriguez, J.L.; Heredia Diaz, O.; Horak, M.J.; *et al.* Pollen-mediated gene flow in maize: implications for isolation requirements and coexistence in Mexico, the center of origin of maize. *PLoS ONE* **2015**, *10*, e0131549.

144. Gressel, J. Dealing with transgene flow of crop protection traits from crops to their relatives. *Pest Manag. Sci.* **2014**, *71*, 658–667.

145. Gonzalez, M.D. Screening and Genotyping of Ga1 Gene and Genotype X Environment Interaction of Cross Incompatibility In Maize. Ph.D. Thesis, Iowa State University, Ames, IA, USA, 2011.

146. Lombardo, L. Genetic use restriction technologies: A review. *Plant Biotechnol. J.* **2014**, *12*, 995–1005.

147. Li, J.; Yu, H.; Zhang, F.; Lin, C.; Gao, J.; Fang, J.; Ding, X.; Shen, Z.; Xu, X. A built-in strategy to mitigate transgene spreading from genetically modified corn. *PLoS ONE* **2013**, *8*, e81645.

148. American Medical Association. Genetically Modified Crops and Foods. Available online: http://www.ilsi.org/NorthAmerica/Documents/AMA_2000InterimMeeting. pdf (accessed on 12 May 2016).

149. Committee on Identifying and Assessing Unintended Effects of Genetically Engineered Foods on Human Health; Food and Nutrition Board; Institute of Medicine; Board on Agriculture and Natural Resources; Board on Life Sciences; Division on Earth and Life Studies; National Research Council. *Safety of Genetically Engineered Foods: Approaches to Assessing Unintended Health Effects*; The National Academies Press: Washington, DC, USA, 2004; p. 256.

150. European Academies Science Advisory Council. Planting the Future: Opportunities and Challenges for Using Crop Genetic Improvement Technologies for Sustainable Agriculture. Available online: http://www.easac.eu/fileadmin/Reports/Planting_the_ Future/EASAC_Planting_the_Future_FULL_REPORT.pdf (accessed on 12 May 2016).

151. The Royal Society. Genetically Modified Plants for Food Use and Human Health—An Update. Available online: https://royalsociety.org/~{}/media/royal_society_content/ policy/publications/2002/9960.pdf (accessed on 12 May 2016).

152. The Royal Society. *Reaping the Benefits: Science and the Sustainable Intensification of Global Agriculture*; Royal Society: Terrace, London, UK, 2009.

153. Hollingworth, R.M.; Bjeldanes, L.F.; Bolger, M.; Kimber, I.; Meade, B.J.; Taylor, S.L.; Wallace, K.B. Society of Toxicology position paper: The safety of genetically modified foods produced through biotechnology. *Toxicol. Sci.* **2003**, *71*, 2–8.

154. American Association for the Advancement of Science. Statement by the AAAS Board of Directors on Labeling of Genetically Modified Foods. Available online: http://www. aaas.org/sites/default/files/AAAS_GM_statement.pdf (accessed on 12 May 2016).

155. American Phytopathological Society Council. Compulsory Labeling of Plants and Plant Products Derived from Biotechnology. Available online: http://www.apsnet.org/members/outreach/ppb/positionstatements/pages/biotechnologypositionstatement.aspx (accessed on 12 May 2016).

156. International Union of Nutritional Sciences. Statement on Benefits and Risks of Genetically Modified Foods for Human Health and Nutrition. Available online: http://www.iuns.org/statement-on-benefits-and-risks-of-genetically-modified-foods-for-human-health-and-nutrition/ (accessed on 12 May 2016).

157. American Medical Association. H-480.958 Bioengineered (Genetically Engineered) Crops and Foods. Available online: https://www.ama-assn.org/ssl3/ecomm/PolicyFinderForm.pl?site=www.ama-assn.org&uri=/resources/html/PolicyFinder/policyfiles/HnE/H-480.958.HTM (accessed on 12 May 2016).

158. Board of Science and Education, Britsh Medical Association. Genetically Modified Foods and Health: A Second Interim Statement. Available online: http://www.argenbio.org/adc/uploads/pdf/bma.pdf (accessed on 12 May 2016).

159. Bruhn, C.; Earl, R.; American Dietetic Association. Position of the American dietetic association: Agricultural and food biotechnology. *J. Am. Diet. Assoc.* **2006**, *106*, 285–93.

160. Biochemical Society. Genetically Modified Crops, Feed and Food. Available online: http://www.biochemistry.org/Portals/0/SciencePolicy/Docs/GM%20Position%20Statement%202011%20Final.pdf (accessed on 12 May 2016).

161. Pramer, D. Statement of the American Society for Microbiology on Genetically Modified Organisms. Available online: http://www.asm.org/index.php?option=com_content&view=article&id=3656&Itemid=341 (accessed on 12 May 2016).

162. Crop Science Society of America. Researchers and Farmers Utilize GM Technology to Address Society's Growing Global Food Production, Security, and Safety Needs. Available online: https://www.crops.org/files/science-policy/issues/reports/cssa-gmo-statement.pdf (accessed on 12 May 2016).

163. Federation of Animal Science Societies. FASS Facts: On Biotech Crops – Impact on Meat, Milk and Eggs. Savoy, IL. Available online: http://www.fass.org/geneticcrops.pdf (accessed on 12 May 2016).

164. Key, S.; Ma, J.K.; Drake, P.M. Genetically modified plants and human health. *J. R. Soc. Med.* **2008**, *101*, 290–298.

165. Nicolia, A.; Manzo, A.; Veronesi, F.; Rosellini, D. An overview of the last 10 years of genetically engineered crop safety research. *Crit. Rev. Biotechnol.* **2014**, *34*, 77–88.

166. European Commission. A Decade of EU-Funded GMO Research (2001–2010). Available online: https://ec.europa.eu/research/biosociety/pdf/a_decade_of_eu-funded_gmo_research.pdf (accessed on 12 May 2016).

167. Van Eenennaam, A.L.; Young, A.E. Prevalence and impacts of genetically engineered feedstuffs on livestock populations. *J. Anim. Sci.* **2014**, *92*, 4255–4278.

168. Delaney, B. Safety assessment of foods from genetically modified crops in countries with developing economies. *Food Chem. Toxicol.* **2015**, *86*, 132–143.

169. Snell, C.; Bernheim, A.; Berge, J.B.; Kuntz, M.; Pascal, G.; Paris, A.; Ricroch, A.E. Assessment of the health impact of GM plant diets in long-term and multigenerational animal feeding trials: A literature review. *Food Chem. Toxicol.* **2012**, *50*, 1134–1148.

170. Ricroch, A.E. Assessment of GE food safety using '-omics' techniques and long-term animal feeding studies. *N. Biotechnol.* **2013**, *30*, 349–354.

171. Herman, R.A.; Price, W.D. Unintended compositional changes in genetically modified (GM) crops: 20 years of research. *J. Agric. Food Chem.* **2013**, *61*, 11695–11701.

172. Committee on Genetically Modified Pest-Protected Plants; Board on Agriculture and Natural Resources; Division on Earth and Life Studies; National Research Council. *Genetically Modified Pest-Protected Plants: Science and Regulation*; The National Academies Press: Washington, DC, USA, 2000; p. 292.

173. Hartung, F.; Schiemann, J. Precise plant breeding using new genome editing techniques: opportunities, safety and regulation in the EU. *Plant J.* **2014**, *78*, 742–752.

174. Kolseth, A.K.; D'Hertefeldt, T.; Emmerich, M.; Forabosco, F.; Marklund, S.; Cheeke, T.E.; Hallin, S.; Weih, M. Influence of genetically modified organisms on agro-ecosystem processes. *Agric. Ecosyst. Environ.* **2015**, *214*, 96–106.

175. Conko, G.; Kershen, D.L.; Miller, H.; Parrott, W.A. A risk-based approach to the regulation of genetically engineered organisms. *Nat. Biotechnol.* **2016**, *34*, 493–503.

176. Diaz-Gomez, J.; Marin, S.; Capell, T.; Sanchis, V.; Ramos, A.J. The impact of *Bacillus thuringiensis* technology on the occurrence of fumonisins and other mycotoxins in maize. *World Mycotoxin J.* **2015**.

177. GMOs and Corn Mycotoxins. Available online: http://graincrops.blogspot.com/2013/08/gmos-and-corn-mycotoxins.html (accessed on 12 May 2016).

178. Tarrio, R.; Ayala, F.J.; Rodriguez-Trelles, F. The Vein Patterning 1 (VEP1) gene family laterally spread through an ecological network. *PLoS ONE* **2011**, *6*, e22279.

179. Yang, Z.; Zhou, Y.; Huang, J.; Hu, Y.; Zhang, E.; Xie, Z.; Ma, S.; Gao, Y.; Song, S.; Xu, C.; Liang, G. Ancient horizontal transfer of transaldolase-like protein gene and its role in plant vascular development. *N. Phytol.* **2015**, *206*, 807–816.

180. Emiliani, G.; Fondi, M.; Fani, R.; Gribaldo, S. A horizontal gene transfer at the origin of phenylpropanoid metabolism: A key adaptation of plants to land. *Biol. Direct* **2009**, *4*, 7.

181. Kyndt, T.; Quispe, D.; Zhai, H.; Jarret, R.; Ghislain, M.; Liu, Q.; Gheysen, G.; Kreuze, J.F. The genome of cultivated sweet potato contains *Agrobacterium* T-DNAs with expressed genes: An example of a naturally transgenic food crop. *Proc. Natl. Acad. Sci. USA* **2015**, *112*, 5844–5849.

182. Bock, R. The give-and-take of DNA: horizontal gene transfer in plants. *Trends Plant Sci.* **2010**, *15*, 11–22.

183. Wang, Q.; Sun, H.; Huang, J. The evolution of land plants: a perspective from horizontal gene transfer. *Acta Soc. Bot. Pol.* **2014**, *83*, 363–368.

184. El Baidouri, M.; Carpentier, M.C.; Cooke, R.; Gao, D.; Lasserre, E.; Llauro, C.; Mirouze, M.; Picault, N.; Jackson, S.A.; Panaud, O. Widespread and frequent horizontal transfers of transposable elements in plants. *Genome Res.* **2014**, *24*, 831–838.

185. Fortune, P.M.; Roulin, A.; Panaud, O. Horizontal transfer of transposable elements in plants. *Commun. Integr. Biol.* **2008**, *1*, 74–77.

186. Yue, J.; Hu, X.; Huang, J. Horizontal gene transfer in the innovation and adaptation of land plants. *Plant Signal Behav.* **2013**, *8*, e24130.

187. Bergthorsson, U.; Richardson, A.O.; Young, G.J.; Goertzen, L.R.; Palmer, J.D. Massive horizontal transfer of mitochondrial genes from diverse land plant donors to the basal angiosperm Amborella. *Proc. Natl. Acad. Sci. USA* **2004**, *101*, 17747–17752.

188. Geering, A.D.; Maumus, F.; Copetti, D.; Choisne, N.; Zwickl, D.J.; Zytnicki, M.; McTaggart, A.R.; Scalabrin, S.; Vezzulli, S.; Wing, R.A.; *et al.* Endogenous florendoviruses are major components of plant genomes and hallmarks of virus evolution. *Nat. Commun.* **2014**, *5*, 5269.

189. Yang, Z.; Wang, Y.; Zhou, Y.; Gao, Q.; Zhang, E.; Zhu, L.; Hu, Y.; Xu, C. Evolution of land plant genes encoding L-Ala-D/L-Glu epimerases (AEEs) via horizontal gene transfer and positive selection. *BMC Plant Biol.* **2013**, *13*, 34.

190. Matveeva, T.V.; Lutova, L.A. Horizontal gene transfer from *Agrobacterium* to plants. *Front. Plant Sci.* **2014**, *5*, 326.

191. Huang, J.; Yue, J. Horizontal gene transfer in the evolution of photosynthetic eukaryotes. *J. Syst. Evol.* **2013**, *51*, 13–29.

192. Diao, X.; Freeling, M.; Lisch, D. Horizontal transfer of a plant transposon. *PLoS Biol.* **2005**.

193. Stegemann, S.; Keuthe, M.; Greiner, S.; Bock, R. Horizontal transfer of chloroplast genomes between plant species. *PNAS* **2012**, *109*, 2434–2438.

194. Markova, D.N.; Mason-Gamer, R.J. The role of vertical and horizontal transfer in the evolutionary dynamics of PIF-like transposable elements in Triticeae. *PLoS ONE* **2015**, *10*, e0137648.

195. Schnell, J.; Steele, M.; Bean, J.; Neuspiel, M.; Girard, C.; Dormann, N.; Pearson, C.; Savoie, A.; Bourbonniere, L.; Macdonald, P. A comparative analysis of insertional effects in genetically engineered plants: Considerations for pre-market assessments. *Transgenic Res.* **2015**, *24*, 1–17.

196. Gao, L.; Cao, Y.; Xia, Z.; Jiang, G.; Liu, G.; Zhang, W.; Zhai, W. Do transgenesis and marker-assisted backcross breeding produce substantially equivalent plants? A comparative study of transgenic and backcross rice carrying bacterial blight resistant gene Xa21. *BMC Genom.* **2013**, *14*, 738.

197. Batista, R.; Saibo, N.; Lourenco, T.; Oliveira, M.M. Microarray analyses reveal that plant mutagenesis may induce more transcriptomic changes than transgene insertion. *Proc. Natl. Acad. Sci. USA* **2008**, *105*, 3640–3645.

198. Lehesranta, S.J.; Davies, H.V.; Shepherd, L.V.; Nunan, N.; McNicol, J.W.; Auriola, S.; Koistinen, K.M.; Suomalainen, S.; Kokko, H.I.; Karenlampi, S.O. Comparison of tuber proteomes of potato varieties, landraces, and genetically modified lines. *Plant Physiol.* **2005**, *138*, 1690–1699.

199. Ladics, G.S.; Bartholomaeus, A.; Bregitzer, P.; Doerrer, N.G.; Gray, A.; Holzhauser, T.; Jordan, M.; Keese, P.; Kok, E.; Macdonald, P.; Parrott, W.; Privalle, L.; Raybould, A.; Rhee, S.Y.; Rice, E.; Romeis, J.; Vaughn, J.; Wal, J.M.; Glenn, K. Genetic basis and detection of unintended effects in genetically modified crop plants. *Transgenic Res.* **2015**, *24*, 587–603.

200. El Ouakfaoui, S.; Miki, B. The stability of the *Arabidopsis* transcriptome in transgenic plants expressing the marker genes nptII and uidA. *Plant J.* **2005**, *41*, 791–800.

201. Kok, E.J.; Keijer, J.; Kleter, G.A.; Kuiper, H.A. Comparative safety assessment of plant-derived foods. *Regul. Toxicol. Pharmacol.* **2008**, *50*, 98–113.

202. Shi, G.; Chavas, J.P.; Lauer, J.; Nolan, E. An analysis of selectivity in the productivity evaluation of biotechnology: an application to corn. *Am. J. Agric. Econ.* **2013**, *95*, 739–754.

203. Howard, P.H. Visualizing consolidation in the global seed industry: 1996–2008. *Sustainability* **2009**, *1*, 1266–1287.

204. Heinemann, J.A.; Massaro, M.; Coray, D.S.; Agapito-Tenfen, Z.F.; Wen, J.D. Sustainability and innovation in staple crop production in the US Midwest. *Int. J. Agric. Sustain.* **2014**, *12*, 71–88.

205. Stein, H. Intellectual property and genetically modified seeds: The United States, trade, and the developing world. *Northwest. J. Technol. Intell. Prop.* **2005**, *3*, 160–178.

206. Mascarenhas, M.; Busch, L. Seeds of change: intellectual property rights, genetically modified soybeans and seed saving in the United States. *Eur. Soc. R. Soc.* **2006**, *46*, 122–138.

207. McGuire, S.; Sperling, L. Seed systems smallholder farmers use. *Food Security* **2016**, *8*, 179–195.

208. Eisenstein, M. Biotechnology: Against the grain. *Nature* **2014**, *514*, S55–S57.

209. What is C4 Rice? Available online:. Available online: http://c4rice.irri.org/index.php/component/content/article/19-about/56-what-is-c4-rice (accessed on 12 May 2016).

210. Sneller, C.H. Impact of transgenic genotypes and subdivision on diversity within elite North American soybean germplasm. *Crop Sci.* **2003**, *43*, 409–414.

211. Biddle, J.B. Can patents prohibit research? On the social epistemology of patenting and licensing in science. *Stud. Hist. Philos. Sci. Part A* **2014**, *45*, 14–23.

212. Plant *vs.* Utility Patents. Available online: http://perennialpatents.com/plantpatent-v-utility-patents/ (accessed on 12 May 2016).

213. Hamilton, L.M. Linux for Lettuce. Available online: http://www.vqronline.org/reporting-articles/2014/05/linux-lettuce (accessed on 12 May 2016).

214. Glenna, L.L.; Tooker, J.; Welsh, J.R.; Ervin, D. Intellectual property, scientific independence, and the efficacy and environmental impacts of genetically engineered crops. *Rural Sociol.* **2015**, *80*, 147–172.

215. Prado, J.R.; Segers, G.; Voelker, T.; Carson, D.; Dobert, R.; Phillips, J.; Cook, K.; Cornejo, C.; Monken, J.; Grapes, L.; Reynolds, T.; Martino-Catt, S. Genetically engineered crops: From idea to product. *Annu. Rev. Plant Biol.* **2014**, *65*, 769–790.

216. Open Source Seed Initiative. Available online: http://osseeds.org/about/ (accessed on 12 May 2016).

217. Krishna, V.; Qaim, M.; Zilberman, D. Transgenic crops, production risk and agrobiodiversity. *Eur. Rev. Agric. Econ.* **2015**, 1–28.

218. Sander, J.D.; Joung, J.K. CRISPR-Cas systems for editing, regulating and targeting genomes. *Nat. Biotechnol.* **2014**, *32*, 347–355.

219. Belhaj, K.; Chaparro-Garcia, A.; Kamoun, S.; Nekrasov, V. Plant genome editing made easy: targeted mutagenesis in model and crop plants using the CRISPR/Cas system. *Plant Methods* **2013**, *9*, 39.

220. Doudna, J.A.; Charpentier, E. The new frontier of genome engineering with CRISPR-Cas9. *Science* **2014**, *346*, 1258096.

221. Woo, J.W.; Kim, J.; Kwon, S.I.; Corvalan, C.; Cho, S.W.; Kim, H.; Kim, S.G.; Kim, S.T.; Choe, S.; Kim, J.S. DNA-free genome editing in plants with preassembled CRISPR-Cas9 ribonucleoproteins. *Nat. Biotechnol.* **2015**, *33*, 1162–1164.

222. Hallerman, E.; Grabau, E. Crop biotechnology: A pivotal moment for global acceptance. *Food Energy Secur.* **2016**, *5*, 3–17.

223. Voytas, D.F.; Gao, C. Precision genome engineering and agriculture: opportunities and regulatory challenges. *PLoS Biol.* **2014**, *12*, e1001877.

224. Huang, S.; Weigel, D.; Beachy, R.N.; Li, J. A proposed regulatory framework for genome-edited crops. *Nat. Genet.* **2016**, *48*, 109–111.

225. Schiemann, J.; Hartung, F. EU perspectives on new plant-breeding techniques. In *New DNA-Editing Approaches: Methods, Applications and Policy for Agriculture*; Eaglesham, A., Hardy, R.W.F., Eds.; North American Agricultural Biotechnology Council: Ithaca, NY, USA, 2015; p. 276.

226. Li, K.; Wang, G.; Andersen, T.; Zhou, P.; Pu, W.T. Optimization of genome engineering approaches with the CRISPR/Cas9 system. *PLoS ONE* **2014**, *9*, e105779.

227. Gaj, T.; Gersbach, C.A.; Barbas, C.F., III. ZFN, TALEN, and CRISPR/Cas-based methods for genome engineering. *Trends Biotechnol.* **2013**, *31*, 397–405.

MDPI AG

St. Alban-Anlage 66

4052 Basel, Switzerland

Tel. +41 61 683 77 34

Fax +41 61 302 89 18

http://www.mdpi.com

Sustainability Editorial Office

E-mail: sustainability@mdpi.com

http://www.mdpi.com/journal/sustainability

www.ingramcontent.com/pod-product-compliance
Lightning Source LLC
Chambersburg PA
CBHW051924190326
41458CB00026B/6404